DYNAMOMETER

DYNAMOMETER

Theory and Application to Engine Testing

Jyotindra S. Killedar

Copyright © 2012 by Jyotindra S. Killedar.

Library of Congress Control Number: 2012909642
ISBN: Hardcover   978-1-4771-2007-1
              Softcover   978-1-4771-2006-4
              Ebook   978-1-4771-2008-8

All rights reserved. No part of this book may be reproduced or transmitted in any form or by any means, electronic or mechanical, including photocopying, recording, or by any information storage and retrieval system, without permission in writing from the copyright owner.

This book was printed in the United States of America.

To order additional copies of this book, contact:
Xlibris Corporation
1-888-795-4274
www.Xlibris.com
Orders@Xlibris.com
114085

Dedication

To My Guru:

सदाशिव समारंभां । शंकराचार्य मध्यमां ।
अस्मदाचार्य पर्यंताम । वन्दे गुरु परंपरां ॥

sadà shiva samàrambhàm | Shankaràcàrya madhyamàm |
asmada àcàrya paryantam | vande guru paramparàm | |

To My parents:

My mother, Umadevi Killedar and my Father, Shankarrao Killedar

My in laws:

Sadanand Desai and Tulasi Desai

To My Family:

For my wife, Smruti Jyotindra Killedar

For my son: Amogh Killedar

My beloved sister Gita Bajirao Patil

Contents

Preface 23

Part 1: Dynamometers

Chapter 1: What Is Dynamometer? 29
1.1 Introduction 29
1.2 Definition of the Dynamometer 29
1.3 History of the Dynamometer 30
1.4 Dynamometer—A Chronology of Innovation 31
1.5 The Word Dynamometer 32
1.6 Essential Features of Dynamometer 32
1.7 Speed Measurement of Dynamometer 32
1.8 Torque Measurement 32
1.9 Torque—turning force 33
1.10 Dynamometer Constant 33
1.11 Mounting of Dynamometer 34
1.11.1 Trunnion bearing 34
1.11.2 Flexural Support 36
1.11.3 Fixed Mounting 36
1.12 Scientific Accuracy of Dynamometers 37
1.13 BHP Concepts 37
Closure 40

Chapter 2: Dynamometer—Classification and Types 41
2.1 Introduction 41
2.2 Prony Brake Dynamometer 43
2.3 The Power Curve of the Prony Brake 45
2.4 Rope Brake Dynamometer 45
2.5 Fan Brake Dynamometer 50
2.5.1 Power Curve of Fan Brake 54
2.6 Hydraulic Dynamometer 54

2.6.1 Hydraulic Dynamometer—Sluice gate Controlled (Constant Fill) 55
2.6.2 Hydraulic Dynamometer—Outlet Valve controlled (Variable Filled) 57
2.7 Eddy-Current Dynamometer—Dry Gap 58
2.8 Eddy-current Dynamometer—Wet Gap......................59
2.9 DC Dynamometer 59
2.10 AC Dynamometer 59
2.11 Transmission Dynamometer..61
2.11.1 Epicyclical Train Dynamometer...................................62
2.11.2 Belt Transmission Dynamometer 64
2.11.2.1 Tatham Dynamometer 64
2.11.2.2 Von Hefner Belt Transmission Dynamometer 65
Closure 67

Chapter 3: Hydraulic Dynamometers 68
3.1 Introduction 68
3.2 Construction of Hydraulic Dynamometer.......................69
3.2.1 Construction of Sluice Gate Machine 69
3.2.2 Pressure-controlled Hydraulic Dynamometer................70
3.3 Water Movement Inside the Dynamometer....................72
3.4 Inertia of Hydraulic Dynamometer and Its Effect 73
3.5 Power and Torque Envelopes...74
3.7 Method of Load Control..79
3.7.1 Sluice Gate Control 79
3.7.2 Pressure-controlled Dynamometer................................80
3.7.2.1 Electric Servo 80
3.7.2.2 Electrohydraulic Servo Control.....................................81
3.8 Cavitation in Hydraulic Dynamometer 82
3.9 Selection of Dynamometer...86
3.10 Application of Hydraulic Dynamometer........................88
Closure 89

Chapter 4: Eddy-current Dynamometer..................................91
4.1 Introduction 91
4.2 History of Eddy-current Dynamometer..........................92
4.3 Principle of Operation 92

4.4	Electrical Laws Associated with the Eddy-current Dynamometer	93
4.5	Construction of Eddy-current Dynamometer	95
4.5.1	Rotor and Shaft Assembly	96
4.5.2	Casing/Carcass Assembly	97
4.5.3	Baseplate Assembly	98
4.6	Inertia of Eddy-current Dynamometer	98
4.7	Power and Torque Envelopes	99
4.7.1	Power Curve of Eddy-current Dynamometer	99
4.7.2	Torque Curve of Eddy-current Dynamometer	101
4.7.3	Power Curve Comparison	102
4.8	Wet Gap Eddy-current Dynamometer	103
4.8.1	Advantages of the Wet Gap Dynamometer	103
4.8.2	Disadvantages of the Wet Gap Dynamometer	103
4.9	Air-cooled Eddy-current Dynamometer	104
4.10	Applications of Air-cooled Eddy-current Dynamometer	105
4.11	Dry Gap Twin Coil Eddy-current Dynamometers	105
	Closure	106

Chapter 5: Magnetic Powder and Hysteresis Dynamometer 107

5.1	Introduction	107
5.2	Magnetic Powder Dynamometer	107
5.3	Power Curves of Magnetic Powder Dynamometer	108
5.4	Torque Curves of Magnetic Powder Dynamometer	109
5.5	Hysteresis Dynamometers	110
5.5.1	Operating Principles	110
5.5.2	Construction of Hysteresis Dynamometer	111
5.5.3	Power Curve of Hysteresis Dynamometer	112
5.5.4	Torque Envelope of Hysteresis Dynamometer	112
5.6	Comparison of Hysteresis, Eddy-current, and Powder Dynamometers	113
5.7	Typical Accuracies of Magnetic Powder and Hysteresis Dynamometer.	114
5.8	Applications of Magnetic Powder and Hysteresis Dynamometers	114
	Closure	114

Chapter 6: Portable Dynamometers 115
- 6.1 Introduction 115
- 6.2 Portable Dynamometer 115
- 6.3 Construction of Portable Dynamometer 116
- 6.4 Cross Section of Portable Dynamometer 117
- 6.5 Power Curve of Portable Dynamometer 118
- 6.6 Typical Applications 121
- 6.7 Advantages of Portable Dynamometer 122
- 6.8 Applications of Portable Dynamometer 122
- 6.8.1 Engine Testing 122
- 6.8.2 Testing of the Engines Insitu Condition 123
- 6.8.3 Farm Tractor PTO Testing 124
- Closure 124

Chapter 7: Direct Current (DC) Dynamometer 126
- 7.1 Introduction 126
- 7.2 Electrical Dynamometers 126
- 7.3 Principle of Operation 127
- 7.4 Rating of the DC 129
- 7.5 Construction of DC Machine 130
- 7.5.1 Trunnion Mounting 131
- 7.5.2 Fixed Mounting 132
- 7.6 Power and Torque Envelope 133
- 7.7 Inertia and Its Effect 133
- 7.8 Typical Accuracies of DC Machine/Dynamometer 134
- 7.9 Control System 134
- 7.9.1 Thyristor Drive Basics 134
- 7.9.2 Four-quadrant DC Drive 136
- 7.10 Basic Drive Operation 139
- 7.11 Changing the Direction of Rotation of DC Dynamometer 141
- 7.12 Stopping a Motor 142
- 7.13 Regeneration—Four-quadrant Operation 142
- 7.14 Emergency Breaking 143
- 7.16 Regeneration—Feedback to Mains 144
- 7.17 DC Dynamometer Application in Engine Testing 145
- 7.18 Types of Load Banks 146
- Closure 147

Chapter 8: Alternating Current (AC) Dynamometer 149
8.1 Introduction 149
8.2 AC Electrical Dynamometers 149
8.3 Principle of Operation 150
8.4 Rating of AC Dynamometer 150
8.5 Construction of AC Dynamometer 150
8.5.1 Trunnion-mounted AC Dynamometer 151
8.5.2 Foot-mounted 152
8.5.2.1 Advantage of Foot-mounted Machines 153
8.6 Liquid-cooled AC Dynamometer 154
8.6.1 The Advantages of Liquid-cooled AC
       Dynamometer 154
8.7 Mounting of In-line Torque Transducer 154
8.8 Comparison of AC and DC dynamometers
       (Bare Machines) 155
8.9 Power and Torque Envelopes 156
8.10 Inertia of AC Dynamometer 157
8.11 Typical Accuracies of AC Machine 157
8.12 AC Drive Principles of Operation 158
8.12.1 AC Drive Basics 160
8.12.2 Converter and DC Link 161
8.12.3 Insulated Gate Bipolar Transistor (IGBT) 162
8.12.4 Converter and Control Logic 163
8.12.5 Pulse Width Modulation (PWM) 164
8.12.6 Control Panel 165
8.12.7 Dynamometer Application in Engine Testing 166
       Closure 167

Chapter 9: Economics of Comparison 168
9.1 Introduction 168
9.2 Parameters Governing Selection of Dynamometer 168
9.3 Important Factors Influencing Selection of
       Dynamometer 169
9.4 Comparison of Hydraulic and Eddy-current
       Dynamometers 170
9.5 Water Supply System and Its Requirement 171
9.5.1 Influence of Water System on Working of
       the Hydraulic Dynamometer 172
9.7 Comparison of AC and DC Dynamometers 175

9.7.1	System Price	176
9.7.2	Drive Price and Maintenance Costs	176
9.7.3	Motor Price	176
9.7.4	Installation Cost	177
9.7.5	Efficiency	177
9.7.6	Power Factor	178
9.7.7	Harmonics	178
9.7.8	Performance	178
9.7.9	Degree of Protection for Motors	178
9.7.10	Modernizing Existing DC Drives	179
9.8	Control Accuracy and Response	180
	Closure	180

Chapter 10: Control Modes—What Do We Control? 182

10.1	Introduction	182
10.2	Open-loop mode	183
10.2.1	Open-loop Characteristics—Hydraulic Dynamometer	184
10.2.2	Open-loop Characteristics—Eddy Dynamometer	184
10.3	Closed Loop	185
10.3.1	Speed Constant Mode	185
10.3.2	Torque Constant Mode	186
10.3.3	Power Law Mode	186
10.4	Application of Modes for Engines	187
10.5	Dynamometer Modes with Relation to Torque Envelope	187
10.5.1	Speed Constant Mode (N = Constant)	187
10.5.2	Torque Constant Mode (M = Constant)	188
10.5.3	Power Law Mode $M(N^2)$	189
10.6	Selection of Right Dynamometer Mode	189
10.7	Manifold Vacuum/MAP Option Control Modes	191
10.8	Bump-less Mode Transfer	191
10.9	Role of PID Controller	192
10.10	The Methodology to Set Up a PI Controller	194
	Closure	195

Chapter 11: Dynamometer: Torque and Power Measurement Accuracy 196

11.3	Calibration of Measured Parameters	202

11.4	What Do We Understand by Traceability?	204
11.5	Torque Measurement	205
11.6	Torque Transmission Process	207
11.7	Errors Associated with Torque Measurement	211
	Closure	216

Chapter 12: Correction Factor and Horsepower 217

12.1	Introduction	217
12.2	How correction factor and Horse power is related?	218
12.3	The STP and NTP conditions	219
12.3.1	STP—Standard Temperature and Pressure	219
12.3.2	NTP—Normal Temperature and Pressure	219
12.4	Use of correction factors	220
12.5	Horsepower and Torque:	221
12.6	Effect of altitude on the power	221
12.7	Effect of Humidity on the power.	222
12.8	Effect of Temperature on the Power	223
12.9	Power Correction Factors	223
12.10	Society of Automotive Engineers	224
12.10.1	SAE J1349 Update:	224
12.10.2	Derivation of SAE Correction Factor Formula	226
12.11	DIN—70012 Method.	229
12.12	JIS D 1001 Method	230
12.13	ISO 1585 Method	231
12.14	Correct sensing of parameters—Ambient Air temperature, Pressure and RH in Test cell.	232
	Closure:	233

Part II: Data Acquisition and Control System

Chapter 13: The Sensors and Transducers 237

13.1	Introduction	237
13.2	Thermocouples	239
13.2.1	Theory of Operation	239
13.2.2	Factors Affecting the Accuracy of the Thermocouple.	240
13.2.3	Identification for Insulated Thermocouple Wire	240
13.2.4	Thermocouple Wire	241
13.2.5	Accuracy of Thermocouple	241

13.2.6	Thermocouple Grade and Extension Grade Wire	241
13.2.7	Types of Thermocouples Commonly Used	241
13.2.8	Cold Junction Compensation (CJC)	243
13.2.9	Precautions and Considerations for Using Thermocouples	244
13.2.10	Connection Problems	244
13.2.11	Lead Resistance	244
13.2.12	Decalibration	245
13.2.13	Noise	245
13.2.14	Common Mode Voltage	245
13.2.15	Thermal Shunting	246
13.3	Resistance Temperature Detector (RTD)	246
13.3.1	Theory of Operation	246
13.3.2	Linearization of RTD	248
13.3.3	Advantages of RTD	249
13.2.4	Disadvantages of RTD	249
13.3.5	Tolerance and Accuracy	249
13.3.6	Comparison of Thermocouple and RTD	250
13.4	Strain Gauge	253
13.4.1	Theory of Operation	253
13.4.2	Factors Influencing Selection of Strain Gauges	255
13.4.3	How Strain Gauge Is Attached to Specimen?	256
13.4.4	Strain Measurement Using a Wheatstone Bridge Circuit	258
13.3.5	Signal Conditioning for Strain Gauges	259
13.5	Load Cell	259
13.5.1	Theory of Operation	260
13.5.2	Load Cell Operating Principles	260
13.5.3	Load Cell Classification Based on Working Principle	261
13.5.4	Types of Load Cells Classified Based on Construction	261
13.5.5	The Important Terminology of the Load Cell	265
13.6	Pressure Transducer	274
13.6.1	Theory of Operation	274
13.6.2	Gauge Pressure	275
13.6.3	Atmospheric Pressure	275
13.6.4	Absolute Pressure	276
13.6.5	Standard Atmospheric Pressure	276

13.6.6 Pressure Measurement Devices/Instruments 276
13.6.6.1 Manometers 276
13.6.6.2 Pressure Gauges 279
13.7   Rotary Torque Transducer 282
13.7.1 Theory of Operation 282
13.7.2 Mechanical Design of Torque Transducer 283
13.7.3 Electrical Design 284
13.7.5 Coupling Requirements for Torque Transducer 286
13.7.6 Mechanical Calibration 287
13.7.7 Electrical Calibration 287
13.8   Magnetic Pickup 289
13.8.1 Magnetic Pickup with 60-toothed Wheel 289
13.8.2 Why 60-toothed Wheel Is Used? 289
13.8.3 Tachogenerator 292
13.8.4 Encoder 292
13.9   Environmental Measurements 294
13.9.1 Ambient Air Intake Temperature 294
13.9.2 Barometric Pressure 296
13.9.3 Relative Humidity 298
13.9.3.1 Measurement of Relative Humidity 299
   Closure 302

Chapter 14: Automation of Testing Process and
          Measurements 303
14.1   Introduction 303
14.2   Different Levels of Automation 304
14.3   Strategies for Automation 306
14.4   Classification of Test Cell Automation 306
14.5   Automation of Control System 307
14.5.1 Engine Throttle Controller 307
14.5.2 Engine Shutdown Actuator 308
14.5.3 Computerized Automation of Testing 308
14.6   Automation of Engine Handling 312
14.6.2 Automatic Driveshaft Connection 312
14.6.3 Automated Guided Vehicles 314
14.7   Docking Trolleys and Carts 317
14.8   Overhead Conveyor and Pallets 320
14.9   Automation of Auxiliary Equipment 324
14.9.1 Why Do We Need Temperature Controllers? 324

14.9.2	Elements of Temperature Controller	325
14.9.3	How Controller Works	327
14.10	Engine Oil Temperature Controller (EOTC)	327
14.14	Engine Water Temperature Controller (EWTC)	330
14.12	Engine Fuel Conditioning Unit	331
	Closure	334

Chapter 15: Basics of Dynamometer Data Acquisition and Control System 335

15.1	Introduction	335
15.2	What Is Data Acquisition?	336
15.3	What Data Acquisition Does?	336
15.4	Where Do We Use Data Acquisition and Control System?	336
15.5	The Elements of Data Acquisition Systems	337
15.5.1	Sensor	338
15.5.2	Signals	338
15.5.3	Data Acquisition Hardware	339
15.5.4	Data Acquisition Software	340
15.6	Data Acquisition Terminology	340
15.7	Types of Data Acquisition Systems	344
15.7.1	Wireless Data Acquisition Systems	344
15.7.2	Serial Communication Data Acquisition Systems	345
15.7.3	USB Data Acquisition Systems	346
15.7.4	Data Acquisition Plug-in Boards	347
15.8	Acquisition Channel Definitions	348
15.9	Alarm Annunciation	349
15.10	Configuration of Alarms as Failsafe and Nonfailsafe	352
15.11	Alarm Trip Choice—Hard or Soft?	353
15.12	Types of Relays	354
15.13	Audiovisual Indications for Tripping	356
15.14	Engine Data Acquisition and Control System (EDACS)	364
15.15	Executing the Automated Test as Programmed	369
15.16	Display of Data—Analog and Digital Indicators	369
15.17	Recording and Printing of Data	369
15.18	Management and Analysis of Data	370
	Closure	370

Part III: Modern Test Cell Concepts

Chapter 16: Engine Test Cell and Its Evolution 375
16.1 Introduction 375
16.2 Types of Test Cells 376
16.3 Noise in Test Cell 376
16.3.1 What Is Noise? 377
16.3.2 What Is dBA? 377
16.3.3 What Is NRC? 377
16.3.4 What Is Sound Transmission Class? 377
16.3.5 How to Interpret the dBA? 379
16.4 Test Cell Ventilation 379
16.5 Test Cell—Engine Exhaust Handling 386
16.6 Water Supply Systems 387
16.6.1 pH Value of Water 389
16.6.2 Purity of Water 389
16.6.3 Water Quantity 389
16.6.4 Cooling Towers 392
16.7 Compressed Air System 394
16.7.1 Air Quality 394
16.7.2 Air Quantity 396
16.7.3 Air Pressure 396
16.8 Fuel System 397
16.9 Engine Oil Supply System 398
16.10 Engine and Dynamometer Foundation 399
16.10.1 Foundation 399
16.10.2 Resonance 401
16.10.3 Concrete 401
16.10.4 Foundation Isolation 402
16.10.5 Foundation Bolts 402
16.10.6 Holding Down Bolts 403
16.10.7 Typical Foundation Hole 403
16.10.8 Dynamometer Baseplate Mounting 404
16.11 Lighting System 405
16.12 Communication System 406
16.13 Fire Fighting System 406
Closure 408

Chapter 17: Measurements of Modern Engine 409
17.1 Introduction 409
17.2 Fuel Consumption Measurement 409
17.2.1 Flow Meter 410
17.2.2 Types of Flow Meter Used in Engine Testing 410
17.2.3 Selection Criteria of a Flow Meter 410
17.2.4 Flow Measurement Orientation.................................. 411
17.2.5 Types of Fuel Consumption Meters 412
17.2.5.1 Orifice-type Flow Meter 412
17.2.5.2 Rotameter 415
17.2.5.3 Volumetric Fuel Consumption Meter
            (Pipette Type) 419
17.2.5.4 Gravimetric Fuel Measurement 424
17.3 Air Flow Measurement.. 426
17.3.1 Intake Air Measurement-Hot Wire Anemometer
         Method 426
17.3.2 Intake Air Measurement—Pitot Tube and
         Differential Pressure Transmitter 428
17.3.3 Intake Air Measurement—Air Box Method................. 431
17.4 Blow-by Measurement 439
17.4.1 What Is Blow-by?... 440
17.4.2 How to Measure Blow-by? 441
17.4.3 How Much Blow-by Is Normal for an Engine? 443
17.4.5 Interpretation of Blow-by Readings........................... 443
17.5 Oil Consumption Measurement 444
17.5.1 What Causes Excessive Oil Consumption? 444
17.5.2 Measuring Oil Consumption 444
17.5.3 Oil Consumption by Slow Flow Meter 445
       Closure 448

Chapter 18: Basics of Engine Testing 449
18.1 Introduction 449
18.2 Definition of an Engine 449
18.3 Classification of the Engines...................................... 449
18.4 Major Components of an IC Engine 450
18.6 Two-stroke Engines 458
18.7 The Four-stroke Engines 460
18.8 Comparison of Four-stroke and Two-stroke Engines.... 464
18.9 Testing Classification 466

| 18.10 | Testing Procedure | 466 |

18.11 What Do We Test Using Dynamometer and Associated Instrumentation? 468
18.11.1 Engine Power and Torque Curves 469
18.11.2 Indicated Horsepower 471
18.11.3 Brake Horsepower 471
18.11.4 Mechanical Efficiency 471
18.11.5 Frictional Horsepower 472
18.11.5.1 Methods to Establish FHP 472
18.11.5.1.1 Willan's Line Method 472
18.11.5.1.2 Morse Test 473
18.11.5.1.3 Motoring test 475
18.11.6 Fuel Consumption 476
18.11.6.1 Volumetric Fuel Measurement 476
18.11.6.2 Gravimetric Fuel Measurement 477
18.11.7 Air Consumption 477
18.11.8 Brake Thermal Efficiency 478
18.11.9 Indicated Thermal Efficiency 479
18.11.10 Brake Mean Effective Pressure 479
18.12 Heat Balance of the Engine 479
18.13 Engine Testing In Industry 480
18.13.1 Research and development 481
18.13.2 Production Testing 481
18.13.3 Quality Audit Testing of Engines 482
18.13.4 Type Testing 482
    Closure 486

Chapter 19: Test Cell Essentials 487
19.1 Introduction 487
19.2 Dynamometer Accessories 487
19.2.1 Dynamometer Water Inlet Filter 487
19.2.1.1 Y-type Strainer 487
19.2.1.2 Bucket Type Filter 489
19.2.1.3 Magnetic Filter 489
19.3 Universal Engine Mounting Test Beds 490
19.3.1 T-slotted Bedplates (CI and MS) 491
19.4 Cardan Shaft 491
19.5 Shaft Guard 492
19.6 Transducer Box 493

19.7	In-cell Control Panel	493
19.8.1	In-Line Mounted Electric Starting Motor	494
19.8.2	Piggyback-Mounted Starting Motor	495
19.8.3	Starter Motor Mounted on Dynamometer Nondrive but Offset to Center	496
19.8.4	Starting Motor Mounted Beneath the Dynamometer	497
19.9	Conventional Method	498
19.10	Nonelectrical Starting Systems	498
19.11	Load Throw Off Valve	499
19.12	Throttle Actuator and Controller	499
19.13	Diesel Engine Shutdown Actuator	501
19.14	Calibration Weights and Arm	502
19.15	Speed and Torque Indicator	502
19.16	Weather Station	503
	Closure	503

Chapter 20: Applications of Dynamometer 504

20.1	Introduction	504
20.2	Tractor PTO Testing	504
20.3	Draw Bar Pull Testing—Towing Dynamometer	507
20.4	Vertical Motor and Vertical Turbine Testing	511
20.4.1	Vertical Dynamometer Calibration	513
20.4.2	Vertical Alignment	514
20.5	Locked Rotor Test	516
20.6	Tandem Dynamometer	518
20.6.1	Single Engine Testing	518
20.6.2	Dual Engine Testing	519
20.7	Outboard Motor Testing	521
	Closure	522

Chapter 21: Driveshaft and Vibrations 524

21.1	Introduction	524
21.2	Terminology in Drive shaft selection	526
21.3	Selection procedure	528
21.3.1	Cardan Shaft Selection	529
21.3.2	Torque rating of Cardan Shaft	529
21.3.3	Speed rating of the cardan shaft	532
21.4	Calculation of torsional vibrations	534

21.4.1 Engine Orders..535
21.4.2 Natural Frequencies of Torsional Vibration 538
21.5   Holzer method for torsional vibration analysis:............540
21.5.1 Three Mass Systems..540
21.5.2 Two Inertia systems...544
21.6   Engine and dynamometer foundation: 552
21.6.1 Schematic 1—Engine and dynamometer firm on
       hard foundation 553
21.6.2 Schematic 2—Engine flexible mounted on hard
       foundation and dynamometer firm on hard
       foundation...553
21.6.3 Schematic 3—Engine Hard mounted on flexible
       Engine stand and dynamometer firm on hard
       foundation...554
21.6.4 Schematic 4:—Engine and Dynamometer hard
       on flexible foundation. 554
21.6.5 Schematic 5:—Engine flexible mounting and
       Dynamometer hard and both flexible mounted on
       common foundation..555
21.6.6 Schematic 6:—Engine stand flexible mounting
       and engine hard on engine stand. Dynamometer
       hard on its own foundation. Engine stand and
       dynamometer foundation both flexible mounted on
       foundation...555
21.6.7 Schematic 7:—Engine flexible on engine stand
       which is flexible on foundation. Dynamometer firm
       on its own foundation...556
   Closure: 556

Chapter 22: Air Intake and Exhaust Extraction Systems...........558
I. Air Intake System 558
22.1   Introduction 558
22.2   The Purpose of Combustion Air Handling
       Unit—CAHU 559
22.3   Combustion Air Handling Unit—CAHU........................560
22.4   Air-Fuel Ratio...562
22.5   Lambda..563
22.6   What Is the Effect of Changing the Air-Fuel Ratio?.....564
II Engine Exhaust Extraction 564

22.7	Introduction	564
22.8	What Is the Exhaust?	564
22.9	Exhaust Blowdown	565
22.10	Exhaust Stroke	565
22.11	The Exhaust Backflow	566
22.12	The Importance of Correct Back Pressure	567
22.13	Exhaust Extraction from the Engine Test Cell	567
22.13.1	Underground Extraction System	567
22.13.2	Overhead Extraction Systems	568
22.14	Exhaust Extraction from the Vehicle Test Cell	569
22.14.1	Telescopic System	569
22.14.2	Common Rail System	570
22.15	The Essential Elements of Exhaust Extraction	571
22.15.1	Exhaust Capture Nozzles	571
22.16	Exhaust Crush-proof Hoses	576
22.17	Fan Motor Assemblies	577
22.18	Hose Reel	577
22.19	Spring Balancer for Hose	578
22.20	Multipoint System	580
22.21	Design Considerations in Exhaust Extraction System	580
22.21.1	Exhaust Back Pressure	580
22.21.2	Pipes and Flexible Pipes	581
22.21.3	Insulation for Pipes	581
22.21.4	Expansion Joints and Bellows	582
22.21.5	Exhaust to Atmosphere	582
22.21.6	Condensate Extraction	583
	Closure	583

Appendix -1:	Power calculation formulae	585
Appendix II:	Engine terminology	587
Appendix III:	Pressure Units	591
Appendix IV:	Temperature Conversion	595
Appendix V:	Torque Conversion	597
Appendix VI:	Permissions and Approvals	599

Index 601

PREFACE

The subject of dynamometer and engine testing is complex, and engines are getting more and more complicated with the involvement of modern technology. The low fuel consumption and low exhaust emissions without compromising the performance are the driving factors for the most modern engines. The testing of these modern engines is becoming more complex in nature as technology advances.

In olden days, the engines were tested in open shed probably at the back of the assembly line. The modern test cells are complex and full of complex electronics and dedicated instrumentation assigned to measure targeted parameters. Computers and robotic mechanisms have taken the place of manual engine testers. More sophisticated test cell management is now in place to evaluate the performance of modern engines.

I started my career in dynamometer field way back in 1984 and continued till 2003. My total experience of thirty-two years reinforced my knowledge in industrial products such as compressors, industrial pumps, dynamometers, and material handling equipment and as software consultant.

I encountered a number of difficulties while I was new in dynamometer field. Aspiring new technology was a challenge as there were very few publications dedicated to dynamometers and engine testing. Moreover, I noticed that an incumbent from the technical college entering the engine and dynamometer field as a novice had to face many challenges in acquiring required knowledge to understand the complex instrumentation and mechanisms. Even today, many engineering and technical schools do not teach the subject of engine testing in required depth.

I realized the need for a proper book that will cater to the needs of the commissioning engineers, service engineers, sales and application engineers, engineers in automotive field, as well as new incumbents in this field.

I decided to put forward my experience and my thoughts on dynamometers in the form of a small book. I dreamed of writing the book during my tenure as engineer in the field of dynamometer in 1984 when I initiated my career in this field. Today, I feel satisfied to fulfill this dream.

Recently, I came across a couple of books that are more oriented toward engine testing. I decided to write my book more oriented toward the dynamometers, their various types, and their application in engine testing.

This book is purely based on my experience and my thoughts. Since 1984, I worked with a few world-class companies manufacturing dynamometers and most modern instrumentation. While writing the book, I realized that this is a herculean task. Preparing the manuscript, drawings, and images was a time-consuming work. This task would not have been completed without the support and inspiration of my wife, Dr. Smruti Killedar (MSc, PhD), and my son, Amogh. They practically relieved me from my domestic duties.

My other family members Dr. Laxmi Desai (MSc, PhD) and Yogesh Meghrajani (ME in electronic engineering) contributed by way of valued suggestions to make this treatise happen. Ms. Dolly Sudhir (MSc) and Mr. Sudhir (BCom, MBA) helped and shared my responsibilities. Mrs. Sujata Joshi (MA, LLM) and Mr. Madhukar Joshi (BA, LLM), successful lawyers, helped me in taking care of legal matters. My niece, Nisha Patil (DEE, BE in electronics), and nephew, Kamlesh Patil (DERE, BE in electronics), helped me in compiling my data in the right order.

Many of my friends who worked in dynamometer field were kind enough to advice and guide me through their active support and suggestions. It is worth mentioning their names for their suggestions and their dedication to dynamometer and engine testing field for more than ten years. They are Rajendra Phadtare, Sanjay Vhawal, Arvind Bawkar, Baban Gaikwad, Vijay Vaidya, Ajit Raibagi, and Sanjeev Keskar, who have shared of my experiences in the dynamometer and engine testing field over the last ten years and kept me up to date. I thank Mr. Volker Leismann and Rohit Nath from Schenck Avery (former name). I am also thankful to my ex-colleagues at AVL India who shared their knowledge during my tenure in AVL India.

I worked in the world-class dynamometer-manufacturing companies, namely, Saj Test Plant Pvt. Ltd. India, Schecnk Avery Ltd., and AVL India. I am especially thankful to Mr. Prakash Jagtap, chairman and managing director, who gave me the opportunity to learn about various dynamometers and instrumentations. His inspiration lighted a lamp in me to learn new things.

I am also thankful to my friend Berteau Joisil (MSME) for correcting the electrical and instrumentation related information in the book.

The author has made an attempt to cover the required topics for the entry level personnel in the field of dynamometer and engine testing. It is impracticable to cover every aspect of engine testing in the one single book like this. For the benefit of the readers a certain books and technical papers are recommended at the end of each chapter under the heading of "Further reading". The few chapters have the web sites mentioned under the bibliography. The names of the web sites are mentioned without mentioning the actual URLs as they are dynamic in nature and keep changing.

I welcome suggestions and comments from readers.

PART 1

DYNAMOMETERS

CHAPTER 1

What Is Dynamometer?

1.1 Introduction

In today's modern world, the IC engines or any prime movers are tested for its performances such as power and torque developed. To test the performance of these prime movers, dynamometers and associated instrumentation are used. The installations of dynamometers have become the important part of the automotive industry. The subject of dynamometer seems to cause more concern, many misunderstandings, and notions. There are three methods of testing an automotive engine in general, and they are

1. testing an engine using dynamometer
2. testing a vehicle with engine under consideration on chassis dynamometer
3. running the vehicle on test track.

1.2 Definition of the Dynamometer

1. A device for measuring the torque, force, or power available from a rotating shaft. The shaft speed is measured with a tachometer, while the turning force or torque of the shaft is measured with a scale or by another method. Power may be read from the instrumentation or calculated from the shaft speed and torque.
2. It is an apparatus for measuring force or power, especially one for measuring mechanical power, as of an engine.
3. It is an instrument for measuring force exerted by men, animals, and machines. The name has been applied generally

to all kinds of instruments used in the measurement of a force, as for example, electric dynamometers, but the term specially denotes apparatus used in connection with the measurement of work or in the measurement of the horsepower of engines and motors.

The most common use of the dynamometer is in determining the power of an electric motor or engine of a car, truck, or other vehicle. A dynamometer that connects to the engine crankshaft is an engine dynamometer. One that has rollers turned by the vehicle drive wheels is a chassis dynamometer; this type is widely used in the automotive industry for mileage accumulation, emissions, fuel economy, and performance testing of cars and trucks.

In this book, our discussions are mainly concentrated on engine dynamometer used for engine/motor testing.

1.3 History of the Dynamometer

The dynamometers are being used to measure the power since long. During the eighteenth century, James Watt introduced a unit of power to compare the power of his steam engines with a more familiar source of work. This unit of power became known as horsepower. It was defined as the amount of power required to move a 550-pound weight up to one foot in one second.

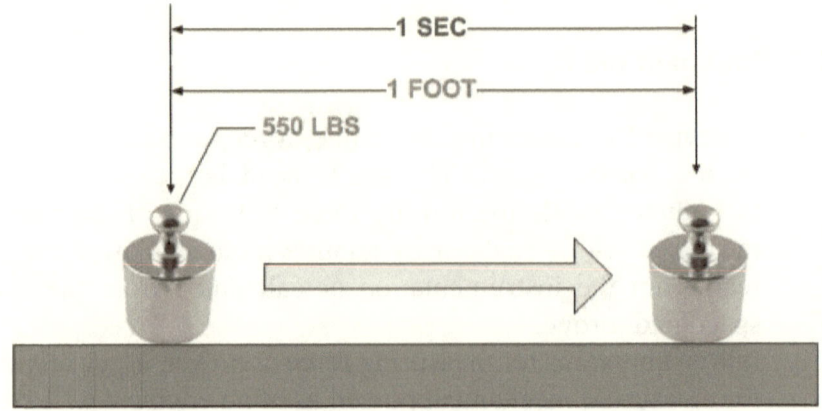

Figure 1.1 Work done.

The first device used probably date back when Gaspard de Prony invented the Prony brake circa 1821. The de Prony brake (or Prony brake) is considered to be one of the earliest dynamometers. Over the next two hundred years, the Prony brake dynamometer and variations of same were developed to measure engine horsepower. Modern day versions of these brake dynamometers are still in use today.

1.4 Dynamometer—A Chronology of Innovation

1. Gaspard de Prony invented the de Prony brake in 1821 in Paris. The de Prony brake (or Prony brake) is considered to be one of the earliest dynamometers.
2. In 1838, Charles Babbage, known to historians as the father of the computer, introduces a dynamometer car to measure the pulling power of English railroad locomotives.
3. William Froude with the invention of the hydraulic dynamometer in 1877, and first commercial dynamometers were produced in 1881 by their predecessor company, Heenan & Froude.
4. In 1921, Professor E. V. Collins of Iowa State College develops a draft horse dynamometer, used to measure a horse's capability to pull the era's heavy metal farm implements.
5. In 1928, the German company "Carl Schenck Eisengießerei & Waagenfabrik" built the first vehicle dynamometers for brake tests with the basic design of the today's vehicle test stands.
6. In 1930, using designs pioneered through collaboration with Rudolph Diesel, John Taylor forms the Taylor Dynamometer and Machine Company to produce engine dynamometers.
7. The eddy-current dynamometer was invented by Martin and Anthony Winther in about 1931. At that time, DC motor/generator dynamometers had been in use for many years. A company founded by the Winthers, Dynamatic Corporation, manufactured dynamometers in Kenosha, Wisconsin, until 2002.
8. In 2002, Dyne Systems of Jackson, Wisconsin, acquired the Dynamatic dynamometer product line. Starting in 1938, Heenan and Froude manufactured eddy-current dynamometers for many years.

9. The first popular, true high speed, computer-controlled, eddy-current motor cycle chassis dynamometer was produced by Factory Pro Dynamometer of San Rafael, CA, USA, in 1990.

1.5 The Word *Dynamometer*

The word *dynamometer* is derived from a Greek word *dunamis* meaning power and *meter* means measure, that is, *dunamis* + *metron* = *dynamometer.*

1.6 Essential Features of Dynamometer

If we think about a good dynamometer which serves the purpose of the engine testing, then the following four essential features are important:

- means of controlling torque
- means of measuring torque
- means of measuring speed
- means of dissipating power.

1.7 Speed Measurement of Dynamometer

The speed in dynamometer is measured by either a mechanical tachometer or an electronic device. In case of an electronic device, the speed measurement consists of a magnetic pulse sensor working in conjunction with a geared wheel generally having sixty teeth. The pulses generated are processed and displayed by electronic digital indicator. In some cases, where accuracy is of utmost importance, an optical encoder is used.

1.8 Torque Measurement

The concept of torque is important enough to be clarified. Actually, it is the direct result of the load of the spring or weight. Its distance from the axis of rotation is also responsible for determining the torque. In reality, dynamometers are used to calculate the production of torque by an engine.

1.9 Torque—turning force

This is also called moment of a force, in physics—the tendency of a force to rotate the body to which it is applied. The torque, specified with regard to the axis of rotation, is equal to the magnitude of the component of the force vector lying in the plane perpendicular to the axis, multiplied by the shortest distance between the axis and the direction of the force component, regardless of its orientation. The following figure shows how the torque is understood.

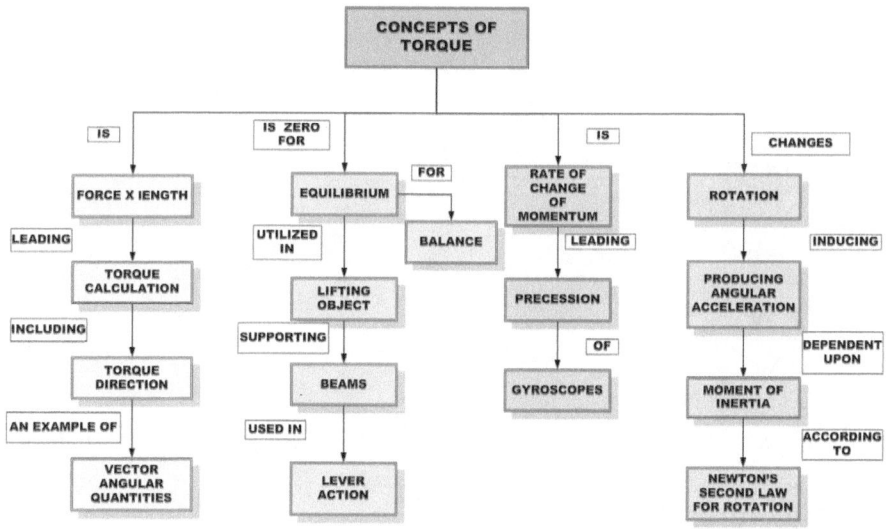

Figure 1.2—Torque
Courtesy of HyperPhysics by Rod Nave, Georgia State University".

Dynamometers, aka "dynos," are brakes used to measure the power of an engine at a given speed. The torque of an engine is determined by a complex measuring mechanism and reaction transferred by the dynamometer to measuring mechanism. Dynamometer manufacturers construct their products using basic components, namely, frame, trunnion bearings, absorption unit, and torque measuring device.

1.10 Dynamometer Constant

As discussed earlier in this chapter by definition, torque is derived quantity. It is a product of force applied and distance from the center to the point of application of the force. Generally, if the force is applied at the circumference of the rotating disk, then the distance between the center of disk and the point of application is radius of the disk under consideration.

Thus, torque T = force (f) × distance (l)

Or torque T = force (f) × radius (r)

Or Torque T = Force (f) x Radius (r)

$$HP = \frac{2\pi NT}{4500} \text{ where } N = rpm \text{ and } T = torque(kgm)$$

$$HP = \frac{2\pi N(F.l)}{4500}$$

$$HP = \frac{NF(2\pi l)}{4500}$$

$$HP = \frac{F.N}{K} \text{ where } K = \frac{4500}{2\pi l} \quad \text{...............1.1}$$

The constant "K" in the equation is known as dynamometer constant.

1.11 Mounting of Dynamometer

The main casing of dynamometer consists of a power absorbing unit. In case of hydraulic dynamometer, it will be rotor and pair of stator and controlling mechanism, and in case of eddy-current dynamometer, it is a rotor, excitation coil, and cooling chamber. This main chamber or casing is also called as cradle.

1.11.1 Trunnion bearing

A pedestal with trunnion bearings is either bolted to baseplate, or they are integral part of baseplate as per the proprietary designs of different manufacturers. The dynamometer cradle is mounted between the pair of the trunnion bearings. This gives the freedom to oscillate when a reaction force acts on cradle as virtue of absorbed power.

Trunnion is nothing but pivot forming one of a pair on which something is supported. Here in this case, a dynamometer carcass is supported which is free to oscillate and transfer the reaction force to the measuring mechanism. The movement is limited by torque arm connected to the side of housing and connected to the torque measuring system. The advantage of trunnion bearing is that it is the simplest type of cradle mounting with freedom of movement for the carcass.

Trunnion bearings located between the ends of the dynamometer housing or carcass, and a set of pedestals do not rotate. They do, however, allow the carcass to rotate slightly for torque measurements. Since bearings and lubricant directly affect performance and accuracy of dynamometer, trunnion bearings should be inspected and rotated frequently. Grease-lubricated trunnion bearings do not require periodic lubrication. However, if grease becomes dry or lumpy, it should be flushed and replaced.

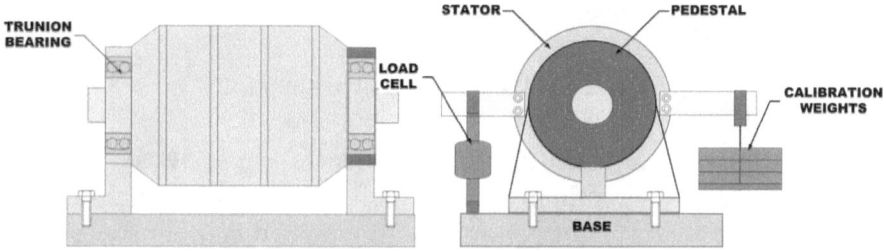

Figure 1.3 Trunnion mounting.

The disadvantage of the trunnion bearing is that after a long use, they try to be sticky. This is mainly because of their minimal movement. This is called the *brinelling effect.*

Some specially designed dynamometers consist of hydrostatic trunnion bearings, referred to as lift trunnion bearings. They are oil pressure lift-type sleeve bearings used to reduce trunnion bearing friction to a negligible value. This in turn typically improves system accuracy. Bearings of this type are oil lubricated with high pressure oil piping system to circulate oil through the bearings and support the carcass on a film of oil as long as the high pressure oil is supplied. Carcass floats during the operation.

1.11.2 Flexural Support

The dynamometer housing is cradle mounted in a flexure support on the frames. The flexure support guarantees absolute maintenance-free operation, optimal measurement accuracy, and minimal hysteresis between loading and unloading. The advantage of the flexure support is simple construction, and no maintenance required. The construction with flexure mount in place of trunnion bearing gives a big advantage of reduced overall weight of dynamometer.

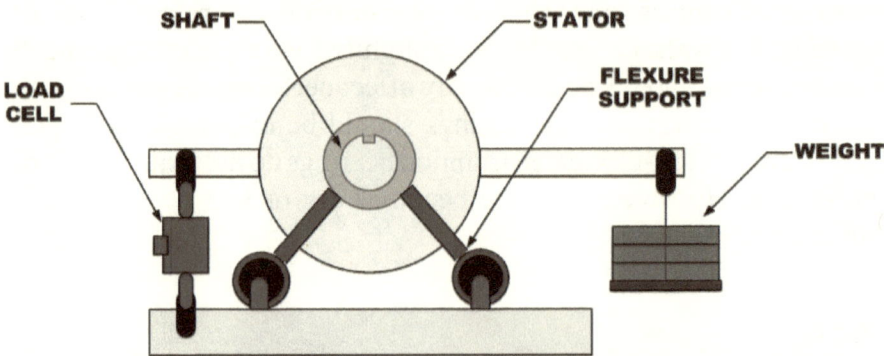

Figure 1.4 Flexure-mounted support.

Flexure supports replace the heavy pedestals and large trunnion bearings. However, a little insight will tell us that while dynamometer is at work, the flexural deflection may move the dynamometer center. However, this movement whatsoever is negligible.

1.11.3 Fixed Mounting

The fixed mounting is simplest type of mounting, and it also eliminates the oscillating cradle assembly. However, this arrangement employs in-line torque transducer for the measurement of the torque.

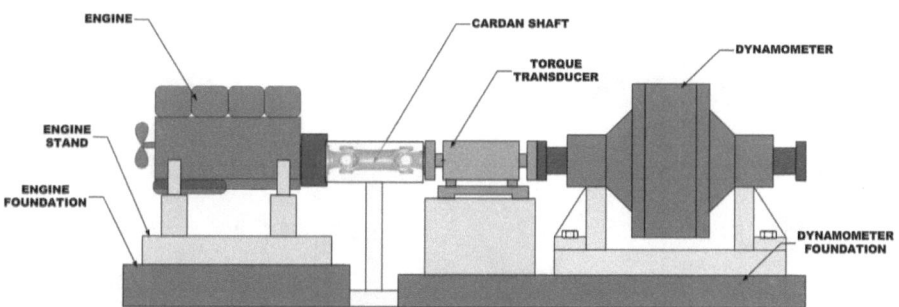

Figure 1.5 Foot-mounted dynamometer with torque transducer.

1.12 Scientific Accuracy of Dynamometers

The accuracy of dynamometer is of prime importance and needs to be evaluated. Many a times, doubts are raised about the absorption and transmission of stator (cradle) reaction to measuring mechanism. The issues raised are about frictional losses in trunnion bearing and measuring mechanism such as load cell mounting swivel joints. This is how it works. The net frictional force will have its reaction force which will act on cradle and ultimately considered in measuring force. Thus, the dynamometer transmits measured reaction to measuring chain faithfully. Hence, dynamometer is considered the hundred percent accurate machine.

1.13 BHP Concepts

Work done is product of force and distance moved in one second Power (energy) is defined as rate of doing work. In FPS system, it is defined as if 550-lb weight is moved one foot in one second, then the work done is 1 HP or 33,000 lbs/min.

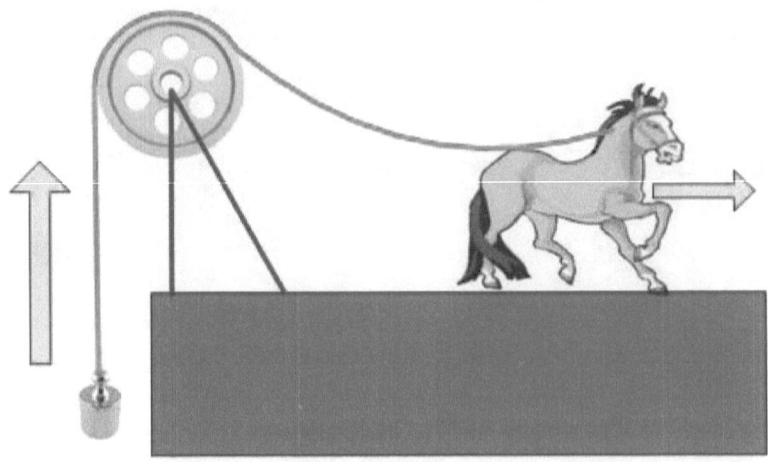

1HP= 550FOOT-POUNDS PER SEC. (FPS)
1HP= 75KG.M PER SEC. (MKS)

Figure 1.6 Horsepower explained.

It was originated by James Watt (1736-1819), the inventor of the steam engine, and the man whose name has been immortalized by the definition of Watt as a unit of power. If F is the force applied at flywheel of radius R, then the work done in on revolution is

Work done = force × distance moved (F = force in lbs and r = feet).

Distance moved in one revolution = perimeter of flywheel = $2\pi r$.

Work done = $F*2\pi r$.

Work done in N revolutions = $\dfrac{F.2.\pi.rN}{60}$, where

N = revolution/minute.

Therefore, $hp = \dfrac{2.\pi.N.F.r}{60 \times 550} =$

$= \dfrac{2.\pi.N.T}{33000}$ where $F.r = T$ lbft– ft

....1.2

Power in MKS System

Power is 1 hp if work is done at the rate of 75 kgm/sec. If a weight of 75 kg is moved to 1 m distance in 1 sec then the work done is 75 kgm/sec = 1 hp.

Work done = force × distance moved.

Distance moved in one revolution = perimeter of flywheel = $2\pi r$.

Work done = $F*2\pi r$ (F = force in kg and r = meter).

Work done in N revolutions = $\dfrac{F.2.\pi.rN}{75}$, where

N = revolution/minute.

Therefore, $hp = \dfrac{2.\pi.N.F.r}{60 \times 75} =$

$= \dfrac{2.\pi.N.T}{4500}$ where $F.r = T lb-ft$

....1.3

Power in SI System

Power is 1 watt if work done is 1 joule per second. It is worth recalling here from physics that 1 joule/sec = 1 Newton-meter/sec.

Work done = force × distance moved.

Distance moved in one revolution = perimeter of flywheel = $2\pi r$.

Work done = $F*2\pi r$ (F = force in Newtons and r = meter).

Work done in N revolutions = $\dfrac{F.2.\pi.rN}{60}$, where

N = revolution/minute.

Therefore, $Watts = \dfrac{2.\pi.N.F.r}{60} \; Nm\,Sec$

$Kw = \dfrac{2.\pi.N.T}{60,000}$ where $F.r = T$ in $N.m$ (1 kW = 1,000 watts)..........1.4

Closure

The dynamometers are basic apparatus used to measure and load the engine or the unit under test. The modern dynamometers, however, are equipped with more sophisticated and complicated instrumentation to suit the today's engine testing requirement. There are various types of dynamometers and modern instrumentation that will be dealt in forthcoming chapters.

Bibliography

1. Timeline of Dynamometer history Internet information.
2. Wikipedia-Dynamometer history.

CHAPTER 2
Dynamometer—Classification and Types

2.1 Introduction

There are various kinds of dynamometers used for different purposes. The dynamometers are mainly classified as absorption type, motoring type, and transmission type. Absorption dynamometer is designed for driving purpose, whereas universal dynamometer is used both for absorption and driving. The absorption type is again subclassified as hydraulic, eddy current, DC, and AC. The following chart shows the detailed classification of all types of dynamometers used in engine or any kind of prime mover testing.

The solid friction type of dynamometer is rarely used nowadays. The rope brake dynamometer is sometimes employed to measure high torque produced at very low rpm, whereas the use of conventional dynamometer like hydraulic or eddy current is not economical. The hydraulic motors having very high torque at corresponding low rpm can be tested using rope brake dynamometer.

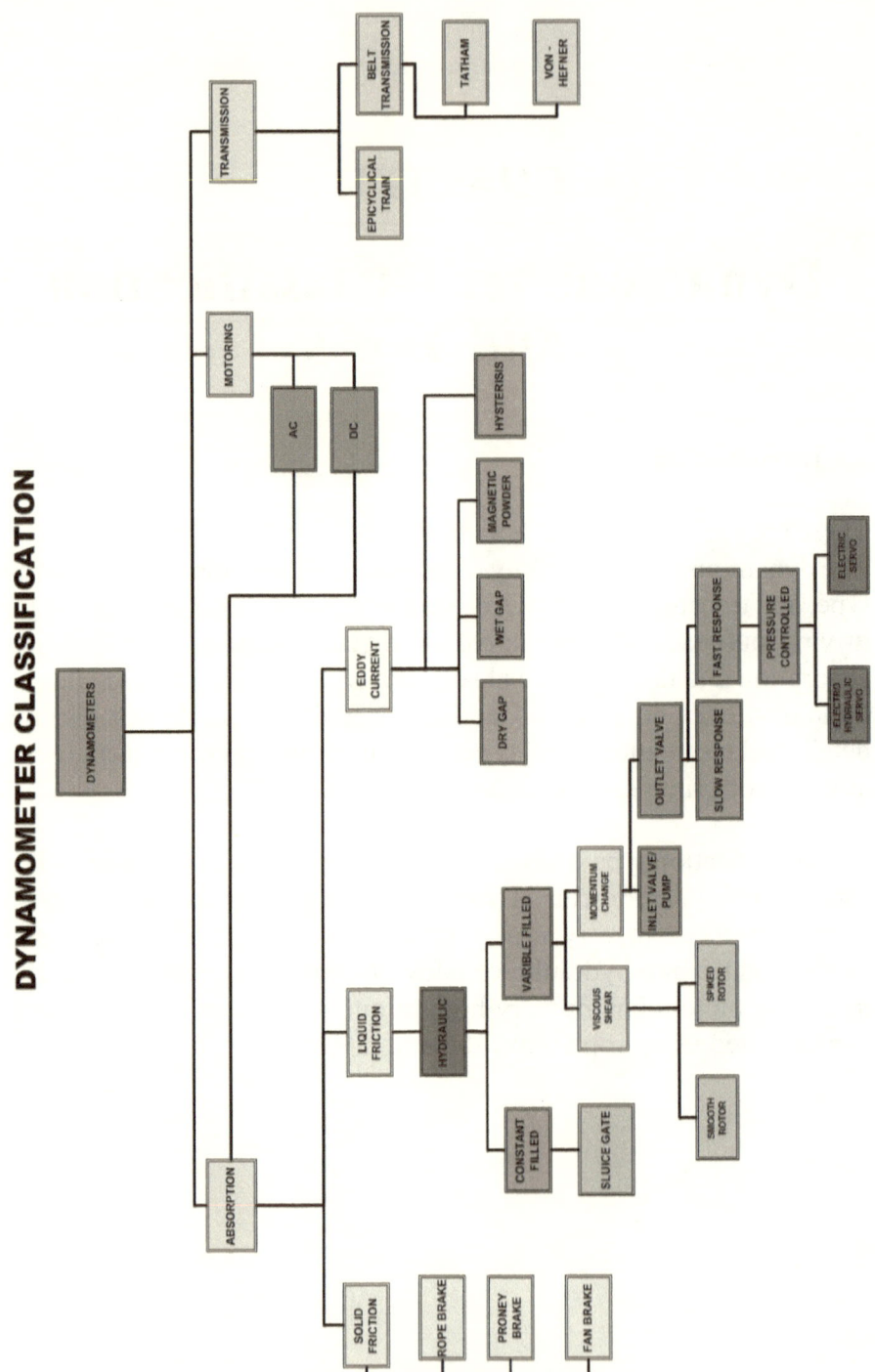

Figure 2.1 Dynamometer classification chart.

2.2 Prony Brake Dynamometer

Gaspard Clair François Marie Riche de Prony (July 22, 1755-July 29, 1839) was a French mathematician and engineer, who worked on hydraulics. He was born at Chamelet, Beaujolais, France. Gaspard de Prony invented the de Prony brake in 1821. The de Prony brake (or Prony brake) is considered to be one of the earliest dynamometers. The term *brake* is referred to the hand brake used in early dynamometers like de Prony brake.

Figure 2.2 Portrait of Prony.

This is a mechanical friction brake in which wooden shoes are pressed against the flywheel mounted on the engine shaft. The braking reaction force is measured by means of deadweights added to the weight pan.

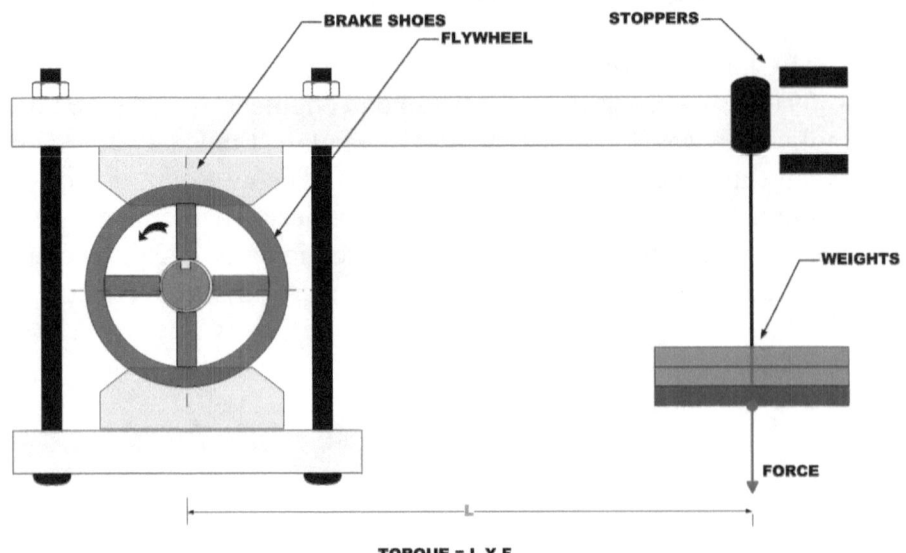

Figure 2.3 Prony brake.

Power is calculated using the standard formula $P = \dfrac{2.\pi.N.T}{4500}$, ...2.1

where
N = RPM
T = torque kgm

2.3 The Power Curve of the Prony Brake

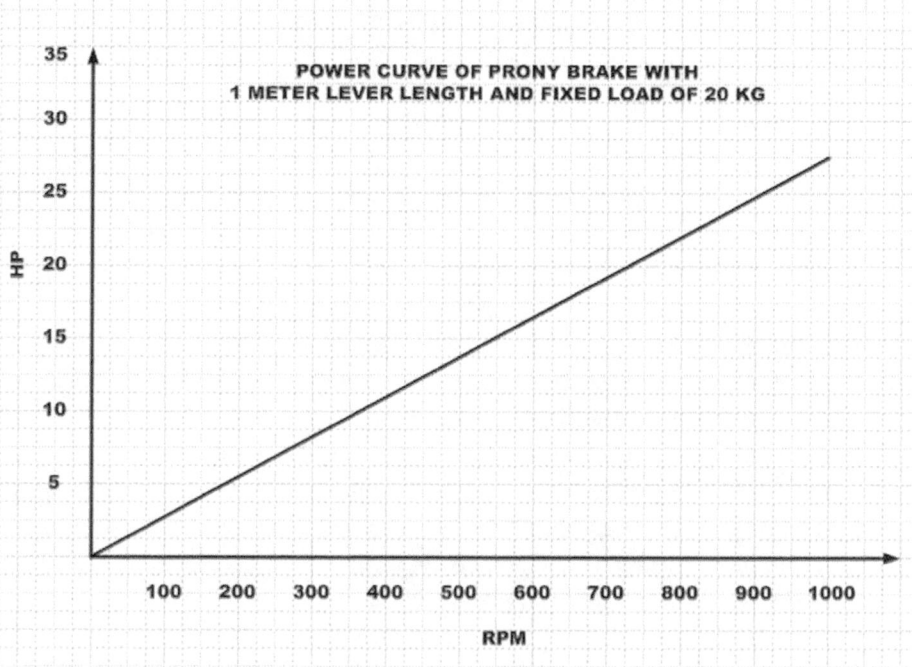

Figure 2.4 Power curve of Proney brake.

2.4 Rope Brake Dynamometer

About 1858, Lord Kelvin developed the "rope" brake, based on the earlier design by Prony, by replacing the wooden friction blocks with a length of rope coiled around the revolving shaft. Variations of these are still used in many engineering school laboratory exercises.

The rope brake dynamometer consists of a rope wound around a drum in the form of flywheel which is rigidly fixed to the shaft of the engine under test whose performance needs to be evaluated. The rope turns on the drum are equispaced by means of the wooden blocks placed on the periphery of the rim of a drum. The upper end of the rope is attached to the spring scale, and the other end is attached to the weight pan which carries known weights.

To measure the power with the help of rope brake dynamometer, engine is run at constant speed. At the constant speed point, the torque transmitted must be balanced by the frictional torque generated by the ropes wound around the drum. While engine is running under load, the drum needs to be constantly cooled. Water is circulated in the drum rim during engine testing to remove the heat generated due to friction.

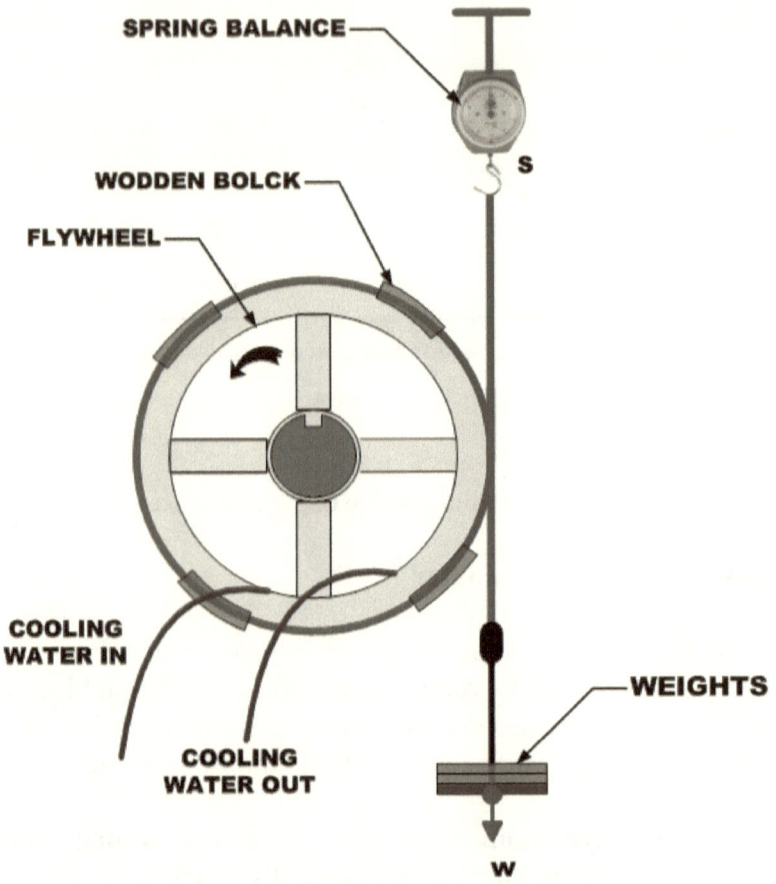

Figure 2.5 Rope brake dynamometer.

Let us derive the formula for torque power absorbed by the rope brake dynamometer.

Let
N = constant RPM of the flywheel (engine)

W = deadweight
S = spring balance reading
D = diameter of the flywheel
d = diameter of the rope

The net load on the shaft = $(W - S)$

Therefore, torque due to ropes = (net load on ropes) × distance of the load = $(W - S).\left(\frac{D+d}{2}\right)$.

But torque transmitted by the engine at constant speed = frictional torque due to ropes = $(W - S).\left(\frac{D+d}{2}\right)$.

Power of the engine = torque transmitted by the engine × angular speed engine

$$= (W - S).\left(\frac{D+d}{2}\right).\omega$$

$$= (W - S).\left(\frac{D+d}{2}\right).\frac{2.\pi.N}{60}$$

If we neglect the diameter of rope, that is, d, then absorbed power of engine

$$= (W - S).\left(\frac{D}{2}\right).\frac{2.\pi.N}{60}$$

$$= (W - S).(R).\frac{2.\pi.N}{60} \text{ watts.}$$

Please note $(W - S)(R)$ is nothing but torque T, and the formula becomes as usual

$$BHP = \frac{2.\pi.N.T}{60} \quad \text{............2.2}$$

Critical Speed

In order to ensure that the cooling water remains in contact with the flywheel chanel section in any position during rotation, it is necessary that the centrifugal force on water particles must be equal to or greater than the gravitational force acting on it. Therefore,

$$F_c \geq w,$$

where
w = weight of water particle

Or $\quad \dfrac{w}{g} r\omega^2 \geq w,$

where
r = mean radius at which water is revolving with channel section.
Ω = angular speed of drum or pulley.

Therefore,
$$\omega^2 \geq (w)\left(\dfrac{g}{w.r}\right).$$

$$\omega \geq \sqrt{\left(\dfrac{g}{r}\right)}.$$

Therefore, speed of drum, above which water particles remain in contact with

Therefore,
$$\omega_{critical} \geq \sqrt{\left(\dfrac{g}{r}\right)}.$$

$$N_{critical} \geq \dfrac{60}{2\pi}\sqrt{\left(\dfrac{g}{r}\right)} rpm. \quad\quad\quad\quad\quad\quad .2.3$$

Example 1

Compute the output power of the engine which is running at constant speed 500 rpm. The deadweight and spring balance are 500 N and 100 N, respectively. The diameter of the flywheel and rope is 2.0 m and 20 mm, respectively.

Solution:

Given:
$N = 500$ rpm
$W = 500$ N
$S = 100$ N
$D = 2.0$ m
$d = 20$ mm $= 0.020$ m

$$BHP = (500 - 100).\left(\frac{2+0.020}{2}\right).\frac{2\pi.500}{60} \text{ watts}$$

$$BHP = (400).(1.01).\frac{2\pi.500}{60} \text{ watts}$$

$$BHP = (404).(52.766) watts$$

$$BHP = 21156 watts \text{ or } 21.156 \ Kw.$$

Example 2

The rim of the flywheel of an IC engine is channel section, and its internal diameter is 500 mm. Find the minimum speed in r.p.m. at which the flywheel will hold a layer of water 25 mm deep at the top of the rim.

Solution:

Given:
Flywheel radius = 500 mm
Water layer thickness = 25 mm

Effective radius of water which can be held in channel:

$$r = \left(\frac{500}{2}\right) - 50 = 200$$

$$N_{critical} \geq \frac{60}{2\pi}\sqrt{\left(\frac{g}{r}\right)}\, rpm.$$

$$N_{critical} \geq \frac{60}{2\pi}\sqrt{\left(\frac{9.81}{0.2}\right)}\, rpm. = 66.87\ rpm.$$

2.5 Fan Brake Dynamometer[2]

Mr. W. G. Walker invented the fan brake dynamometer. The fan dynamometer is one of the choices because of its adaptability, cheapness, and simplicity. Its great disadvantage lies in the fact that it is very susceptible to changes in atmospheric conditions, such as air pressure, temperature, humidity in test environment. In its simplest form, the fan dynamometer is a fan whose blades are placed at right angles to the plane of rotation so that they will offer resistance to the air as they revolve.

The fan dynamometer is simply a paddle fan which is driven by the engine. The faster that the fan turns, the more power has to be put into it to keep it turning. The HP required to drive the fan is in direct relationship to the speed, so as long as the speed is known, the HP can be calculated.

Fan is used as loading media to the engine shaft. The fan brake will provide a fixed load, which is predetermined, and will be applied to the engine shaft. The fan brake is calibrated using a DC or AC dynamometer. The only measuring parameter will be a RPM. If fan brake is calibrated for 2 kW at 1,500 rpm and if it is connected to engine shaft, then tester will see that a 1,500 rpm is achieved or not. If engine speed reaches 1,500, then it is presumed that the power absorbed is 2 kW at 1,500 rpm. Generally, it is used for small two wheel scooter/motorcycle engine in production. It is more of like to have a go or no-go gauge-type testing. Fan brakes can be calibrated for different speed (rpm) range.

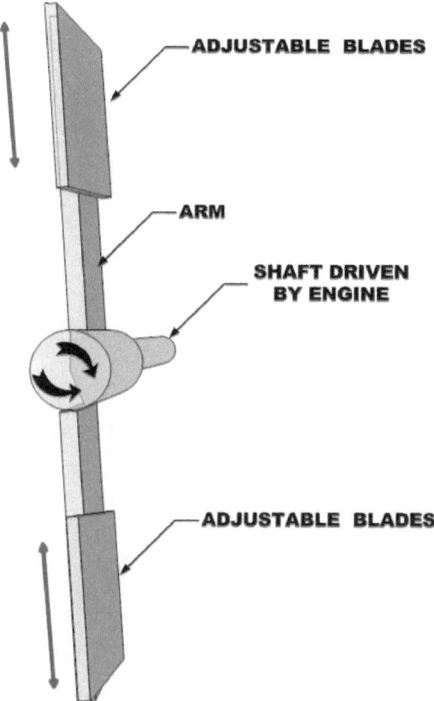

Figure 2.6 Fan brake dynamometer.

This measurement method produces results those are of good approximation of HP, but it is neither exact nor hundred percent repeatable. Simple mechanical differences such as air pressure and the amount of water present in the air can affect the measurement.

Disadvantage

Fan brakes calibrated at one measuring point used as absorption dynamometers in testing internal combustion engines have the disadvantage that a given fan will run only at one speed when the engine is delivering full power.

The power absorbed is proportional to volume swept time cube of speed of rotation. The power absorbed is also proportional to the size of blades (plates), plate distance from the center of the rotation, and upon the cube of speed. The torque absorbed in this dynamometer varies as the square:

$$T = \alpha N^2$$

$$P = \alpha N^3$$

$$P = F_d \cdot V$$

$$P = \frac{1}{2} \rho v^3 A . C_d, \quad \ldots\ldots\ldots\ldots 2.4$$

where
F_d = the force vector of drag
ρ = density of the fluid
v = velocity of the object relative to the fluid
A = reference area of the blade
C_d = drag coefficient

It can be assumed that the power absorbed is proportional to width of the plate (w). Let N represent the RPS of the fan and C a constant denoting the power absorbed by the air resistance on a unit is located at a unit distance from the axis of rotation at unit speed of revolution.

The following figure shows the blade of the fan brake dynamometer. Let us consider a small element of width "W" and height "dr."

Let
w = width of blade
dr = width of an element
N = rpm of the blade
R = distance of the element from the axis of rotation.

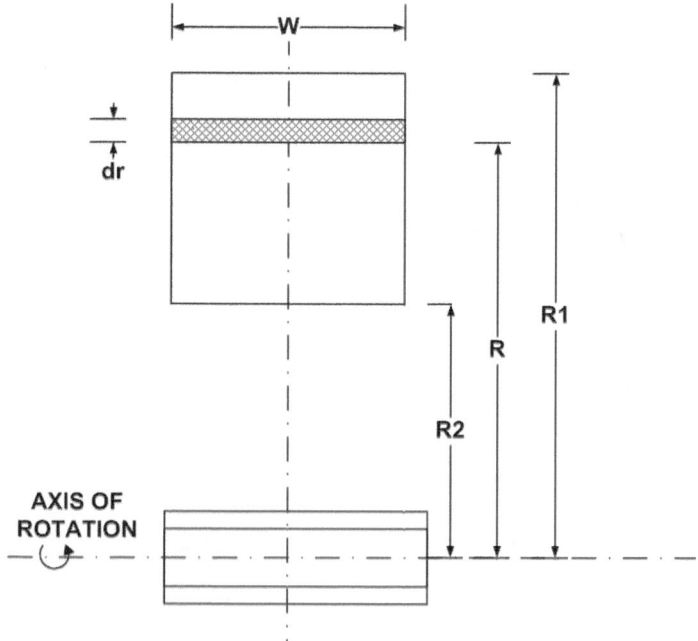

Figure 2.7 Fan blade.

Then the power absorbed by the element dr is

$$HP = N^3.C.W.R^{2..5} dr \quad \text{..................................2.5}$$

The power absorbed by the whole plate is

$$\int_{R2}^{R1} HP = N^3.C.W.R^{2..5} dr \quad \text{..............................2.6}$$

$$HP = \frac{N^3.C.W}{3.5}(R1^{3.5} - R2^{3.5}) \quad \text{......................2.7}$$

Since there are always two plates, power absorbed is

$$HP = \frac{2C}{3.5} N^3.W(R1^{3.5} - R2^{3.5}).$$

Empirically, $\dfrac{2.C}{3.5} = \dfrac{1}{68x10^6}$

$$HP = \dfrac{1}{68x10^6} N^3.W(R1^{3.5} - R2^{3.5}). \quad\quad\quad\quad 2.8$$

This formula can be subsequently applied to the boom carrying the plates.

2.5.1 Power Curve of Fan Brake

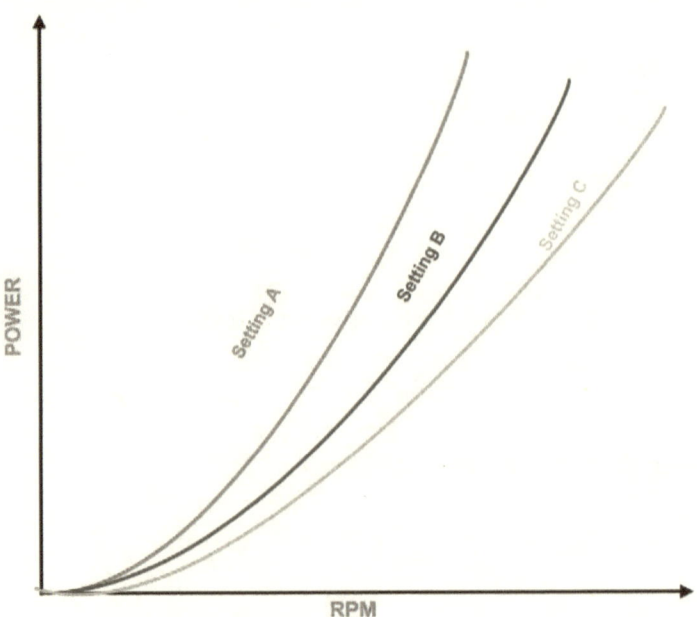

Figure 2.8 Power curve of fan brake.

2.6 Hydraulic Dynamometer

Hydraulic dynamometers are also known as *water brakes*. They are broadly classified as constant filled dynamometers such as sluice gate-type dynamometers and variable filled dynamometers such as pressure-controlled dynamometers.

2.6.1 Hydraulic Dynamometer—Sluice gate Controlled (Constant Fill)

Generally speaking, sluice gate machines are open-loop-controlled dynamometers. These machines dominated the testing of engines since early twentieth century till the widespread of electronics.

The rotor and stator are normally made of bronze, which is known for its corrosion resistance. The pair of rotor and stator can be made of chrome steel for corrosion resistance and long life. The economic way of achieving the corrosion resistance and long life will be to have the rotor and stator made of two percent Ni cast iron which has strong resistance to corrosion and wear.

The dynamically balanced rotor assembly is mounted and supported in ball bearings. The bearings are lubricated by grease. The rotor may be keyed or oil injection fitted on shaft. The low speed application will normally employ key-fitted rotor. Stators are fixed to the left-half and right-half split casings. The stators are vertically split and support the mechanism for the movement of the sluice gate. When sluice gates are fully closed, the rotor vanes are hidden.

The water leakage form the vortex chamber is prohibited by multirope graphite-based rope. The construction resembles somewhat like the stuffing box of the steam engine. The casing assembly is supported on a baseplate by means of lubricated self-aligning ball bearings housed in large trunnion bearings.

The power absorbed by the dynamometer is controlled by the movement of sluice gate. The back-and-forth movement of the sluice gates exposes the number of rotor vanes which forms the vortex in conjunction with facing stator vanes. It is worth to point out here that the dynamometer is constantly filled, and movement of the sluice gate merely exposes the number of rotor vanes to stator vanes, and in turn, power absorbed depends upon the circulating quantity of water in the number of exposed vortices.

The movement of sluice gates can be controlled by manually rotating a handwheel or by a motorized control mechanism which can be accessed from remote push buttons located on control panel in the test cell.

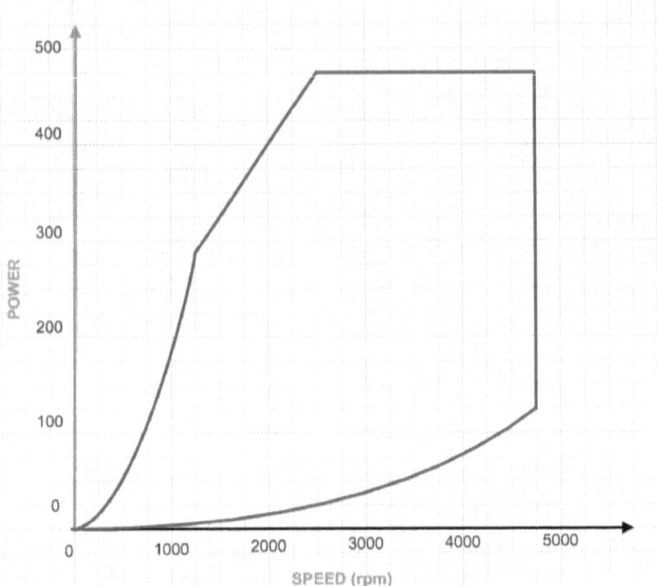

Figure 2.9 Power envelope of hydraulic dynamometer—Speed versus BHP curve.

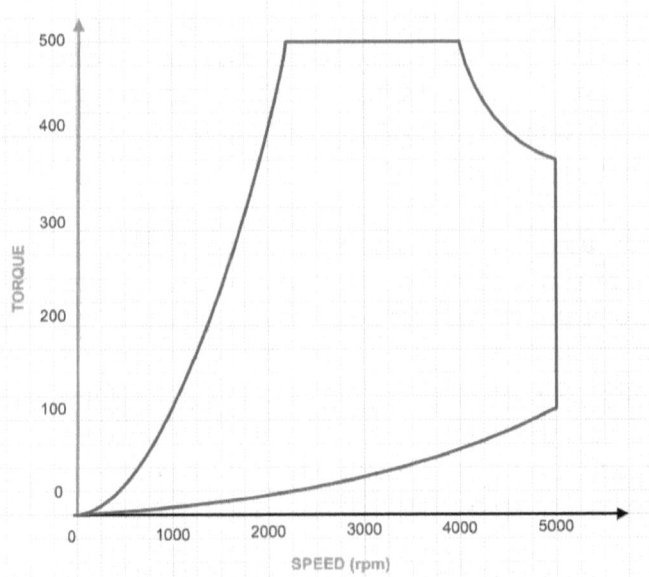

Figure 2.10 Torque envelope of hydraulic dynamometer.

2.6.2 Hydraulic Dynamometer—Outlet Valve controlled (Variable Filled)

The dynamometer working compartment consists of special semicircular-shaped vanes cast into stainless steel rotor and stators. Water flowing in a toroidal vortex pattern around these vanes creates a torque reaction through the dynamometer casing which is resisted and measured by a precision load cell. The dynamometer load is controlled by a "butterfly" water outlet valve. The power absorbed by the dynamometer is carried away by the water in the form of heat.

The loading mechanism could be a butterfly valve, whih is controlled either by a hydraulic rotary actuator powered by hydraulic power pack or by a DC servo motor.

Sluice gates are absent in this type of dynamometer. These machines can be used in closed loop with suitable feedback-controlled electronic controller.

Figure 2.11 Hydraulic dynamometer. Courtesy of Power Test.

2.7 Eddy-Current Dynamometer—Dry Gap

Jean B. L. Foucault in 1855 demonstrated the conversion of mechanical work into heat by rotating a copper disk between the poles of an electromagnet. Since 1935, this simple means of developing drag torque based on eddy current was widely exploited in dynamometer.

Eddy-current dynamometers are further classified as dry gap—and wet gap-type dynamometers. Dry gap machine has indirect cooling method. In this type of machine, the rotor rotating in the magnetic field is isolated from direct contact with water. The dynamometer casing houses twin or single magnetizing coil/coils which produces a retarding controllable magnetic field that resists the applied torque. Heat generated in this process is dissipated by cooling water which is circulated in isolated cooling passages from the rotor chamber.

Rotation of the casing is resisted by a precision strain gauge load cell that gives accurate measurement of total input torque. Detailed discussion on eddy-current dynamometer can be found in Chapter 5.

Figure 2.12 Dry gap eddy-current dynamometer. Courtesy of Horiba.

2.8 Eddy-current Dynamometer—Wet Gap

In wet gap machine, as the name implies, the gap between rotor and loss plate is wet. The rotor is partially submerged in water.

2.9 DC Dynamometer

Fundamentally, a DC dynamometer is a DC motor aka DC machine. For the purpose of engine testing, a DC machine is converted into dynamometer by simply mounting it on trunnion for the measurement of torque.

Advantages

- Motors and absorbs.
- Fast response.
- Air cooled (no water required).
- Line regenerative.

Disadvantages

- High cost (dynamometer and power amplifier).
- High inertia.
- Requires high amperage electrical service.

2.10 AC Dynamometer

An AC dynamometer is simply AC motor or AC machine mounted in trunnion for the ease of measurement of torque. Both AC and DC machines can be directly coupled to test specimen under testing with the help of transmission-type torque transducer.

Figure 2.13 AC dynamometer.

Advantages

- Motors and absorbs.
- Very low inertia.
- Very fast response.
- Line regenerative.

Disadvantages

- High cost.
- Requires high amperage electrical service.

Typical power curve of AC machine is as shown below.

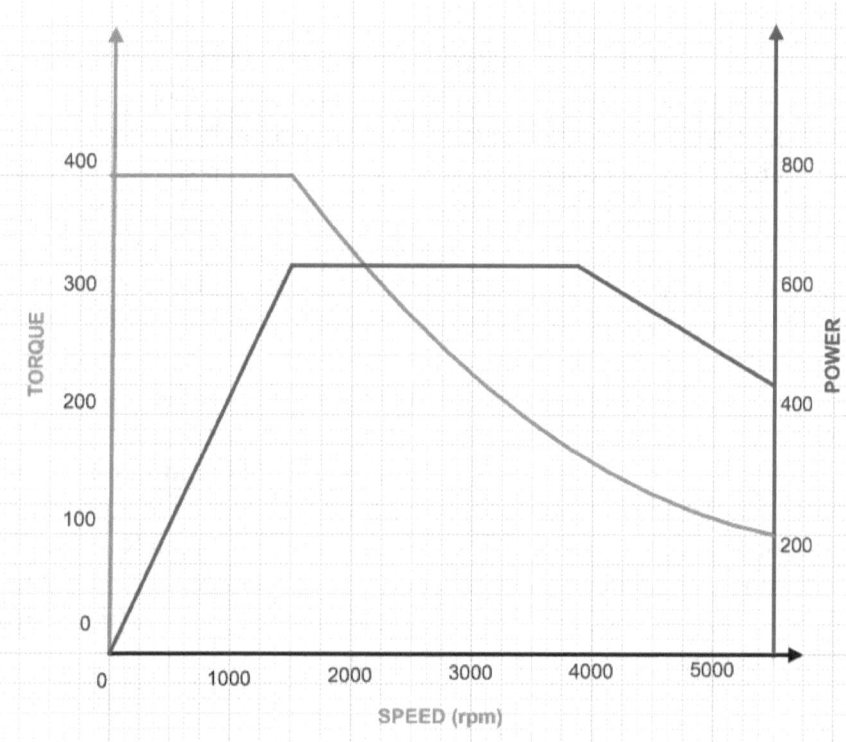

Figure 2.14 Power envelope of AC machine.

2.11 Transmission Dynamometer

Transmission dynamometer is one in which power is measured, without being absorbed or used up, during transmission. In transmission dynamometer, the work done or the energy of the prime mover is measured before it is utilized to drive the machine. In other words, the work is not absorbed while being measured. Examples are belt transmission, epicyclical, and torsion dynamometers. These dynamometers are suitable for measuring the large power produced by various prime movers.

2.11.1 Epicyclical Train Dynamometer[1]

An epicyclical gear train placed between the prime mover and the machine is used to measure the power transmitted. Consider the sun and planet type of epicyclical gear train as shown in the figure below. The sun gear B is connected to the prime mover shaft and rotates in one direction, say counterclockwise direction. The internal gear D is connected to the driven machine shaft and rotates in clockwise direction. The power is transmitted from gear B to D through the intermediate gear C revolves freely on a pin fixed to the arm A. While transmitting power, the tangential force is exerted by the gear B on the gear C and the reaction of the internal gear D on the gear C. If the friction of the pin on which gear C revolves is neglected, these two tangential forces act in upward direction and are equal in magnitude. Therefore, the total upward force on the lever arm A is equal to $2F$. The moment due to force $2F$ about the center of gear B causes the lever arm A to rotate in counterclockwise direction. This moment is balanced by suspending a deadweight W from the arm, which causes the arm to float between the stops.

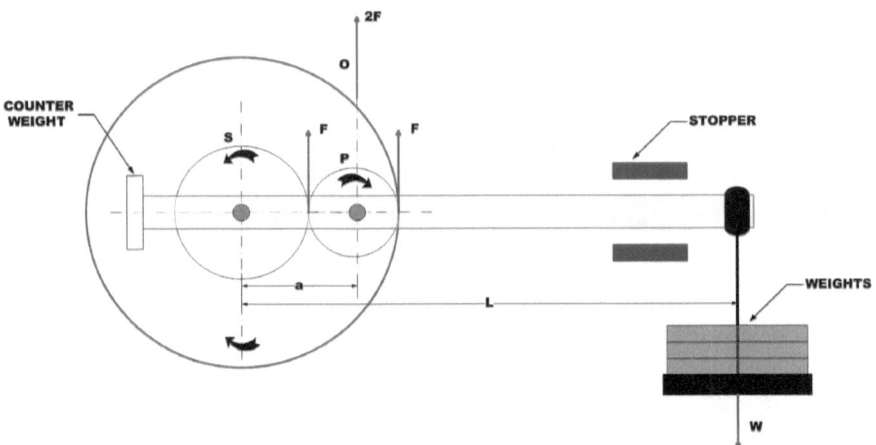

Figure 2.15 Epicyclical dynamometer.

Taking the moments about fulcrum O,

$$w.l = 2(F).a$$

or

$$F = \frac{w.l}{2.a}.$$

Therefore, power

$$P = \frac{F.2.\pi.r.N}{4500} \ HP, \ \dots\dots\dots\dots\dots\dots\dots\dots 2.9$$

where
F = tangential force in kgf
R = radius of wheel S in m
N = speed of rotation of S in rpm

In SI

$$P = \frac{F.2.\pi.r.N}{60.(1000)} \ Kw,$$

where
F = tangential force in N
R = radius of wheel S in m
N = speed of rotation of S in rpm

Substituting for F

$$P = \frac{w.l}{2\ a} \frac{2.\pi.r.N}{4500} \ HP \ \dots\dots\dots\dots 2.10$$

$$P = \frac{w.l}{2\ a} \frac{2.\pi.r.N}{60.(1000)} \ Kw \ \dots\dots 2.11$$

2.11.2 Belt Transmission Dynamometer[1]

The belt transmission dynamometer is most widely used among transmission dynamometers. It works on the principle that when belt transmits power, there exists a difference in tension between the two sides of the belt. A belt transmission dynamometer measures this difference in tensions, and let us know the amount of power being transmitted. Depending upon the kinematic arrangement, there are two types of belt transmission dynamometers:

1. Tatham belt transmission dynamometer
2. Von Hafner belt transmission dynamometer.

2.11.2.1 Tatham Dynamometer

In Tatham dynamometer, an endless belt passes over the driving and driven pulleys through two intermediate pulleys as shown in figure below.

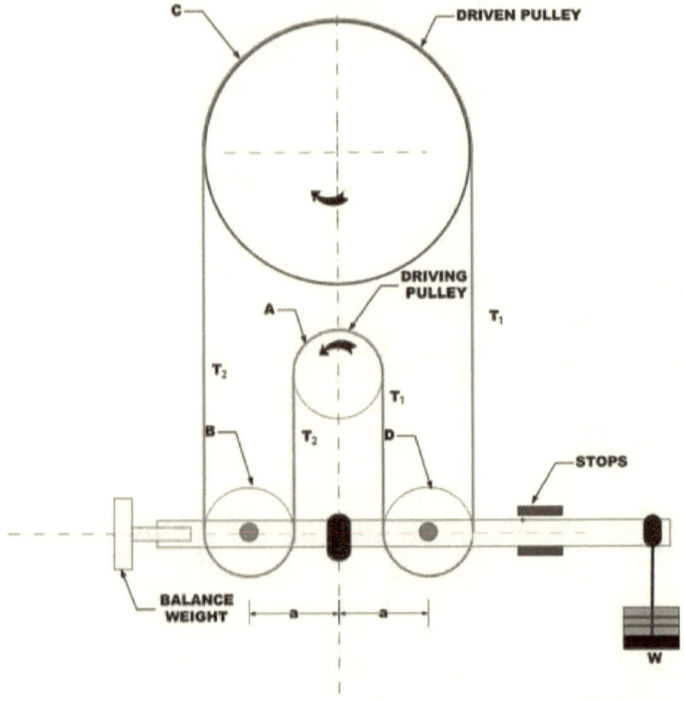

Figure 2.16 Tatham belt transmission dynamometer.

The intermediate pulleys B and D revolve about pins mounted on the lever. The lever is provided about the fulcrum O on the fixed frame. When driving pulley A rotates in counterclockwise direction, the tight and slack sides of the belt are shown in the *figure above*. The total upward forces acting on the pins of intermediate pulleys B and D are $2T_2$ and $2T_1$, respectively. The moment caused by these forces about the fulcrum O is balanced by suspending a weight W at a distance l from the fulcrum.

Taking moments about the fulcrum O, we get

$$W \times l = 2T_1 \times a - 2T_2 \times a$$

Therefore, $T_2 - T_1 = Wl/2a$

$$\text{Power } P = 2\pi N(T_1 - T_2)r/60, \quad\quad\quad\quad\quad\quad 2.12$$

where
r = radius of driving pulley A
N = speed driving pulley A in rpm

2.11.2.2 Von Hefner Belt Transmission Dynamometer

The Von Hefner transmission dynamometer is generally used for a horizontal belt drive. In this dynamometer, a continuous belt runs over driving and driven pulls through two idler pulleys mounted on a triangular-shaped frame as shown in *figure below*. The frame is free to turn about a fixed axis through point O on the line joining the centers of driving and driven pulleys. The triangular frame is connected to lever arm which is pivoted at fulcrum Ot.

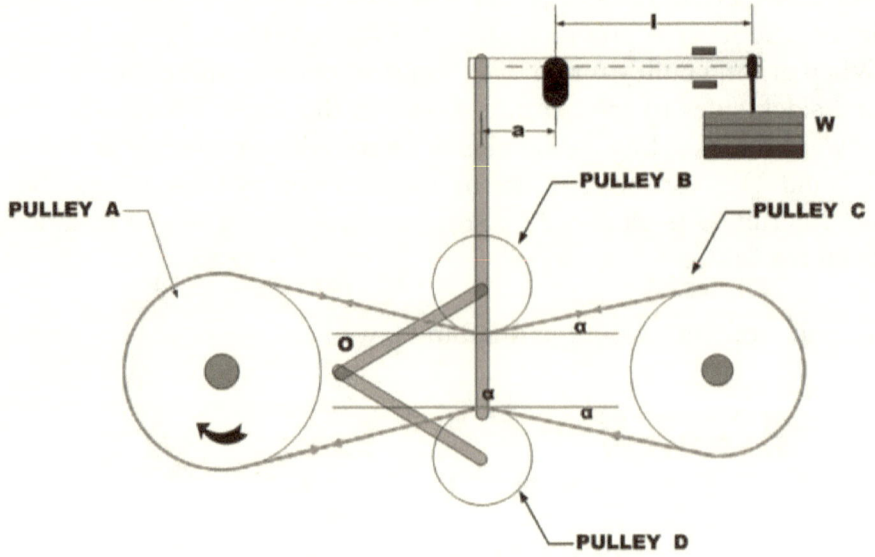

Figure 2.17 Von Hefner transmission dynamometer.

The net downward force on the idler pulley caused by the difference between the belt tension is transmitted to one end of the lever. The lever is balanced through balanced weight suspending at the other end of the lever. The dynamometer is so adjusted that lever floats horizontally. Generally, the diameters of driving and driven pulleys are of equal size so that four straight portions of the belt are equally inclined at the angle α(alpha) to the line joining the centers of driving and driven pulleys. The downward force on the pulley D due to tight side tension T_1 is $2T_1 \sin\alpha$. Similarly, the upward force on the pulley B due to slack side tension $2T_2 \sin\alpha$.

Therefore, net force

$$F = 2(T_1 - T_2)\sin\alpha$$

This net force F is acting downward on the lever arm at point P. For the equilibrium of the lever arm, taking moments about fulcrum O1, we get

$$F = 2(T_1 - T_2)\sin\alpha \times a = Wl$$

$$T_1 - T_2 = wl/2a\sin\alpha$$

$$\text{Power } P = 2\pi N(T_1 - T_2)r/60, \dots\dots\dots\dots\dots\dots 2.13$$

where
r = radius of the pulley
N = speed of the driving pulley

Closure

The dynamometers are of different types and are selected based on requirement of engine testing. The type of dynamometer to be used for a particular application depnds upon the accuracy and the type of testing to be done. The dynamometer for the production testing of an engine will be different for Research and development testing in terms of accuracy, cost, and type of instrumentation used.

Bibliography:

1. *Theory of Machines* by R. S. Khurmi, J. K. gupta. Published by S Chand & Co Ltd, 14th Edition 2005
2. E. S Eschlin, J. W. Willette, *Influence of Environment on Fan Dynamometer*.

CHAPTER 3
Hydraulic Dynamometers

3.1 Introduction

It is a well-known fact that William Froude invented the hydraulic dynamometer in 1877 and produced the first commercial dynamometers in 1881.

Brief History

William Froude (1810-1879): An Oxford graduate, he worked for Isambard Kingdom Brunel on the Great Western Railway and subsequently turned his attention to naval engineering. From his extensive research with ship-hull designs, he established the *"Froude number,"* which is the hydrodynamic equivalent of the Mach number, still used today in the design of all marine vessels. During this period, he invented the hydraulic dynamometer (1877) for testing ship engines and founded the company *Heenan & Froude* (1881). The company progressively extended its product range to include equipment to test engines for ships, cars, and aircraft.

Figure 3.1 William Froude.

3.2 Construction of Hydraulic Dynamometer

The construction of the hydraulic dynamometer essentially consists of a stator and rotor with cups/vanes. The vane angle is normally 45°. The hydraulic dynamometer can be either a sluice gate control (constant fill) or servo-controlled (variable fill) dynamometer.

3.2.1 Construction of Sluice Gate Machine

The hydraulic dynamometer consists of a stator housing carrying stator blade rings or stators. The rotor and stator are of special semicircular-shaped vanes cast into stainless steel rotors and stators. Special designs may involve the bronze stator and impellor/rotor. Water flowing in a toroidal vortex pattern around these vanes creates a torque reaction through the dynamometer casing which is resisted and measured by a precision load cell. This type of dynamometer is also called as constant-filled dynamometer as water level is always full. The load is controlled by the opening of the sluice gate which exposes the more and more rotor vanes as it keeps opening.

The rotation of rotor creates the pumping action, and water is pumped with high velocity and flows over the stator vanes, connected to casing, in a toroidal vortex pattern and creates a torque reaction through the dynamometer casing which is resisted and measured by a precision load cell.

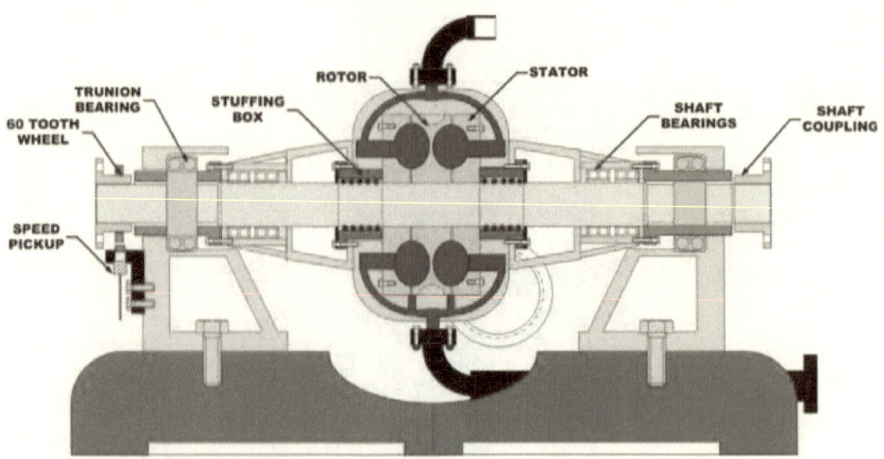

Figure 3.2 Sluice gate-controlled hydraulic dynamometer.

3.2.2 Pressure-controlled Hydraulic Dynamometer

The hydraulic dynamometer consists of a stator housing carrying stator blade rings or stators. The rotor and stator are of special semicircular shaped vanes cast into stainless steel rotors and stators. Special designs may involve the bronze stator and impellor/rotor. Water flowing in a toroidal vortex pattern around these vanes creates a torque reaction through the dynamometer casing which is resisted and measured by a precision load cell. The dynamometer is mounted in large trunnion bearings. The trunnion bearing supports the stator housing like a cradle. The dynamometer load is controlled by a "characterized" water outlet valve and operated by a closed-loop electrohydraulic servo system. Some manufacturers operate this valve by using electric servo-controlled DC motor.

The shaft carrying the rotor is fitted with two half couplings on either side which facilitates connection of test specimen. Dynamometer can be connected from either end. In most of the cases, the couplings and the main rotor/impeller are mounted on shaft by oil injection method. This interference fit method eliminates the stress rising keyways and the mounting holes on the dynamometer rotor and shaft.

Dynamometers are equipped with water pressure switch, bearing temperature monitoring thermocouples, water inlet and outlet temperature monitoring thermocouples to ensure safety against high bearing temperature, and high water temperature.

The dynamometer working compartment consists of special semicircular-shaped vanes cast into stainless steel rotors and stators. Rotor is connected to the test specimen under test, usually an IC engine. The rotation of rotor creates the pumping action, and water is pumped with high velocity and flows over the stator vanes, connected to casing, in a toroidal vortex pattern and creates a torque reaction through the dynamometer casing which is resisted and measured by a precision load cell.

The dynamometer load is controlled by a "butterfly" type water outlet valve and operated by a closed-loop electrohydraulic servo system or by a DC electric servo motor.

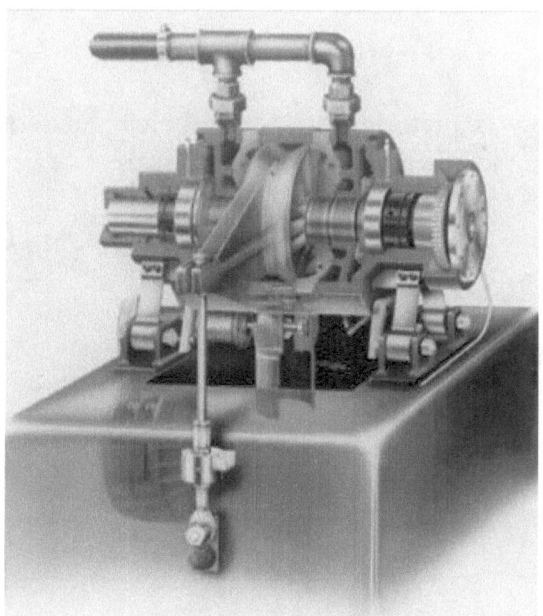

Figure 3.3 Cross section of pressure-controlled hydraulic dynamometer. Courtesy of Horiba

3.3 Water Movement Inside the Dynamometer

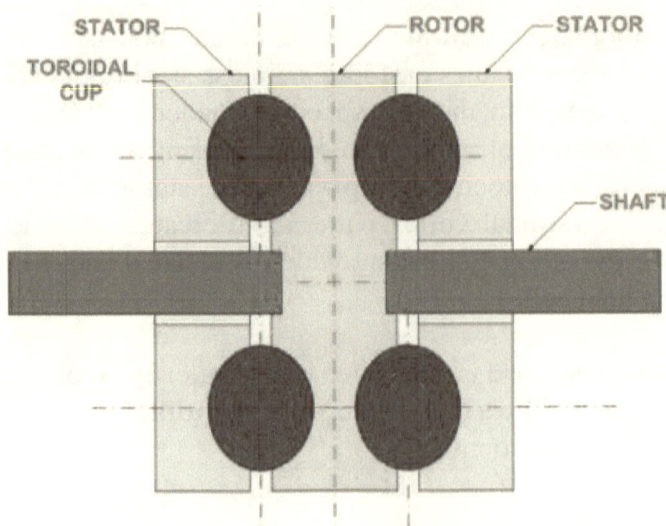

Figure 3.4 Vortex chamber.

The water flow is pictorially depicted below, understand how the water is moving inside the toroidal shape cups.

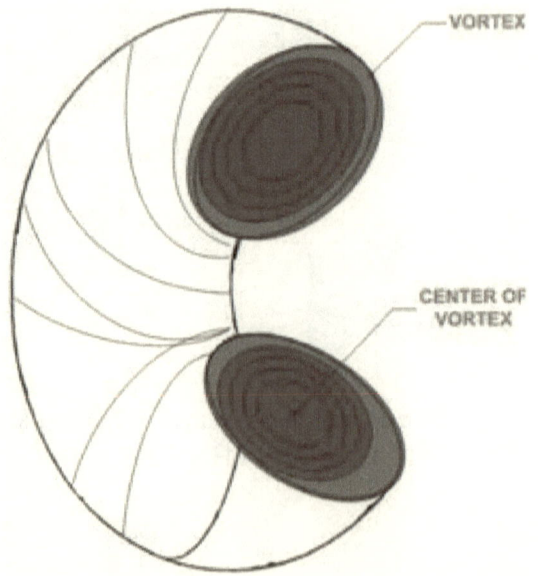

Figure 3.5 Water flow in vortex around the periphery.

The following figure clarifies in more detail about how the water flow takes place inside the cups and around the periphery of the dynamometer. The blue arrow indicates that the water will flow along with rotor in peripheral direction. The red arrow indicates how the water flows inside the toroidal cups.

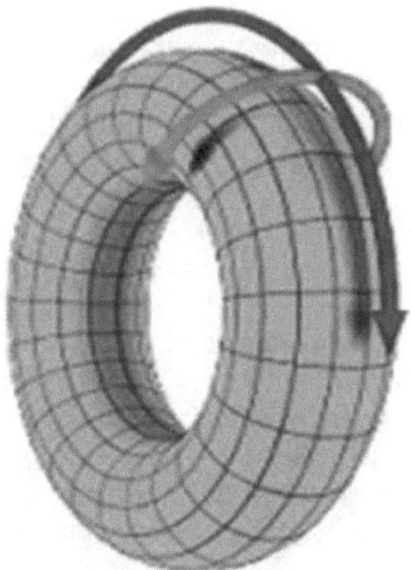

Figure 3.6 Water flow inside the dynamometer.

3.4 Inertia of Hydraulic Dynamometer and Its Effect

Inertia of hydraulic dynamometer is less compared to equivalent size of eddy-current machine. The inertia of dynamometer governs the response time to changes in load. Lower the inertia, higher is the response to load changes. The response of load control valve also plays major role in load control response of the dynamometer.

In case of hydraulic dynamometer, the total inertia of rotating mass included the inertia of rotor, rotor shaft, couplings on end, keys, bolts, and nuts if rotor is bolted to shaft instead of being oil injected. In addition, water entrapped in vane pockets also needs to be accounted for. The mass of fluid contained in the impeller should be added to the mass of the metal of shaft. The polar inertia I of the impeller disk of diameter D and thickness T or the weight W has the inertia:

$$I = \rho \left(\frac{\pi}{32}\right) D^4 T \quad \ldots\ldots\ldots\ldots 3.1$$

$$I = \frac{w.D^2}{8g} \frac{lb}{ft}/sec^2. \quad \ldots\ldots\ldots\ldots 3.2$$

The impeller is rigidly fixed to the shaft, and the compressibility of the liquid contained in the vane pockets of impeller is negligible. Therefore, no shaft windup or compliance needs to be introduced.

Shaft Inertia

$$I = \frac{M.R^2}{4} = \frac{MD^2}{8} \quad \ldots\ldots\ldots\ldots 3.3$$

3.5 Power and Torque Envelopes

Power and torque curves of hydraulic dynamometer are shown below. The following curve indicates the power envelope of the hydraulic dynamometer. The envelope has five segments. Let us see what each segment tells us about the dynamometer characteristics.

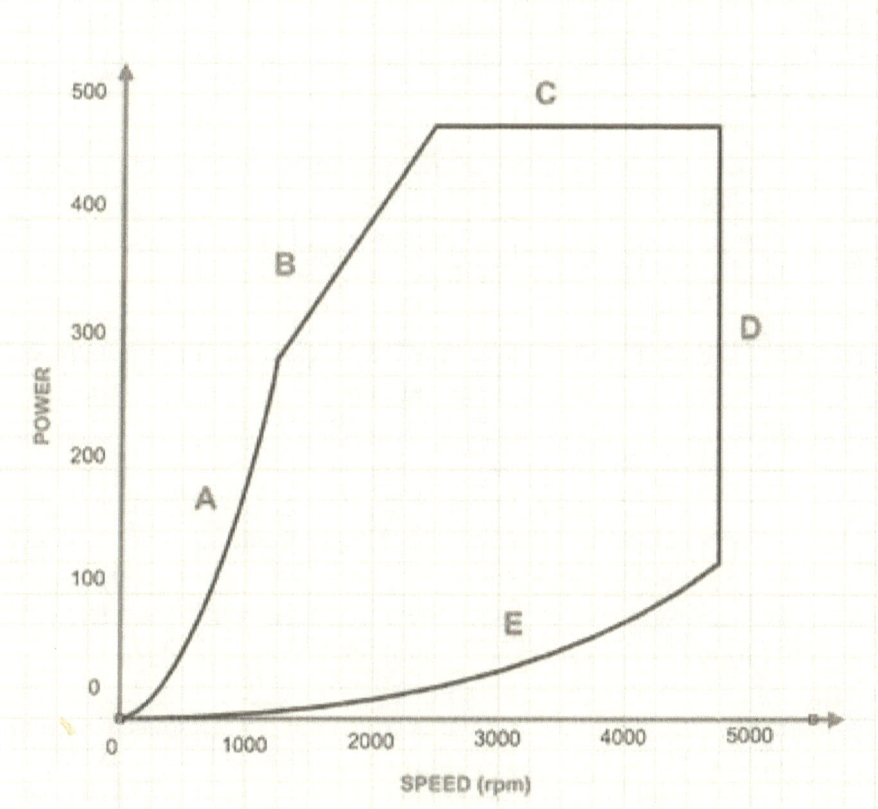

Figure 3.7 Typical power curve of hydraulic dynamometer.

The analysis of above figure, that is, power envelope, conveys the deeper meaning about each segment.

Let us consider the each segment of the power curve.

Segment A: This segment indicates how much maximum power can be achieved by a given dynamometer or the dynamometer under consideration. The nature of the curve is dependent on the following parameters:

1. Amount of water that can be accommodated by a whirl chamber. The whirl chamber is the space formed by each half cup on rotor and stator.
2. Geometric dimensions of the toroidal space.
3. The speed at which the dynamometer rotor is running.

Segment B: This segment of the power curve indicates the maximum torque achievable by the dynamometer. This is dependent on the following factors:

1. Shaft design, maximum permissible load on shaft which carries the rotor.
2. Hoops stress of the whirl chamber or toroidal space.
3. Pressure of the water in the whirl chamber.

Segment C: Rated power, this is what a given dynamometer will produce under installed setup and dependent on the following factors:

1. Maximum cooling water that can be circulated which is ultimate media to carry the heat generated due to power absorbed. A slight increase in power limit can be achieved by increasing the inlet water pressure to the dynamometer.
2. Hoops stress of the whirl chamber or toroidal space.

Segment D: This segment indicates the "maximum speed" limit of dynamometer and is dependent on the following parameters:

1. Bearing types and lubrication method adopted for lubrication of bearings.
2. Critical speed of the shaft or the complete system.
3. Dynamometer mounting arrangement.
4. Hoops stress developed in the whirl chamber.

Segment E: This segment indicates the minimum power to be absorbed by the dynamometer, and it is dependent on the following parameters. In this case, the loading mechanism of the dynamometer is not active.

1. Water inlet pressure and quantity.
2. Speed at which dynamometer is rotating.
3. Rotor inertia of the rotor.

In case of a particular hydraulic dynamometer, certain design parameters are fixed while designing the dynamometer.

The maximum power and the minimum power curves are fixed for a particular model and cannot be changed in practice or during the use of the dynamometer at site. The dynamometer comes with fixed rotor and stator which has predesigned power absorption capacity.

From the above discussion about each segment, the power curve now can be represented as:

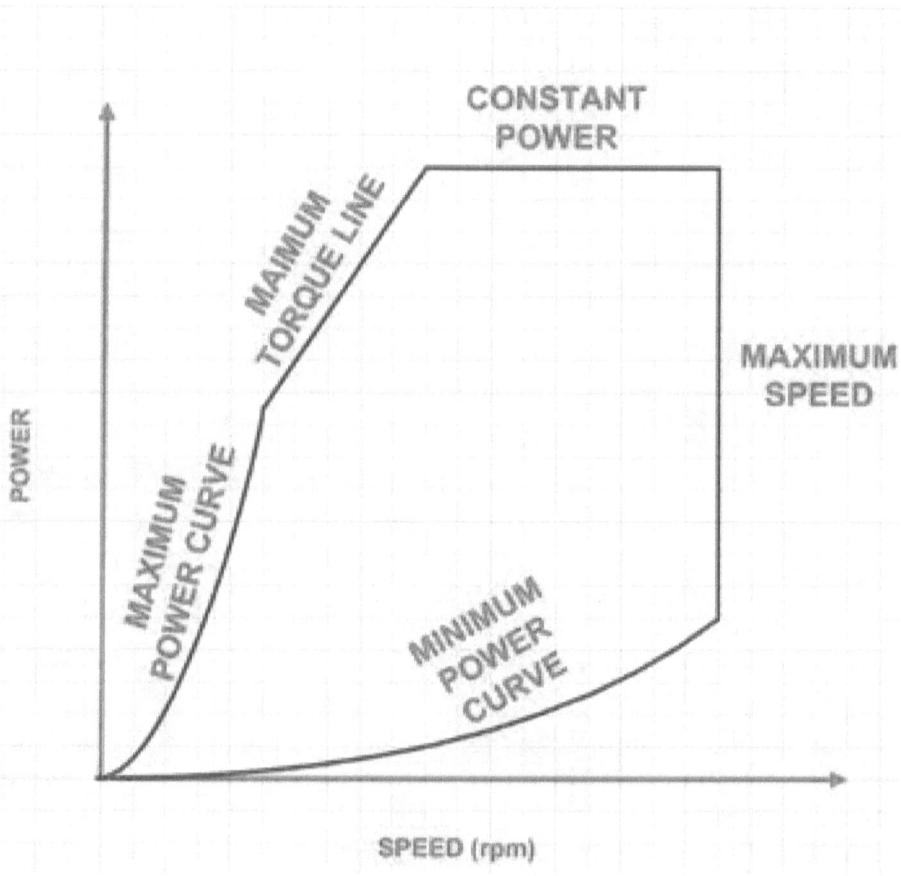

Figure 3.8 Hydraulic dynamometer power curve.

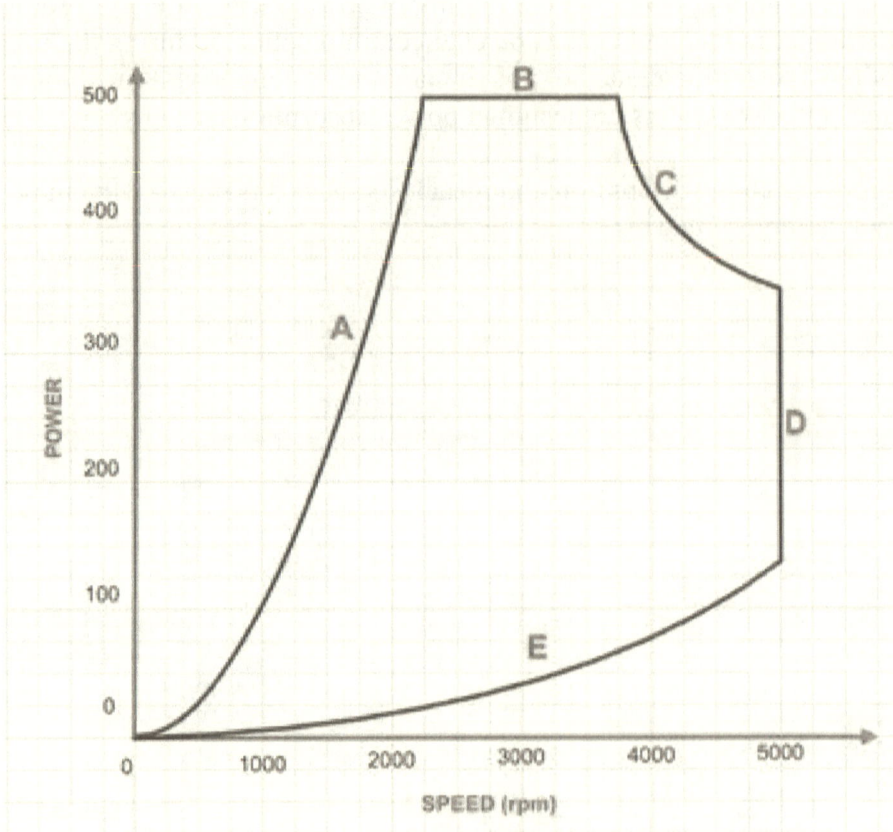

Figure 3.9 Torque curve of hydraulic dynamometer.

3.6 Direction of Rotation

The hydraulic dynamometers are typically unidirectional, meaning they absorb the power in one direction of rotation. The hydraulic dynamometers are not effective if their direction is reversed other than what is specified by the manufacturers. This is due to the angle of the vanes which are oriented in one direction. If one needs to have the bidirectional hydraulic dynamometer, then extra pair of stator and rotor is added. The torque rating of the bidirectional dynamometer is usually lower than that of unidirectional dynamometers.

The dynamometers are available with 90° vane angles for bidirectional use, but their power capacity is much smaller for the same size of the rotor and stator.

3.7 Method of Load Control

The load control mechanism is a device on or inside of the dynamometer to vary the load exerted on the specimen under test which is normally an IC engine. This depends on the type of dynamometer and is described below.

3.7.1 Sluice Gate Control

The method of control in constant fill dynamometer or typically in sluice gate-controlled machine is by the movement of sluice gates. The sluice gates cover the impeller or rotor vanes exposure to stator vanes. When a sluice gate is moved, more and more rotor vanes get exposed to stator vanes, thus forming more number of vane pockets and more water is churn, thus imparting the load/resistance to rotation.

The movement of the sluice gates can be manual or remote controlled from the control panel. The movement of sluice gates inside of dynamometer is either coming closer or moving apart from each other, thus closing the impeller or opening the later.

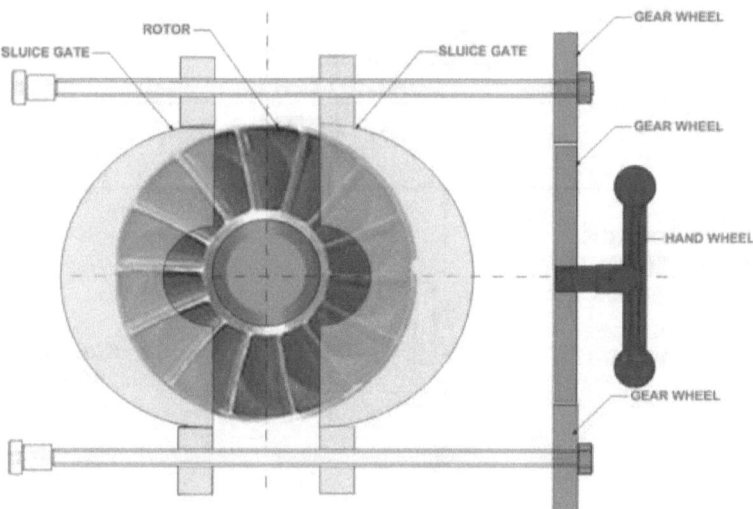

Figure 3.10 Sluice gate control.

3.7.2 Pressure-controlled Dynamometer

The load control is achieved through a specially designed butterfly valve which is either controlled by electrohydraulic servo control or DC electric servo control. In this method of control, a closed-loop operation is possible, and dynamometer load control capability is enhanced from open-loop control to closed-loop control. In closed loop, load control dynamometer can be run in various modes, such as

1. constant speed control
2. constant torque control
3. power law control or $T\alpha N^n$; normally, $n = 2$ but can be varied to simulate propeller or wind resistance approximately.

3.7.2.1 Electric Servo

The variable fill dynamometer is controlled by outlet butterfly valve. The butterfly outlet valve is controlled by electric servo-controlled motor to vary the loading on the test specimen, usually an engine. The following figure shows the schematics of electrical servo-controlled outlet valve.

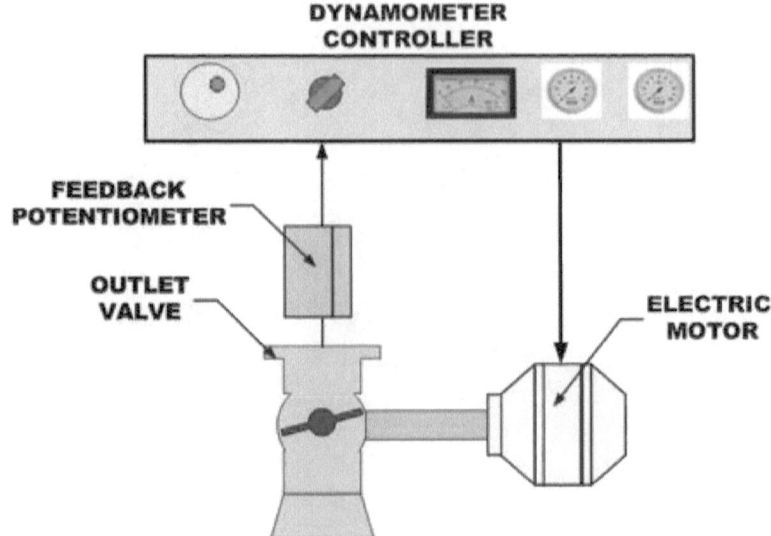

Figure 3.11 Schematics of electrical servo control.

3.7.2.2 Electrohydraulic Servo Control

Some of the hydraulic dynamometers employ electrohydraulic servo control valve to control the outlet of the dynamometer to achieve the desired loading. This involves the hydraulic power pack, servo valve, and hydraulic actuator. The servo valve controls the direction of the outlet butterfly valve of the dynamometer in conjunction with hydraulic actuator. The feedback potentiometer sends the signal about the exact position of outlet butterfly valve to the controller. The following figure shows the schematics of electrohydraulic control of the outlet valve.

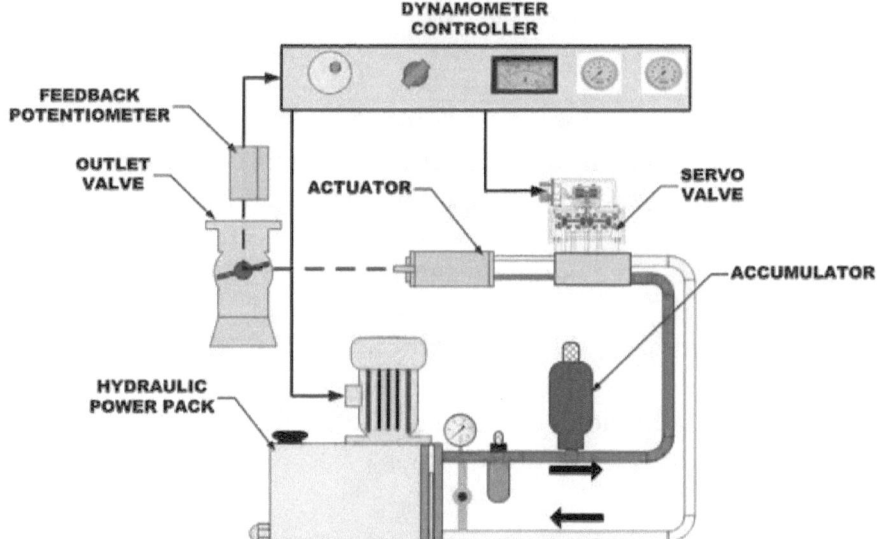

Figure 3.12 Schematics of electrical hydraulic servo control.

The dynamometer in the following figure shows an example of a dynamometer employing electrohydraulic servo control mechanisms. The dynamometer is known for its fast response in controlling the torque and speed demanded by the test engineer.

Figure 3.13 Hydraulic dynamometer with electrohydraulic control valve. Courtesy of SAJ Test Plant Pvt. Ltd., Pune, India.

3.8 Cavitation in Hydraulic Dynamometer

Majority of hydraulic dynamometers are equipped with vane-type rotor and operate at very high speed and are subjected to cavitation phenomenon. Cavitation occurs when the local static pressure in the fluid reaches a level below the vapor pressure of the liquid at the actual temperature.

When a dynamometer is rotating at high speed, the water pressure is developed. If the water pressure drops below vapor pressure or its temperature increases, it begins to vaporize, just like boiling water. The vaporization does not cause the damage. The damage happens when the bubbles can't escape so they implode, causing physical damage to the dynamometer internals or the rotor. A combination of temperature and pressure constraints will result in cavitation in any system. This is undesirable in working condition. The vaporization causes a loud noise because the bubbles are imploding and making the liquid to move faster than the speed of sound. The first appearance of cavitation is called cavitation inception.

Mechanism of Cavitation

Step 1: Formation of Bubbles Inside the Liquid Being Circulated Through Dynamometer

The bubbles form inside the liquid when it vaporizes; this is also called a phase change from liquid to vapor. The vaporization of liquid inside a closed container can occur for two reasons.

1. If either pressure on the liquid surface decreases to a level of the liquid vapor pressure or pressure on the liquid surface falls below the vapor pressure at the operating temperature.
2. The temperature of the liquid rises, raising the vapor pressure such that it becomes equal to or greater than the operating pressure at the liquid surface.

For example: If water at room temperature is kept at 25°C (77°F) in a closed container and the system pressure is reduced to its vapor pressure 0.036 bar (approximately 0.52 psia), the water starts vaporizing, or if the operating pressure remains constant at approximately 0.036 bar (0.52 psia) and the temperature of the water is allowed to rise above 77°F, then the water begins to vaporize.

Similar to closed container, vaporization can occur in the hydraulic dynamometer when the local static pressure reduces below the vapor pressure of the water at operating temperature.

Step 2: Growth of Bubbles

If there is no change in operating conditions, new bubbles will continue to form, and old bubbles grow in size. Due to the rotors rotating action, the bubbles achieve very high velocity.

Step 3: Collapse of Bubble

As the bubbles start to move along the vane, the pressure around the bubbles begins to increase until a point where the pressure outside the bubble is greater than the pressure inside the bubble. The bubble collapses the process is not an explosion but rather an implosion, that is, inward bursting.

As bubbles collapse, the surrounding liquid rushes to fill the void forming a liquid micro jet. This micro jet ruptures the bubble with a great force so that hammering action occurs. The highly localized hammering effect can pit the dynamometer rotor.

After the bubble collapses, a shock wave emanates outward from the point of collapse. This shock wave is what we actually hear and call it as "cavitation." To summaries, the mechanism of cavitation is all about formation, growth, and collapse of the bubbles inside the water being circulated

The definitions of the following terms will enable the clear understanding of the mechanism of cavitation.

Thoma's Number

Thoma's number is nondimensional number. Cavitation will not be present if critical value of Thoma's number σ is exceeded. Thoma's number is given by the following expression:

$$\sigma = \frac{P_S - P_V}{\dfrac{\rho V^2}{2g}} \quad \text{..........................3.4}$$

where
σ = Thoma's cavitation number
P_s = static pressure of the fluid/water, kg/m²
P_v = cavitation inception pressure. This is generally considered as vapor pressure in the liquid, kg/m²
V = velocity of flow, m/s
ρ = desnity of the fluid, kg/m³
g = acceleration due to gravity, m/s²

The critical value of the Thoma's number has been determined to be of order 1.0-1.5 for hydraulic machines, especially for the centrifugal pumps. There is no specific measurement of Thoma's number for hydraulic dynamometer. For any practical purpose, it is better to have highest Thoma's number as possible.

A mere observation of Thoma's formula will tell us that to achieve higher number, one needs to have

1. Static pressure of the water should be as high as possible. However, this is limited by the mechanical design constraints of seals and joints used in hydraulic dynamometer.
2. Fluid (water in case of dynamometer) temperature should be as low as possible to decrease the water vapor pressure.
3. Fluid flow velocity should be as low as possible. This is directly proportional to the speed of the dynamometer (RPM). The peripheral velocity of water is directly proportional to the r.p.m. and increase as r.p.m. increases. Therefore, in case of hydraulic dynamometer, the cavitation chances are at very high rpm.

Avoiding Cavitation

1. Increasing the gap between the actual local static pressure in the liquid and the vapor pressure of the fluid at the actual working temperature and vaporization and cavitation may be avoided.

    The ratio between static pressure and vaporization pressure, which is an indication of the possibility of vaporization, is often expressed by the cavitation number.

2. Reducing the temperature of the working fluid—water

    The vaporization pressure is highly dependable on the fluid temperature.

Effect of Cavitation

Cavitation can accelerate the corrosion, pitting, and erosion of the dynamometer and dynamometer parts, shortening the life of the dynamometer rotor.

3.9 Selection of Dynamometer

The selection of the dynamometer is a scientific process and needs a careful study of the engine capacities to be tested on the dynamometer. The following entities for the engines to be tested should be identified and noted for the purpose of the selection of dynamometer:

1. The engine maximum power and the corresponding speed;
2. The engine maximum torque and the corresponding speed;
3. The engine minimum power and the corresponding speed;
4. The engine minimum torque and the corresponding speed;
5. The lowest test point that test engineer intends to test on the dynamometer.

The above entities should be noted for all the engines that the test engineer intends to test on the dynamometer which he is planning to be installed. If the test engineer has the range of the engines to be tested, it may be difficult to accommodate all the engines on one dynamometer. Sometime, it is difficult to accommodate both test points of high torque at low rpm or speed and the maximum power at high speed of the engine.

Control Accuracies

The dynamometer should be capable to maintain the control accuracies at the lowest test point. This could be a conflicting issue if the test point of a particular engine is outside the control limit. This needs to be carefully studied and accommodated while selecting the dynamometer.

Curve Fitting

It is the best practice to plot the engine curves on the dynamometer power envelope. The superimposing of the engine curves will give the clear understanding to the test engineer about how many engines he is able to accommodate on a particular dynamometer.

The following figure indicates the engine curves superimposed on a typical dynamometer power envelope.

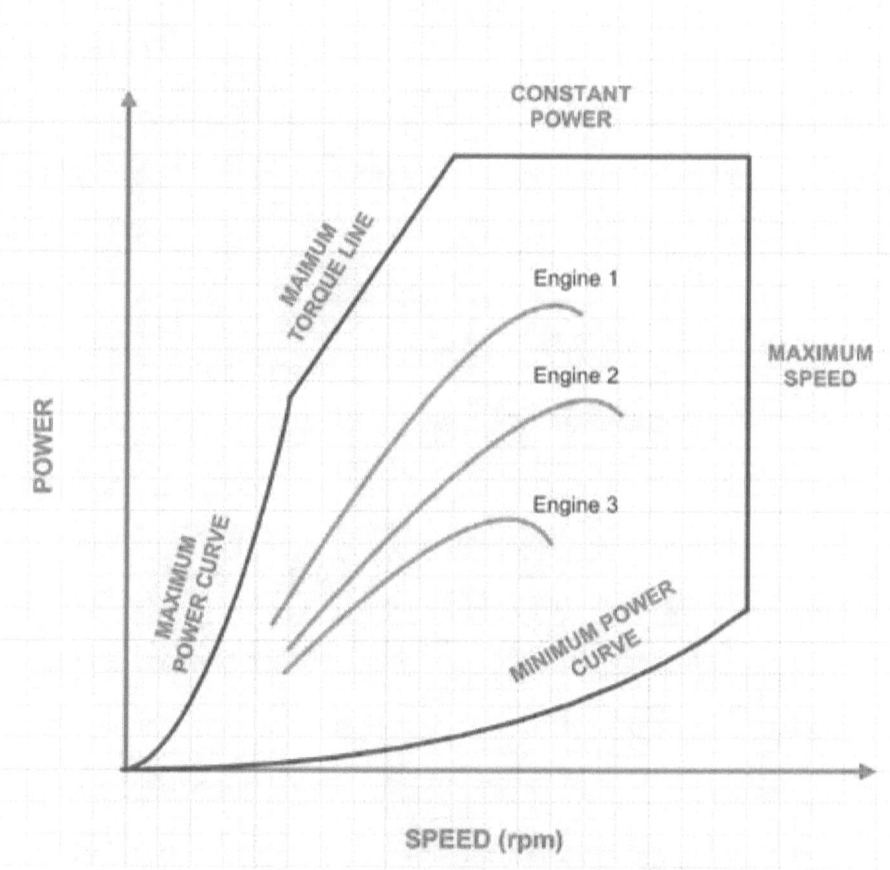

Figure 3.14 Dynamometer selection.

The above exercise of the plotting engine power curve on dynamometer power curve will help the test engineer to select the proper dynamometer for his application.

3.10 Application of Hydraulic Dynamometer

The hydraulic dynamometers are widely used in testing of automotive, industrial engines. The very large marine diesel engines testing is tested on hydraulic dynamometers. The hydraulic dynamometers are preferred for testing of the engines having power more than 3,000 kW due to its good power to weight ratio. The following figure shows a hydraulic dynamometer used for testing of industrial engines.

Figure 3.15 Typical hydraulic dynamometer installations. Courtesy of SAJ Test Plant Pvt. Ltd., Pune, India.

Figure 3.16 Typical Hydraulic Dynamometer for testing diesel engine
Courtesy of SAJ Test Plant Pvt. Ltd., Pune, India.

Closure

The hydraulic dynamometer is one of the most suited means for the testing of the engines and plays very important role in the engine testing for production, R&D, quality, and after-sales market, that is, repairs workshop and garages. The hydraulic dynamometers are known for the trouble-free low maintenance requirement dynamometers. With modern instrumentation and control system, the hydraulic dynamometer becomes a great choice for the test cell engineer.

Bibliography

1. William Froude Image—Wikipedia web site.
2. William Froude information: Britannica encyclopedia web version.

Further Reading

1. Pump handbook _Igor Karasik.—for cavitation information
2. Francois Avellan, *Introduction to cavitation in Hydraulic Machinery*.

CHAPTER 4
Eddy-current Dynamometer

4.1 Introduction

The phenomenon of eddy currents was discovered by French physicist Leon Foucault in 1851, and for this reason, eddy currents are sometimes called Foucault currents. Eddy currents are electric currents induced in conductors when a conductor is exposed to a changing magnetic field due to relative motion of the field source and conductor or due to variations of the field with time. This can cause a circulating flow of electrons, or current, within the conductor. These circulating eddies of current have inductance and thus induce magnetic fields.

The eddy currents generate heat as well as electromagnetic forces (EMFs). The induction furnace uses heat produced for induction heating. The EMFs are used to give a strong braking effect in eddy-current dynamometer.

Michael Faraday discovered that when a magnetic field passes through a conductor or when a conductor passes through a magnetic field, an electric current will flow through the conductor if there is a closed path through which the current can circulate. The term *eddy current* is derived from analogous currents produced when dragging an object in water. The turbulence created by dragging an object is known as eddy currents. The eddy currents have a tendency to oppose the cause producing them.

4.2 History of Eddy-current Dynamometer

The story of the eddy-current dynamometer and its invention is dated around 1920 when Martin and Anthony Winthers were granted patents for almost three hundred mechanical devices. These included the first induction coupling, a magnetic clutch. They are chiefly known for the invention of the eddy-current dynamometer.

4.3 Principle of Operation

The power absorbed by the dynamometer is directly proportional to the input current flowing through the magnetizing coils which are encapsulated and housed within the casing. The varying the amount of current flowing through the coils varies the amount of magnetic field generated by the coils. As shaft of dynamometer is rotated by the engine or test specimen under test, the rotor mounted on shaft cuts the magnetic field, and the eddy currents are generated within the loss plate or heat carrying members. The eddy current acts to oppose the rotation of the shaft, thus applying a load to the test specimen or the engine. The absorbed engine power is converted into heat in heat carrying members and dissipated by cooling water flowing through it.

Eddy currents, also called Foucault currents, are currents induced in conductors, opposing the change in flux that generated them. It is caused when a conductor is exposed to a changing magnetic field due to relative motion of the field source and conductor or due to variations of the field with time. This can cause a circulating flow of electrons, or a current, within the body of the conductor. These circulating eddies of current create induced magnetic fields that oppose the change of the original magnetic field due to Lenz's law, causing repulsive or drag forces between the conductor and the magnet. The stronger the applied magnetic field or the greater the electrical conductivity of the conductor or the faster the field that the conductor is exposed to changes, then the greater the currents that are developed and the greater the opposing field.

Figure 4.1 Eddy-current dynamometer. Courtesy of Horiba

4.4 Electrical Laws Associated with the Eddy-current Dynamometer

Faraday's First Law: Whenever magnetic flux linking with a coil changes with time, an EMF is induced in the coil, or whenever a moving conductor cuts the magnetic flux, an EMF is induced in the conductor.

Faraday's Second Law: The magnitude of the induced EMF is equal to the product of the number of turns of the coil and the rate of change of flux linkage.

Lenz's Law: The direction of the induced EMF by electromagnetic induction is in a direction to oppose the main cause producing it.

Fleming's Right-hand Rule: This rule helps in deciding the direction of the induced EMF. Hold the right-hand thumb, forefinger, and the middle finger set at right angles to each other, and the *thumb* points the direction of the *motion* of the conductor, and the *forefinger* points the direction of the *field*, and the *middle finger* points the direction of the induced *EMF*.

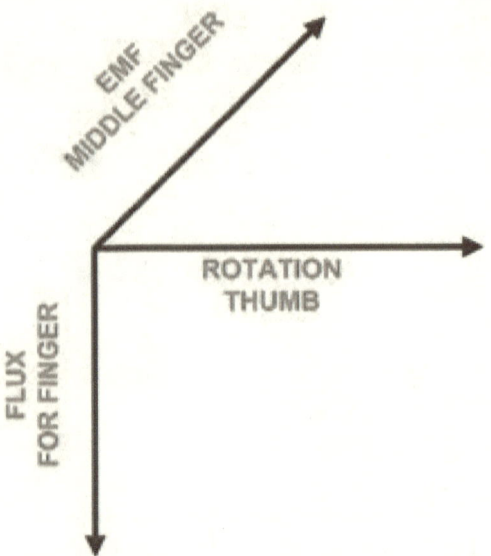

Figure 4.2 Fleming's right-hand rule.

Fleming's Left-hand Rule: This rule helps in deciding the direction of force acting on a conductor. Hold the left-hand thumb, forefinger, and the middle finger set at right angles to each other, and the *thumb* points the direction of the *force* acting on the conductor, and the direction of the *forefinger* points the direction of the *magnetic field*, and the *middle finger* points the direction of the *current* in the conductor.

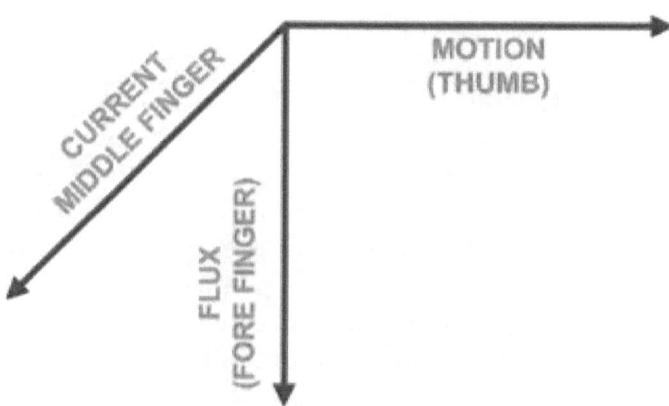

Figure 4.3 Fleming's left-hand rule.

4.5 Construction of Eddy-current Dynamometer

The eddy-current dynamometer consists of the following major components or subassemblies:

1. rotor and shaft assembly
2. casing/carcass assembly
3. bedplate/baseplate assembly.

Figure 4.4 Water-cooled twin coil eddy-current dynamometer. Courtesy of SAJ Test Plant Pvt. Ltd., Pune, India.

4.5.1 Rotor and Shaft Assembly

The simplified structure of rotor assembly provides a low inertia which is mainly responsible to fast response of the eddy-current dynamometer. The finger-shaped/spiked rotor is manufactured from highly permeable magnetic steel plate and attached to shaft. Rotor is attached to shaft by employing one of the following methods:

1. *Friction welding*: Two stub shafts are friction welded to spiked rotor and then machined. Rotor after machining looks like a homogenous single piece.
2. Rotor is mounted on shaft with interference fit by oil injection method. The interference fit is good enough to withstand the maximum designed torque of the dynamometer.
3. Two stubs of shaft are bolted and suitably locked.

Rotor and shaft are manufactured from a nonmagnetic steel to prevent the bearings from becoming magnetized. Half couplings are provided on both ends of the shaft to facilitate the connection to test specimen on either end of the eddy-current dynamometer. Rotor assembly is balanced using stage balancing to reduce the vibrations due to imbalance in rotor assembly. Bearing elements are selected considering the maximum speed of operation, bending load due to overhang of the shaft couplings and maximum torque induced on the shaft.

Figure 4.5 Cross section of WT dynamometer. Courtesy of Horiba.

4.5.2 Casing/Carcass Assembly

Carcass or the casing could be a single piece or two pieces depending on a single coil or twin coil design. Casing houses magnetic coil which is encapsulated with high quality rubber compound, usually silicon based. Casing forms a cooling chamber with heat carrying member. This cooling chamber is designed in such a way that it dissipates the heat to water being circulated and maintains proper working temperature for optimum performance of the dynamometer. It also houses the temperature monitoring sensors to monitor the temperature of magnetic coil as well as heat carrying members. In addition to the water cooling, air ventilation is implemented which circulates the air inside the dynamometer. Fanning action of the rotor also helps circulating the air inside the machine. Different manufacturers employ the ventilation slots either on the periphery or on the side of the casing.

The rotor of eddy-current dynamometers is also sometimes constructed as drum-type construction.

Figure 4.6 Drum-type rotor. Courtesy of Horiba.

4.5.3 Baseplate Assembly

The baseplate assembly supports the casing/carcass assembly that houses rotor assembly. The carcass is connected to bedplate via trunnion bearings or flexure supports. The baseplate also helps in absorbing the vibrations and transferring to the foundation.

4.6 Inertia of Eddy-current Dynamometer

The inertia of the eddy-current dynamometer is total inertia of the rotor shaft and the spiked rotor. The coupling inertia is added to this.

4.7 Power and Torque Envelopes

The power and torque curves differ in hydraulic dynamometer and are discussed below.

4.7.1 Power Curve of Eddy-current Dynamometer

The following figure shows the dynamometer characteristics curve, which is graph of speed versus power.

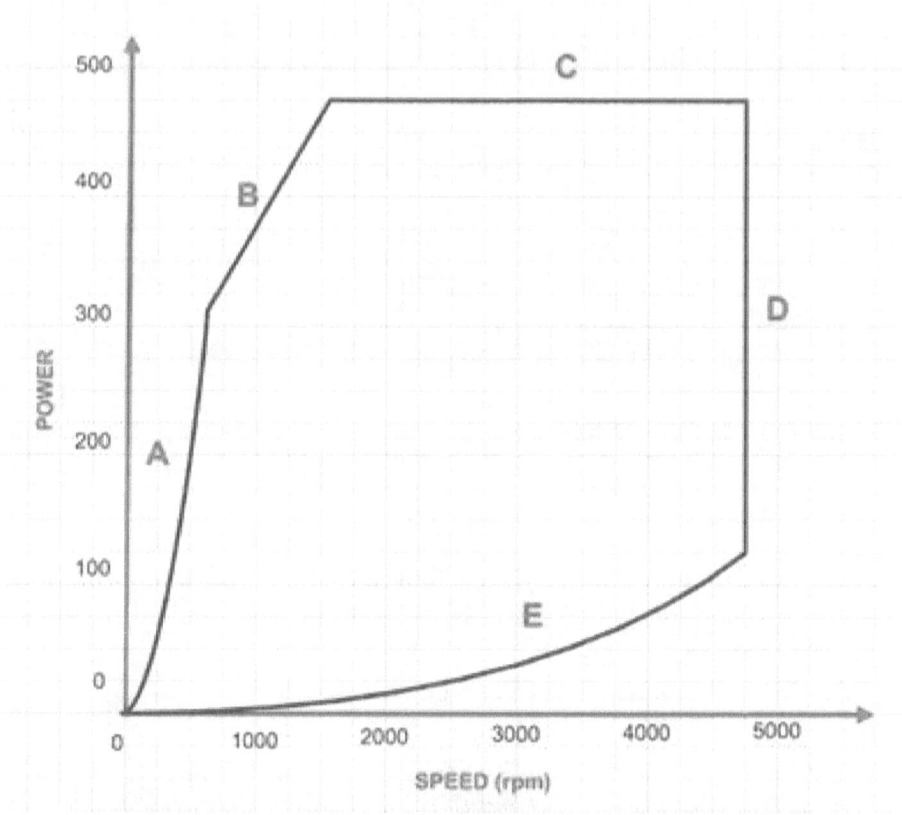

Figure 4.7 Power curve.

The analysis of above figure, that is, power envelope, conveys the deeper meaning about each segment.

Let us consider the each segment of the power curve.

Segment A: This segment indicates how much maximum power can be achieved by a given dynamometer or the dynamometer under consideration. The nature of the curve is dependent on the following parameters:

1. Amount of current circulated in the coils.
2. Geometric dimensions of the rotor.
3. The speed at which the dynamometer rotor is running. This governs the rate of change of magnetic flux responsible for the generation of eddy currents.

Segment B: This segment of the power curve indicates the maximum torque achievable by the dynamometer. This is dependent on the following factors:

1. Shaft design, maximum permissible load on shaft which carries the rotor.
2. Current in the coil and the corresponding speed.

Segment C: Rated power, this is what a given dynamometer will produce under installed setup and dependent on the following factors:

1. Maximum cooling water that can be circulated which is ultimate media to carry the heat generated due to power absorbed.
2. The maximum current can be carried by the coils.

Segment D: This segment indicates the "maximum speed" limit of dynamometer and is dependent on the following parameters:

1. Bearing types and lubrication method adopted for lubrication of bearings.
2. Critical speed of the shaft or the complete system.
3. Dynamometer mounting arrangement.

Segment E: This segment indicates the minimum power to be absorbed by the dynamometer, and it is dependent on the following parameters. In this case, the loading mechanism of the dynamometer is not active.

1. Speed at which dynamometer is rotating.
2. Rotor inertia of the rotor.

In case of a particular eddy-current dynamometer, certain design parameters are fixed while designing the dynamometer.

The maximum power and the minimum power curves are fixed for a particular model and cannot be changed in practice or during the use of the dynamometer at site. The dynamometer comes with fixed rotor and coils which have predesigned power absorption capacity.

4.7.2 Torque Curve of Eddy-current Dynamometer

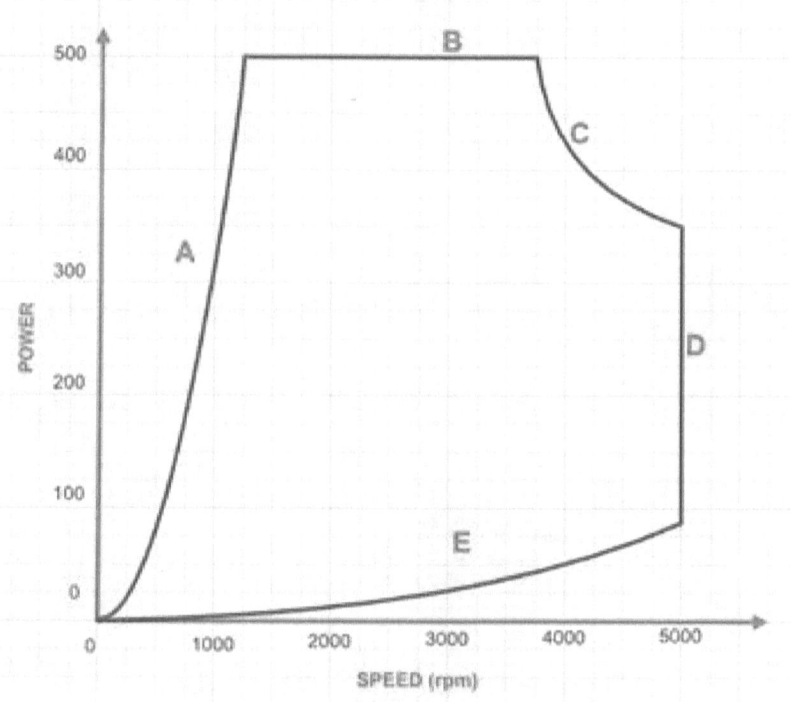

Figure 4.8 Torque curve.

4.7.3 Power Curve Comparison

The following figure shows the power envelopes of the eddy-current and hydraulic dynamometer of the same horsepower capacity. It can be seen that curve A of eddy-current dynamometer is much steeper than the curve of hydraulic dynamometer. The torque capacity of eddy-current dynamometer is a little larger than that of hydraulic dynamometer at the same speed.

Also, the minimum power absorption is lower in case of eddy-current dynamometer than in hydraulic dynamometer.

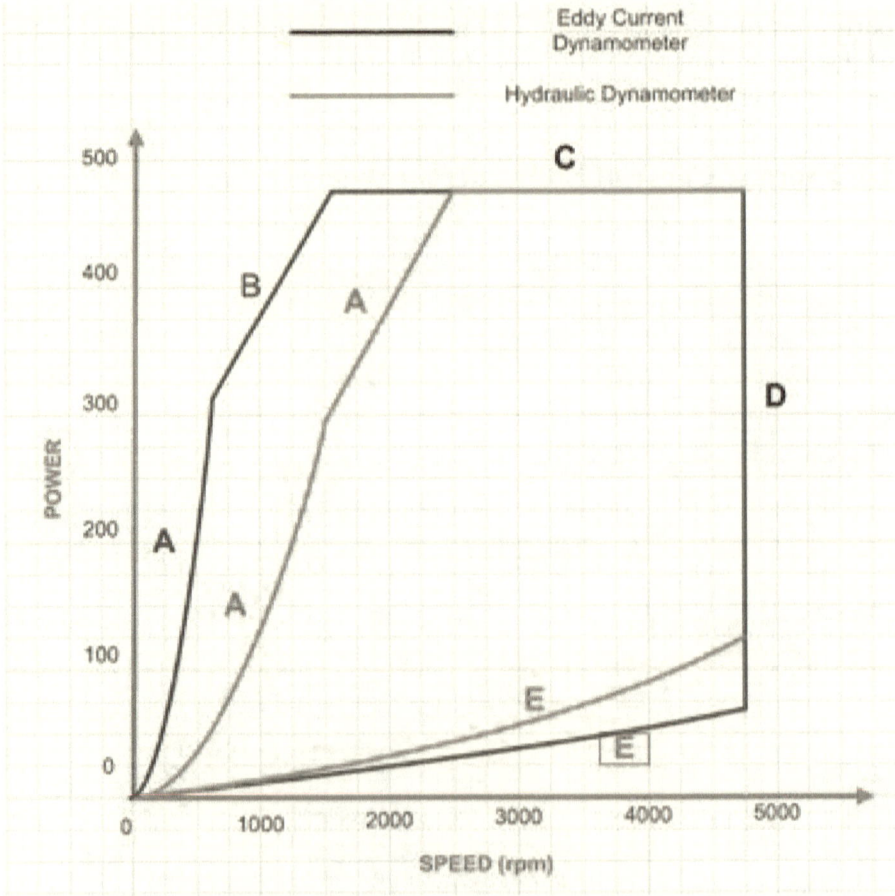

Figure 4.9 Power envelope comparison.

4.8 Wet Gap Eddy-current Dynamometer

In the wet gap eddy-current dynamometer, as the name implies, the rotor or the part of rotor is immersed in the water. The rotor is in contact with the water. The following figure shows the cross section of the wet gap eddy-current dynamometer.

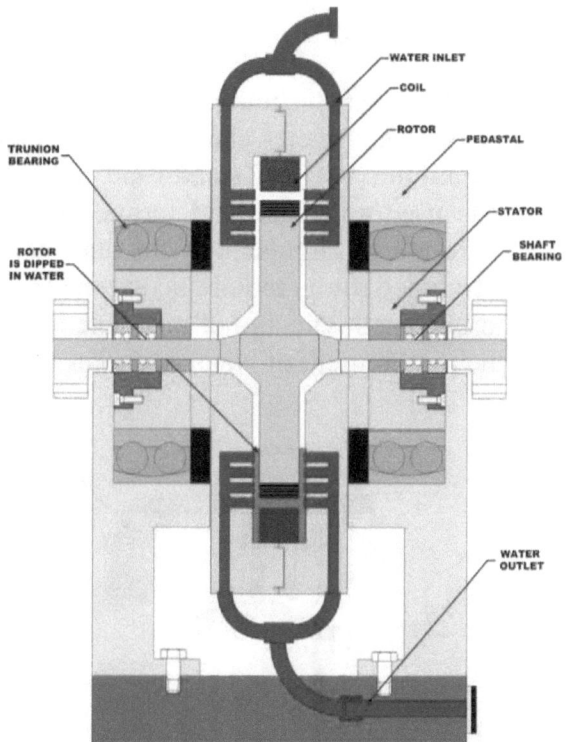

Figure 4.10 Wet gap eddy-current dynamometer.

4.8.1 Advantages of the Wet Gap Dynamometer

1. There is no fear of overheating as water is directly sprinkled in the air gap.
2. Cooling is more effective.

4.8.2 Disadvantages of the Wet Gap Dynamometer

1. Wet gap machines will have higher inertia due to drag of water.

2. Wet gap dynamometers will have higher minimum torque characteristics due to drag of water present in wet gap.
3. The rotor in wet gap dynamometer is prone to corrosion due to contact with water. However, this can be reduced by the proper selection of material for rotor.

4.9 Air-cooled Eddy-current Dynamometer

Air-cooled eddy-current power absorber is mainly made up of stator, rotor, and electromagnetic coils on both sides. Its main application is in the motorcycle chassis dynamometer, chassis dynamometer, emissions loading equipment, automotive test stand, and the presentation of teaching instruments used as a load and torque measurement. The heat dissipation is through air circulated by the fans mounted on rotor assembly. Their self-cooled rotors require no external water supply or resistor banks. The biggest advantage of air-cooled dynamometer is that it eliminates the requirement of water and water piping.

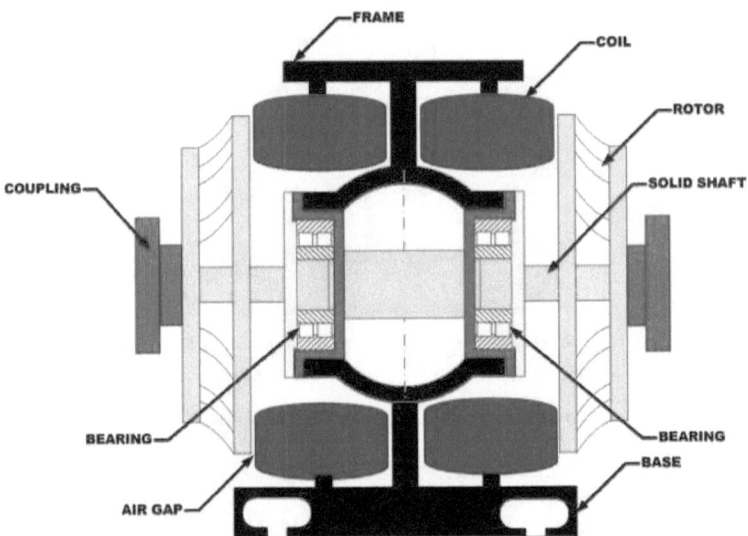

Figure 4.11 Air-cooled eddy-current dynamometer.

4.10 Applications of Air-cooled Eddy-current Dynamometer

The main application is in the end of line motorcycle chassis dynamometer, chassis dynamometer, emissions loading equipment, automotive test stand, and the presentation of teaching instruments used as a load. For the applications that do not require stall-speed loading, an air-cooled eddy-current absorber is usually the most cost-effective electric absorber.

Figure 4.12 Air-cooled dynamometer application.
Courtesy of Dynomite Dynamometers, USA.

4.11 Dry Gap Twin Coil Eddy-current Dynamometers

The eddy-current dynamometers are also available in twin coil design. The coils are placed on either side of the rotor. The advantage claimed by twin-coiled design is elimination of thrust on rotor crated by axial component of magnetic flux. The overall size of the coil diameter also gets reduced as compared to single coil machine as its placement is on side of the rotor. The following figure shows the cross section of the twin coil eddy-current dynamometer.

The following figure also highlights the typical installation with eddy-current dynamometer.

Figure 4.13 Typical installation of twin coil eddy-current dynamometer. Courtesy of Saj Test Plant Pvt. Ltd., Pune, India.

Closure

The eddy-current dynamometers are widely used in for testing of engines in production, R&D, and quality control. Eddy-current dynamometers are preferred over hydraulic dynamometers by the test engineers because of their simple design, accuracy, and control achievable. EC dynamometer is the most commonly used dynamometer in modern engine and chassis dynamometers in the present days. They are capable of changing the load most efficiently at a rapid speed.

Bibliograpghy

1. Time Line Dynamometers Internet Web Site.
2. Images provided by Horiba Dynamometer e.
3. Images provided by Saj Test Plant Pvt. Ltd.
4. Images provided Dynomite Dynamometers.

CHAPTER 5
Magnetic Powder and Hysteresis Dynamometer

5.1 Introduction

In chapter 1, we have seen the classification of the dynamometer and the different types of dynamometer. The magnetic powder and hysteresis dynamometers are absorption-type dynamometers. These dynamometers find applications in testing of variety of motors and small engines.

5.2 Magnetic Powder Dynamometer

A powder dynamometer is similar to an eddy-current dynamometer, but a fine magnetic powder is placed in the air gap between the rotor and the coil. The resulting flux lines create "chains" of metal particulate which are constantly built and broken apart during rotation, creating great torque. Powder dynamometers are typically limited to lower RPM due to heat dissipation issues.

The magnetic powder brake dynamometer, as its name suggests, uses a magnetic powder. The electrical current passing through the coil generates a magnetic field, which changes the property of the powder, thus producing a smooth braking torque through friction. They are ideal for applications operating in the low to middle speed range or when operating in the middle to high torque range. Powder brakes provide full torque at zero speed.

The magnetic powder dynamometers are water cooled and typically allowing for power ratings up to 48 kW. They have accuracy ratings of ±0.3 percent to 0.5 percent full scale, depending on size and system configuration.

Figure 5.1 Magnetic powder dynamometer. Courtesy of Magtrol.

Figure 5.2 Powder dynamometer in tandem with EC dynamometer. Courtesy of Magtrol

5.3 Power Curves of Magnetic Powder Dynamometer

A typical power curve of powder dynamometer resemble as shown below.

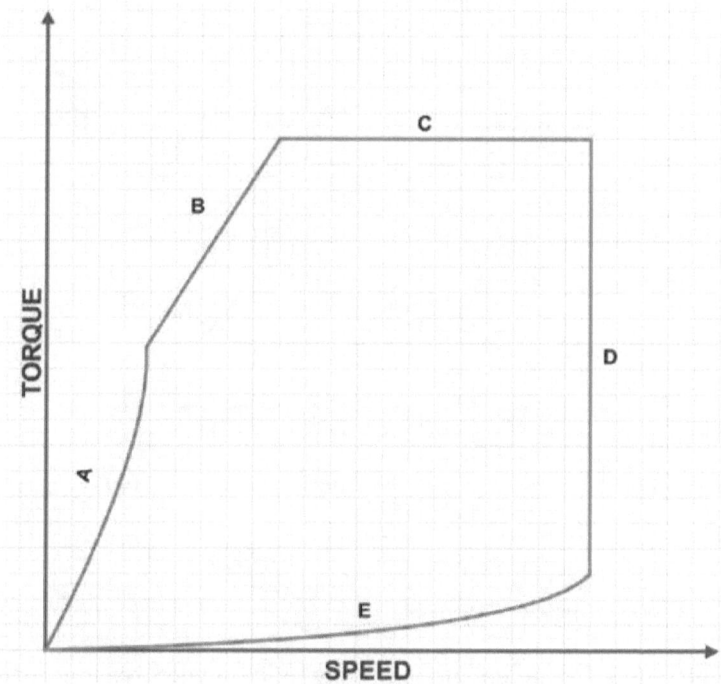

Figure 5.3 Power envelope of magnetic powder dynamometer.

The power envelope is described as follows:

- A: Power rises due to current and the presence of magnetic powder.
- B: Constant torque line and limited by the permissible max torque by the shaft and rotor material.
- C: Maximum power due to maximum current.
- D: Maximum speed—this is limited due to maximum speed bearings can sustain, maximum peripheral velocity of rotor, and critical speed limit.
- E: Minimum power—this is governed by zero coil current and inertia of rotating masses.

5.4 Torque Curves of Magnetic Powder Dynamometer

Magnetic powder dynamometers produce their rated torque at zero speed.

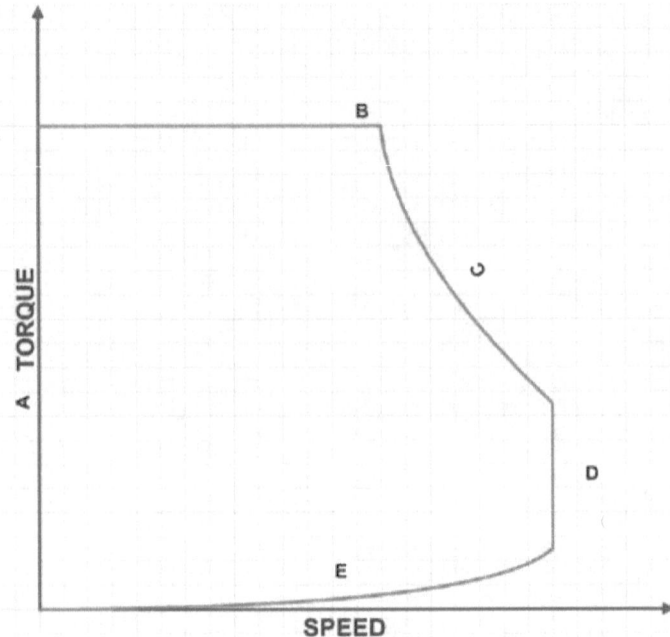

Figure 5.4 Torque envelope of magnetic powder dynamometer.

5.5 Hysteresis Dynamometers

5.5.1 Operating Principles

The hysteresis effect in magnetism is applied to torque control by the use of two basic components—a reticulated pole structure and steel rotor/shaft assembly—fastened together but not in physical contact. Until the field coil is energized, the drag cup and shaft can spin freely on its bearings. When a magnetizing force from either a field coil or magnet is applied to the pole structure, the air gap becomes a flux field. The rotor is magnetically restrained, providing a braking action between the pole structure and the rotor. Because torque is produced strictly through a magnetic air gap, without the use of friction or shear forces, hysteresis brakes provide operating characteristics, such as

- absolute smoothness
- torque independent of speed
- maintenance free
- highest repeatability.

In an electrically operated hsteresis dynamometer, adjustment and control of torque is provided by a field coil. This allows for complete control of torque by adjusting DC current to the field coil. Adjustability from a minimum value to a maximum value of rated torque is possible.

5.5.2 Construction of Hysteresis Dynamometer

Hysteresis dynamometers use a steel rotor that is moved through flux lines generated between magnetic pole pieces. This design, as in the usual "disk-type" eddy-current absorbers, allows for full torque to be produced at zero speed, as well as at full speed. Heat dissipation is assisted by forced air. Hysteresis and "disk-type" EC dynamometers are one of the most efficient technologies in small range dynamometers, such as 200 hp (150 kW) and less. A hysteresis brake is an eddy-current absorber which, unlike most "disk-type" eddy-current absorbers, puts the electromagnet coils inside a vented and ribbed cylinder and rotates the cylinder instead of rotating a disk between electromagnets. The potential benefit for the hysteresis absorber is that the diameter can be decreased and operating rpm of the absorber may be increased.

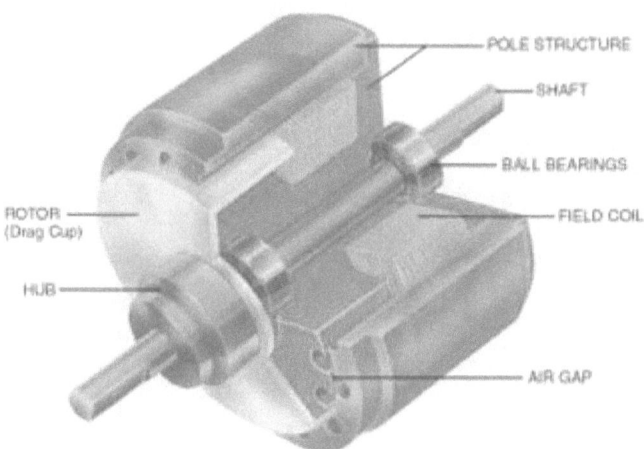

Figure 5.5 Cross section of hysteresis dynamometer.
Courtesy of Magtrol Dynamometers.

5.5.3 Power Curve of Hysteresis Dynamometer

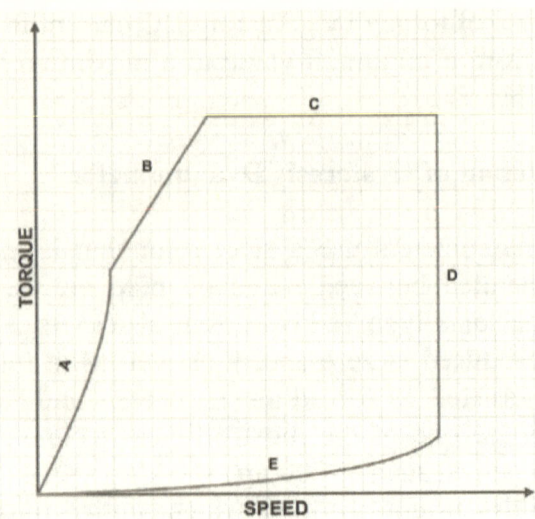

Figure 5.6 Power envelope for hysteresis dynamometer.

5.5.4 Torque Envelope of Hysteresis Dynamometer

The hysteresis dynamometer has high torque capacity at extremely low speed (RPM) as shown in the following torque envelope.

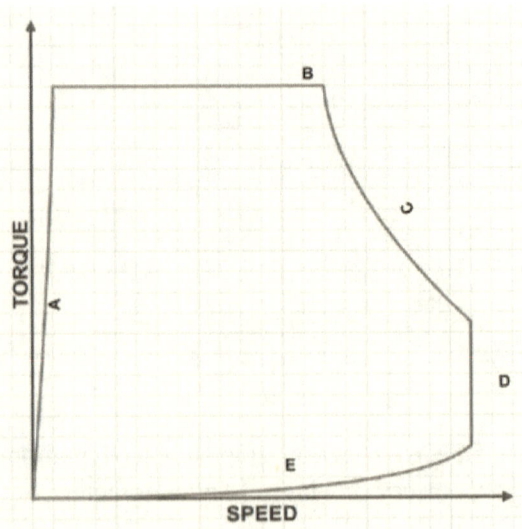

Figure 5.7 Torque envelope for hysteresis dynamometer.

Description of power envelope

- A: Power rises due to current and the presence of magnetic powder.
- B: Constant torque line and limited by the permissible max torque by the shaft and the rotor material.
- C: Maximum power due to maximum current.
- D: Maximum speed—this is limited due to maximum speed bearings can sustain, maximum peripheral velocity of rotor, and critical speed limit.
- E: Minimum power—this is governed by zero coil current and inertia of rotating masses.

5.6 Comparison of Hysteresis, Eddy-current, and Powder Dynamometers

The difference between eddy-current, magnetic powder, and hysteresis brakes is the design of the brake itself. All the three dynamometers have a rotor and stator of which the design is adapted to its technology. A coil (in the stator) produces a magnetic field when supplied by an excitation current.

Powder Brake: It uses a metallic powder between the rotor and stator of the brake. When excitation current is applied, the powder sticks on the rotor (according to the magnetic field line) and brakes by friction. Powder brakes have a strong braking torque at 0 rpm but are limited in speed (up to 1,000-3,000 rpm depending on the size of the dynamometer).

Eddy-current Brakes: These have a splined rotor. The splined part of the rotor in the magnetic field produces the braking effect. Eddy-current dynamometers apply braking power proportional to the speed. There is no power absorption at 0 rpm and is specially adapted for middle to very high speed applications.

Hysteresis Brakes: These have a stator in two parts with a rotor in the form of a cup rotating in the magnetic field between two stator parts. Hysteresis brakes provide braking power from 0 rpm to maximum speed of that machine.

Both hysteresis and eddy-current dynamometers can work in a vertical application.

5.7 Typical Accuracies of Magnetic Powder and Hysteresis Dynamometer.

    A Torque ± 0.25 percent of full-scale torque.
    B Speed ± 1 RPM.

5.8 Applications of Magnetic Powder and Hysteresis Dynamometers

Dynamometers find wide applications in testing the following:

1. Medium-sized electric motors.
2. Fractional horsepower motors.
3. Pneumatic tools.
4. Hydraulic motors.
5. Small IC engines used for agricultural sprayer and tools.
6. Testing of dental tools.

Closure

The powder and hysteresis dynamometers are ideal for applications operating in the low to middle speed range or when operating in the middle to high torque range. These dynamometers provide full torque at zero speed. These dynamometers are used for testing of appliance, automotive, aviation, computer, HVAC, lawn and garden, medical and dental, electric motor, office equipment, and power tools.

Bibliography

1. *Manufacturer of Powder and Hysteresis Dynamometer*, Magtrol Inc: USA.

CHAPTER 6

Portable Dynamometers

6.1 Introduction

The most of the dynamometer installations are fixed, which means that dynamometer is installed in a test cell along with the accessories, and engine is brought to the test cell for the purpose of the testing and evaluating the performance. However, many a times, a need arises that dynamometer to be brought to the engine. These types of situations may arise in practical life. The situations where the dynamometer is ported are listed below.

1. Engine overhauling workshops.
2. Large generating set engines.
3. Truck engine before overhauling is tested by portable dynamometer to evaluate whether or not to go for complete overhaul.

6.2 Portable Dynamometer

Portable dynamometers are units that can be installed in virtually any setup because of their small footprint. Most of the portable dynamometers couple directly to either the bell housing of an automotive engine or the SAE housing of diesel engines. Portable dynamometers can also be used in driveshaft applications and can be fitted for mounting into stationary stands. It also can be mounted within the driveline of heavy-duty trucks for in-vehicle engine testing.

The greatest advantage of a portable dynamometer is that when your engine testing is completed, they can be stored back until they are called up again for service. Portable dynamometers are sometimes called "fit and go" solution for on-site testing needs.

6.3 Construction of Portable Dynamometer

The portable hydraulic dynamometers consist of a pair of power absorbing stator and rotor elements capable of fully bidirectional operation. The power elements are made from corrosion resistant stainless steel. The power elements are sometimes constructed using bronze or Monel metal. The shaft is normally manufactured using stainless steel.

The portable hydraulic dynamometers are usually unidirectional; however, for bidirectional operation, perforated disk power elements are used, or the rotor and stator elements are manufactured with vane angle of 90°.

The rotor assembly is supported by grease-packed precision ball bearings. For high speed application, normally oil mist lubrication is used.

To assure smooth, vibration-free operation, the rotor assembly is dynamically balanced. The mechanical carbon face seals are used to ensure positive sealing between the water and the bearing compartments. Unlike with labyrinth seals, there is no need to drain the dynamometer during shutdowns to prevent water from entering the bearings.

The dynamometer module is supported in the single trunnion mounted by rugged, tapered roller-type trunnion bearings. This results in smooth, vibration-free operation under the most severe operating conditions as well as excellent torque measurement accuracy.

Figure 6.1 Portable dynamometer.
Courtesy of Power Test Dynamometers.

6.4 Cross Section of Portable Dynamometer

The following illustration shows the internal construction of the portable dynamometer. The power absorption elements will be the same as that of stationary dynamometer. The mounting arrangement is normally custom-built to suit the individual engine flywheel.

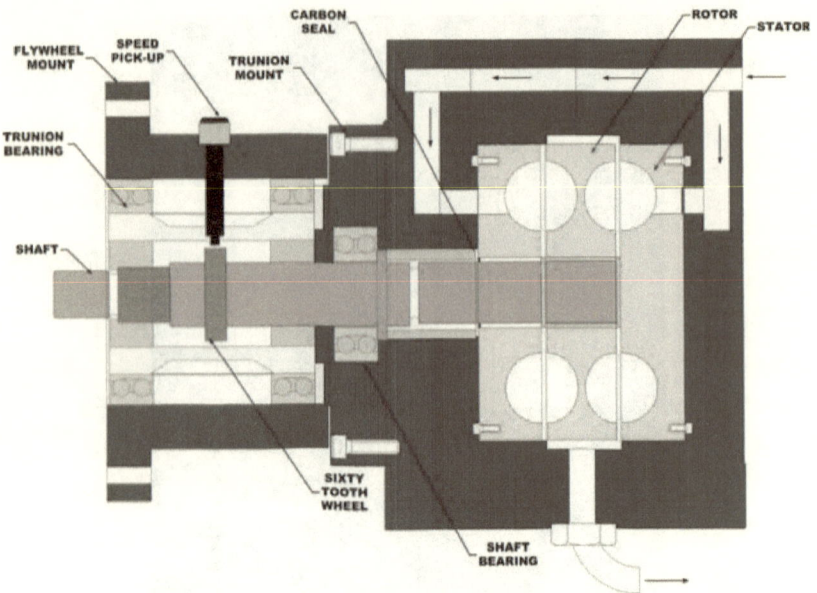

Figure 6.2 Cross section of portable dynamometer.

6.5 Power Curve of Portable Dynamometer

The power envelope will be identical to that of stationary dynamometer in its nature. The typical power curve will be as shown below on natural scale.

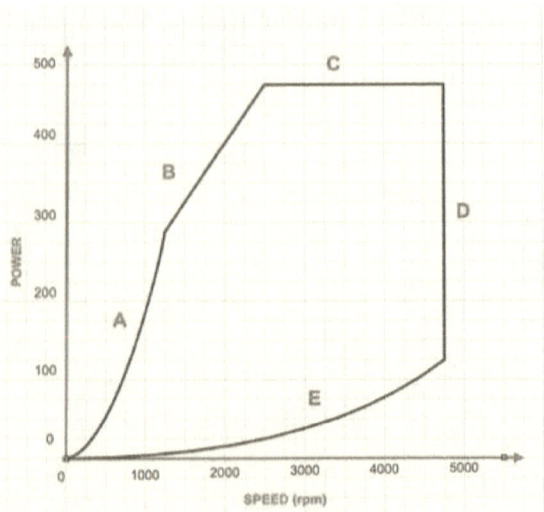

Figure 6.3 Typical power curve of hydraulic dynamometer.

The analysis of above figure, that is, power envelope, conveys the deeper meaning about each segment.

Let us consider the each segment of the power curve.

Segment A: This segment indicates how much maximum power can be achieved by a given dynamometer or the dynamometer under consideration. The nature of the curve is dependent on the following parameters:

1. Amount of water that can be accommodated by a whirl chamber. The whirl chamber is the space formed by each half cup on rotor and stator.
2. Geometric dimensions of the toroidal space.
3. The speed at which the dynamometer rotor is running.

Segment B: This segment of the power curve indicates the maximum torque achievable by the dynamometer. This is dependent on the following factors:

1. Shaft design, maximum permissible load on shaft which carries the rotor.
2. Hoops stress of the whirl chamber or toroidal space.
3. Pressure of the water in the whirl chamber.

Segment C: Rated power, this is what a given dynamometer will produce under installed setup and dependent on the following factors:

1. Maximum cooling water that can be circulated which is ultimate media to carry the heat generated due to power absorbed. A slight increase in power limit can be achieved by increasing the inlet water pressure to the dynamometer.
2. Hoops stress of the whirl chamber or toroidal space.

Segment D: This segment indicates the "maximum speed" limit of dynamometer and is dependent on the following parameters:

1. Bearing types and lubrication method adopted for lubrication of bearings.
2. Critical speed of the shaft or the complete system.
3. Dynamometer mounting arrangement.
4. Hoops stress developed in the whirl chamber.

Segment E: This segment indicates the minimum power to be absorbed by the dynamometer, and it is dependent on the following parameters. In this case, the loading mechanism of the dynamometer is not active.

1. Water inlet pressure and quantity.
2. Speed at which dynamometer is rotating.
3. Rotor inertia of the rotor.

In case of a particular hydraulic dynamometer, certain design parameters are fixed while designing the dynamometer.

The maximum power and the minimum power curves are fixed for a particular model and cannot be changed in practice or during the use of the dynamometer at site. The dynamometer comes with fixed rotor and stator which has predesigned power absorption capacity.

From the above understanding about each segment, the power curve now can be represented as:

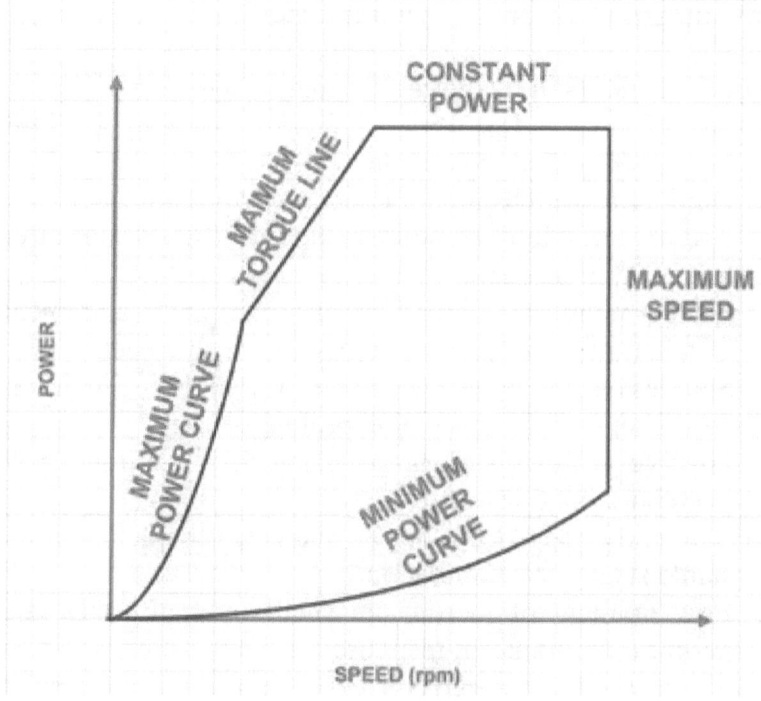

Figure 6.4 Hydraulic dynamometer power curve.

6.6 Typical Applications

Portable dynamometer may be employed for the testing of the engines in the following configurations:

1. Directly mounted on flywheel housing
2. Under the truck
3. Testing of power of tractor at PTO shaft.

The portable dynamometers with modern control system along with electrohydraulic-controlled butterfly outlet valve or fully electrically controlled valve provide the following modes of operation:

1. Manual mode—also called as open loop.
2. Constant speed mode.
3. Constant torque mode.
4. Power law mode T (torque) α N^2.

6.7 Advantages of Portable Dynamometer

Some of the benefits of the portable dynamometer are

1. Easy to install and to operate.
2. Outdoor as well as indoor operation is possible.
3. Use of portable dynamometer does not need special engine test stand or test cell.
4. Flange-mounted configuration permits alignment-free installation directly to the engine or the flywheel.
5. Stainless steel power elements are highly capitation and corrosion resistant providing prolonged life.
6. Positive sealing permits operation at all altitude positions from horizontal to vertical.
7. Inherently steep, open-loop torque speed characteristic assures stable steady-state operation.
8. Low moment of inertia and small internal water volumes permit rapid transient response.
9. Simple modular design concept permits quick overhaul by the user.

6.8 Applications of Portable Dynamometer

The few applications of the portable dynamometer are described below.

6.8.1 Engine Testing

The following figure/photos will clear the views about how portable dynamometers are actually used in practice. They can be mounted directly on the flywheel housing with suitable adapter and coupling specially designed to suite the engine under test. This is of great advantage for the testing in repair shops after the engine is overhauled. The following figure shows the engine coupled with portable dynamometer. This arrangement is highly suitable for repair workshops.

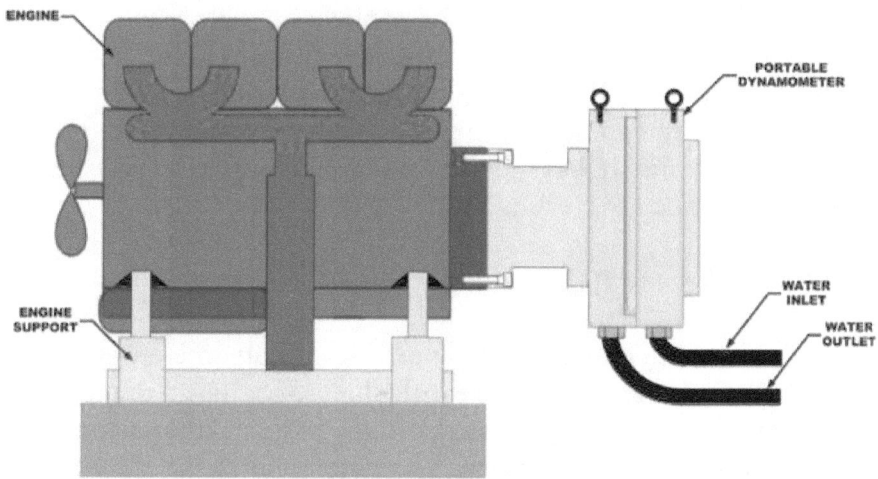

Figure 6.5 Schematic arrangement of engine testing in repair workshops.

6.8.2 Testing of the Engines Insitu Condition

Before removing the engine from the truck chassis for overhauling, it can be tested for diagnosing the problem. This can be achieved by taking the portable dynamometer mounted on trolley under the vehicle (trucks). The propeller shaft from engine to rear axle can be disconnected, and portable dynamometer can be adapted to test the engine using suitable driveshaft to connect it to the engine transmission.

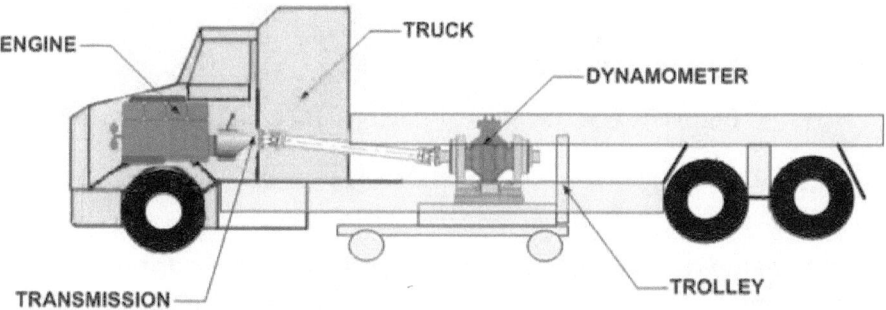

Figure 6.6 Schematics of portable dynamometer application.

This will gain knowledge about the engine health, and decision-making regarding engines overhaul will be justified.

The portable dynamometer finds its application mainly in the repair shop. The main idea is taking dynamometer to the engine, instead of engine coming to the dynamometer test stand as it normally happens in the test cells at manufacturing place. It can be employed to test the farm tractors at site. If it is mounted on mobile trailer, it can be taken where there is need for testing.

6.8.3 Farm Tractor PTO Testing

One of the applications of the portable dynamometer is testing performance of the power take off shaft of a farm tractor. The dynamometer can be brought where the tractor is, and the performance can be evaluated. The following illustration shows the PTO testing using portable dynamometer mounted on trolley. The dynamometer could be hydraulic or eddy current. The air-cooled eddy-current dynamometer could be a good solution where water supply is unavailable.

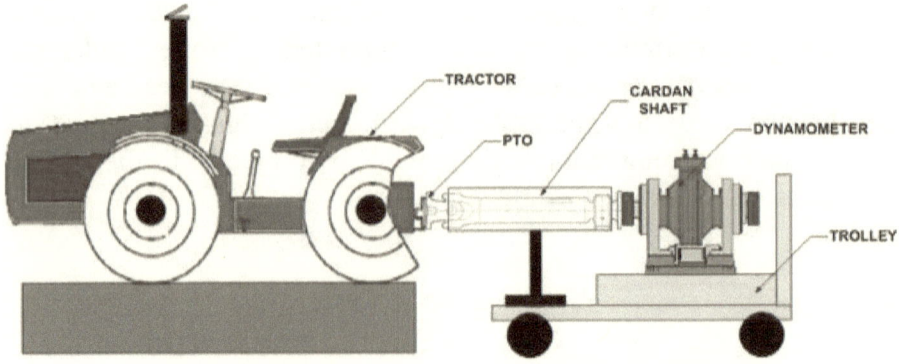

Figure 6.7 Schematics of tractor PTO application.

Closure

The portable dynamometer finds application in various engine testing repair workshops and engine rebuilding shops. The portable dynamometer is an economic solution for the reconditioning workshops where the engine performance before and after the engine overhaul can be established. The installation is economical and affordable to the engine rebuilders.

The portable dynamometers could be hydraulic or eddy current. In case of eddy-current dynamometer, safety needs to be monitored for safe working of the system. Another alternative is to employ air-cooled eddy-current dynamometer; however, hydraulic portable dynamometers are widely accepted for the engine testing.

CHAPTER 7

Direct Current (DC) Dynamometer

7.1 Introduction

A DC dynamometer is a DC motor that is mounted in trunnion bearings as cradle which acts as motor as well as generator depending on which mode it is running. A dynamometer is an apparatus that measures mechanical force, speed, or power. In any engine testing application by dynamometer test cell testing, an engine, motor, or transmission is coupled to the dynamometer via couplings and a driveshaft. When the device under test is running, the dynamometer can exert a braking force on it.

The load cell and speed pick—up sensor mounted on the dynamometer will measure engine (device under testing) torque and speed. Acquiring these values, the dynamometer can be used to calculate engine power output. Dynamometers allow us to reproduce a desired speed or torque for a test under controlled conditions.

7.2 Electrical Dynamometers

As we described in the first chapter, dynamometers use different methods to induce a braking force on the device under test. What exactly are the benefits of different types of dynamometers?

There are certain advantages and disadvantages of using electrical machines, such as DC/AC machine, as dynamometer.

The physical principle of production of mechanical force by the interactions of an electric current and a magnetic field was known as early as 1821. Electric motors of increasing efficiency were constructed throughout the nineteenth century, but commercial exploitation of electric motors on a large scale required efficient electrical generators and electrical distribution networks.

The DC unit is especially suitable for transient test cycles such as the European Transient Cycle (ETC), the European Load Response (ELR), the European Steady-state Cycle (ESC), and the Federal Transient Procedure (FTP) where motoring is required

DC dynamometers when used for engine testing can be employed for

1. engine testing under load
2. motoring of the engine
3. engine stating.

7.3 Principle of Operation

DC dynamometer is a machine to test the mechanical power of the unit under testing. A standard DC dynamometer comprises of a DC motor, a gearbox, a rotational speed sensor, a base frame, and an in-line torque sensor, for example, HBM torque testing ring, which is connected to the machine using the flange joint. It also comprises of a control unit of the DC dynamometer. The working principle is to convert the DC from the DC generator into the AC through the DC convert system.

The electrical dynamometer like DC machine fundamentally is a generator. It can change testing unit's mechanical energy to electricity. As mentioned above, the DC dynamometer consists of DC motor, armature, excitation's control unit, load circuit, and torque and speed measuring sensors. When DC motor is generating mode as show in figure 7.1, The electromotive force equation is

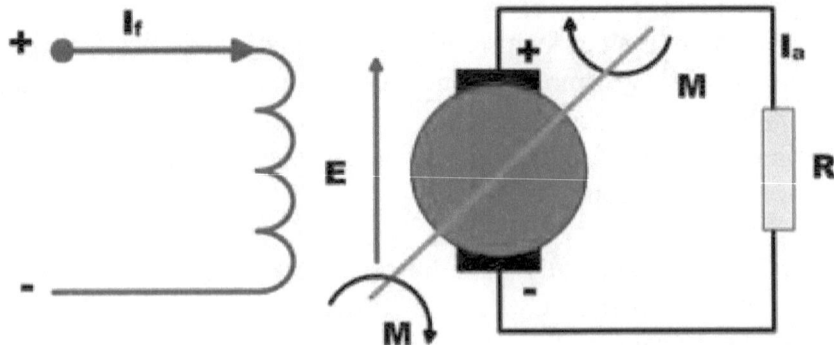

Figure 7.1 DC motor in generating mode.

$$E = C_e . \emptyset . N, \quad\quad\quad\quad\quad\quad\quad\quad\quad\quad\quad\quad (7.1)$$

where
C_e = constant relative to motor's structure
Φ = magnetic pole's magnetic flux, Wb
N = motor RPM

And its armature voltage function is

$$E = U + I_a . R_a = I_a(R_a + R), \quad\quad\quad\quad\quad\quad (7.2)$$

where
E = reserve electromotive force
R_a = armature resistance in Ω
I_a = armature current in ampere (A)
R = load resistor
$U = I_a R_a$

Electromagnetism torque function is

$$M = C_m . \emptyset . I_a, \quad\quad\quad\quad\quad\quad\quad\quad\quad\quad (7.3)$$

where
C_m = constant relative to motor's structure

When a dynamometer is in steady state, that is, at steady speed, the engines' torque (M_e) is equal to total of electromagnetism torque M and friction torque M_f. Therefore,

$$M_e = M + M_f$$

In fact, friction torque is negligible; hence, above equation can be rewritten as:

$$M_e = M + M_f = M$$

From equations (7.2) and (7.1), we can get

$$I_a = \frac{E}{R+R_a} = \frac{C_e \cdot \varnothing \cdot N}{R+R_a} \quad \ldots \ldots \ldots \ldots \ldots \ldots \ldots \ldots \ldots \ldots (7.4)$$

Inserting equation (7.4) into equation (7.3), we get, that is, we are substituting the value of I_a from equation (7.4) into equation (7.3):

$$M = C_m \cdot \varnothing \cdot I_a = \frac{C_m \cdot C_e \cdot \varnothing^2}{R+R_a} \quad \ldots \ldots \ldots \ldots \ldots \ldots \ldots \ldots (7.5)$$

From equation (7.5), electromagnetism torque can be changed by adjusting excitation current (I_a), reserve electromotive force (E), and load resistor (R). In fact, load resistor often adopts active load, which can generate resistance torque through feedback to power. DC motor's power is limited; however, DC motor's electromagnetism max torque (M_{max}) is unlimited.

The DC dynamometer has two control modes:

1. torque constant
2. speed constant mode.

7.4 Rating of the DC

The rating of the DC dynamometer is defined in the following way:

Absorption capacity in kW.
Motoring capacity in kW.
Base speed in RPM.

Maximum speedin RPM.

Indication of speed describes the meaning that continuous operation is possible at the rated capacity from base speed up to maximum speed. This range is called the constant capacity range.

The maximum torque to be measured is lowered in inverse proportion to speed when speed is above the base speed. If the speed is below the base speeds, the measured torque is kept constant. The range is called the constant torque range.

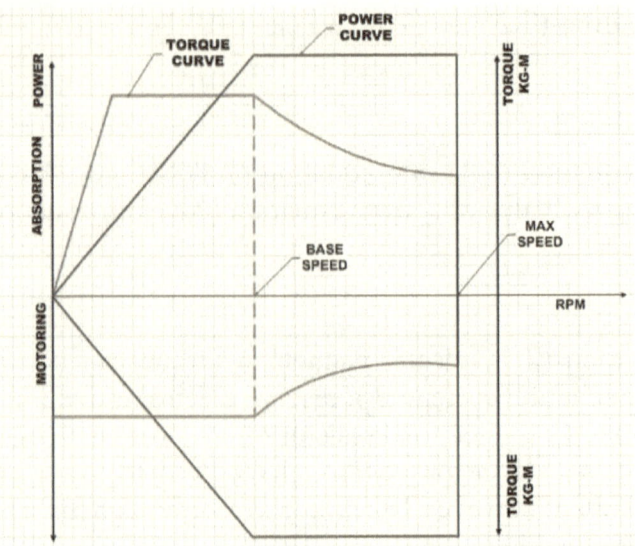

Figure 7.2 DC dynamometer measuring range.

7.5 Construction of DC Machine

The DC machine can be employed in trunnion-mounted or fixed mounting on a baseplate. Both the methods are in practice; however, fixed mounting with in-line torque transducer is more popular.

DC machines are made up of several components that include the following:

1. frame
2. shaft
3. bearings
4. main field windings (stator)
5. armature (rotor)
6. commutator
7. brush assembly.

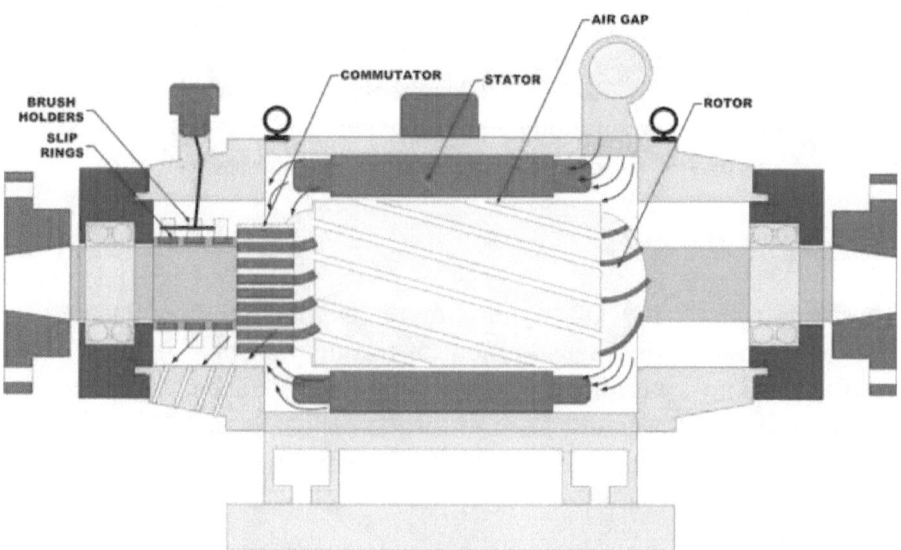

Figure 7.3 Cross section of DC machine.

7.5.1 Trunnion Mounting

The following figure shows the trunnion-mounted DC machine with universal load cell mounted on the DC machine with the torque arm. In this case, DC machine swings due to torque absorbed.

Figure 7.4 shows trunnion-mounted DC machine with calibration arms attached to it.

Figure 7.4 Trunnion-mounted DC machine with calibration arms.
Courtesy of Associated Electrodyne Industries Pvt. Ltd.

7.5.2 Fixed Mounting

In special cases, the DC dynamometer is foot-mounted instead of cradle-mounted. The measurement of torque is critical in such cases as there is no swinging arm and universal load cell. Where DC machine is foot-mounted, in-line torque sensor is employed for the measurement of torque developed by the unit under testing, typically an IC engine.

The foot-mounted DC machine has certain advantages as it eliminates bulky trunnion bearings, pedestals housing trunnion bearings, and swinging arm connecting to load cell. In-line torque transducer is incorporated in the shaft connecting the unit under testing (engine) and the DC dynamometer. Figure 7.5 shows typical application of DC dynamometer in "foot-mounted" state.

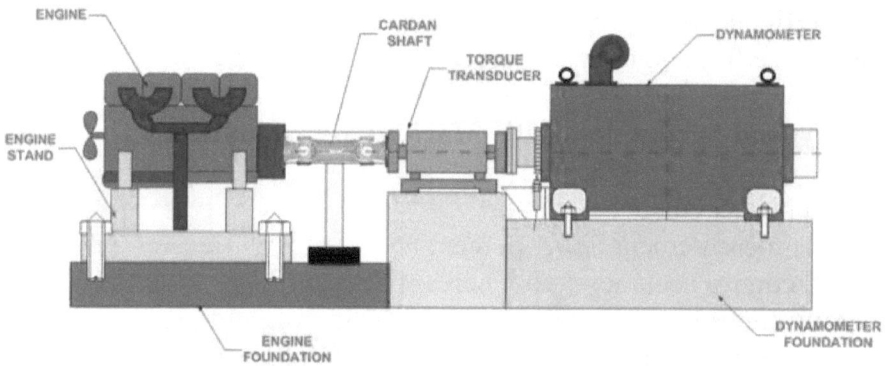

Figure 7.5 Foot-mounted DC dynamometer.

7.6 Power and Torque Envelope

The typical torque and power curves for DC dynamometer are shown in the following figures.

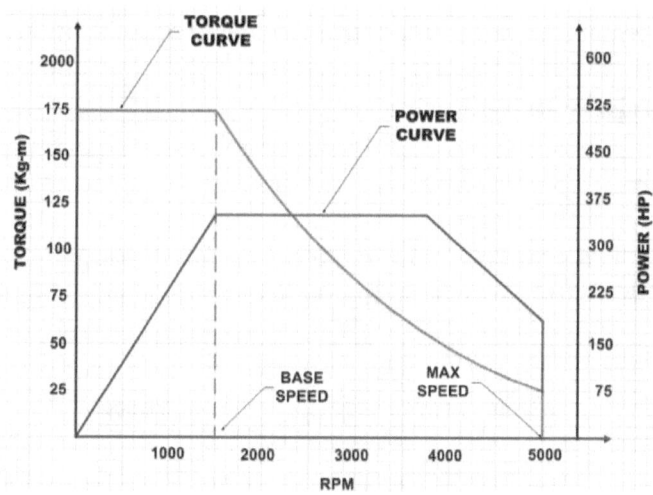

Figure 7.6 Power envelope of DC machine.

7.7 Inertia and Its Effect

The DC machine has a higher inertia compared to AC machines. The inertia plays an important role in the deciding the response time. Higher the inertia, lower is the response time to accelerate or decelerate from a given set point.

7.8 Typical Accuracies of DC Machine/Dynamometer

Speed measurement : ±1r/min.
Torque measurement : ±0.1 percent FS.
Speed control accuracy : ±0.1 percent FS.
Torque control accuracy : ±0.2 percent FS.
Active response time : < 5 m.

7.9 Control System

A thyristor bridge is a technique commonly used to control the speed of a DC motor by varying the DC voltage. Voltage values given in these examples are used for explanation only. The actual values for a given load, speed, and motor vary. A primary function of a DC drive is to convert AC voltage into a variable DC voltage. It is necessary to vary DC voltage in order to control the speed of a DC motor.

7.9.1 Thyristor Drive Basics

Thyristor

A thyristor is a semiconductor device commonly used to convert AC to DC. A thyristor consists of an anode, a cathode, and a gate.

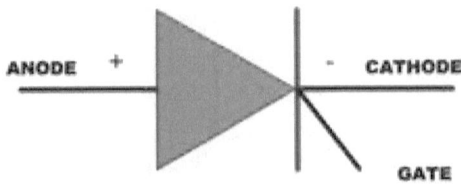

Figure 7.7 Thyristor symbol.

Thyristor Operation

A thyristor acts as a switch. Initially, a thyristor will conduct (switch on) when the anode is positive with respect to the cathode, and a positive gate current is present. The amount of gate current required to switch on a thyristor varies. Smaller devices require only a few milliamps; however, larger devices such as required in the motor circuit of a DC drive may require several hundred milliamps.

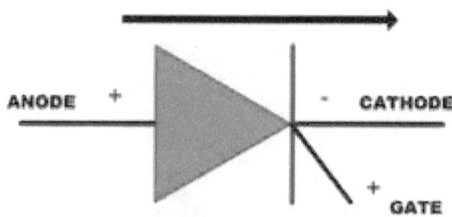

Figure 7.8 Gate current.

Holding current refers to the amount of current flowing from anode to cathode to keep the thyristor turned on. The gate current may be removed once the thyristor has switched on. The thyristor will continue to conduct as long as the anode remains sufficiently positive with respect to the cathode to allow sufficient holding current to flow. Like gate current, the amount of holding current varies from device to device. Smaller devices may require only a few milliamps, and larger devices may require a few hundred milliamps. The thyristor will switch off when the anode is no longer positive with respect to the cathode as shown in below figure.

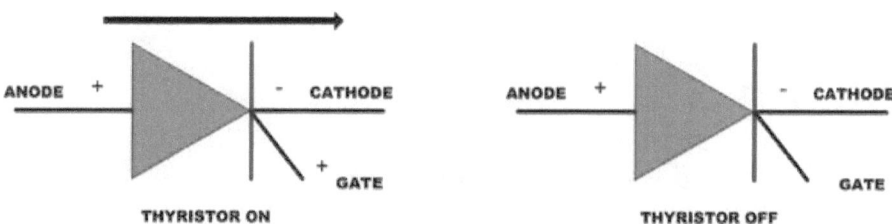

Figure 7.9 Thyristor status.

AC to DC Conversion

The thyristor provides a convenient method of converting AC voltage to a variable DC voltage for use in controlling the speed of a DC motor. In this example, the gate is momentarily applied when AC input voltage is at the top of the sine wave. The thyristor will conduct until the input's sine wave crosses zero. At this point, the anode is no longer positive with respect to the cathode, and the thyristor shuts off. The result is a half-wave rectified DC.

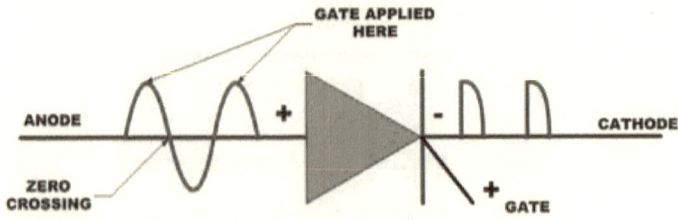

Figure 7.10 Half-wave rectifier.

The amount of rectified DC voltage can be controlled by timing the input to the gate. Applying current on the gate at the beginning of the sine wave results in a higher average voltage applied to the motor. Applying current on the gate later in the sine wave results in a lower average voltage applied to the motor.

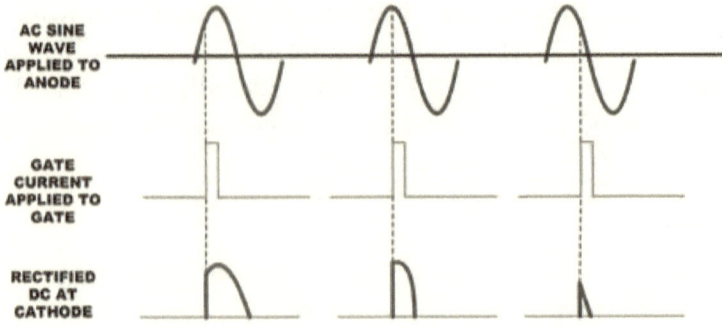

Figure 7.11 Rectified DC voltages.
Courtesy of Siemens AG

7.9.2 Four-quadrant DC Drive

As discussed in Section 7.1, the DC dynamometer application requires application of DC machine in all the four quadrants, that is,

1. clockwise rotation motoring
2. Counterclockwise rotation motoring
3. Clockwise braking/generating
4. Counterclockwise braking/generating.

DYNAMOMETER

The above four modes are shown in the following figure:

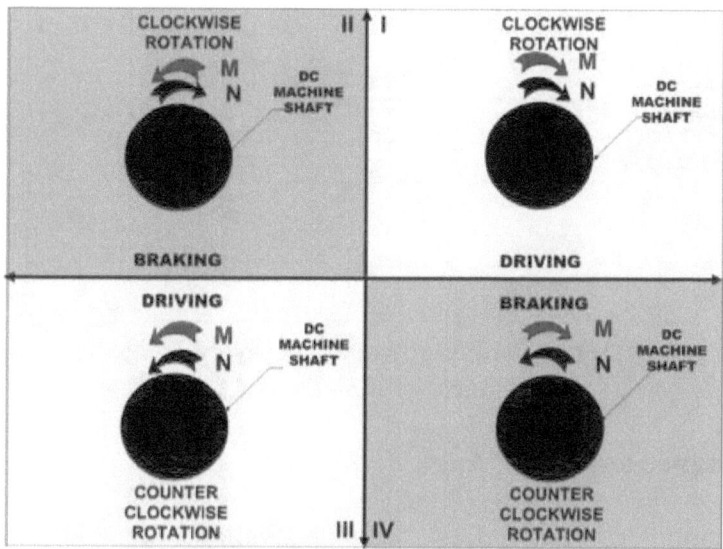

M = Torque, N = Speed.
Figure 7.12 Four-quadrant operation.
Courtesy of Siemens AG

In order for a drive to operate in all four quadrants, a means must exist to deal with the electrical energy returned by the motor. Electrical energy returned by the motor tends to drive the DC voltage up, resulting in excess voltage that can cause damage. One method of getting four-quadrant operations from a DC drive is to add a second bridge connected in reverse of the main bridge. The main bridge drives the motor. The second bridge returns excess energy from the motor to the AC line. This process is commonly referred to as regeneration. This configuration is also referred to as a four-quadrant design.

DC Drive Convertor

The output of a thyristor is with ripple and not smooth enough to control the voltage of drives, those employed in engine testing application. To counter this problem, a three-phase six-thyristor bridge is employed in practice. This is shown in figure 7.13.

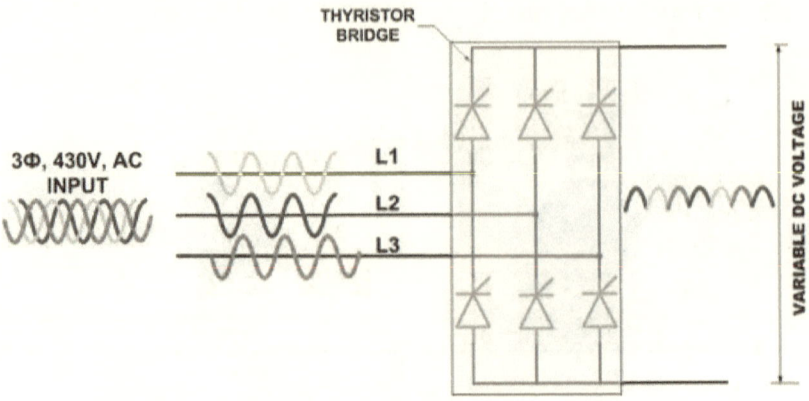

Figure 7.13 Six-thyristor rectifying bridge.
Courtesy of Siemens AG

Importance of Getting Angle

The amount of rectified DC voltage available depends upon the gating angle of thyristor in relationship to the incoming line AC supply voltage. However, the negative and positive value of the AC sine wave must be considered when working with a fully controlled three-phase (3Ø) rectifier. A simple formula can be used to calculate the amount of rectified DC voltage in a three-phase (3Ø) bridge. Converted DC voltage (V_{DC}) is equal to 1.35 times the RMS value of input voltage (V_{RMS}) times the cosine of the phase angle (cos α)

$$V_{DC} = (1.35) \cdot V_{RMS} \cdot Cos\, \alpha \qquad \text{...............................7.6}$$

The amount of DC voltage that can be derived from a 430-volt AC input is –580.5-volt DC to +580.5-volt DC. The following table shows the calculated sample values for various phase angles.

Table 7.1 Output DC voltages

Input AC voltage (V)	A	Cosine	Output DC voltage (V)
430	0	1.00	580.50
430	30	0.87	505.04

430	60	0.50	290.25
430	90	0.00	0
430	120	−0.50	−290.25
430	150	−0.87	−505.035
430	180	−1.00	−580.50

The following illustration approximates the output waveform of a fully controlled thyristor bridge rectifier for 0°, 60°, and 90°. The DC value is indicated by the heavy horizontal line. It is important to note that when thyristors are gated at 90°, the DC voltage is equal to zero. This is because thyristors conduct for the same amount of time in the positive and negative bridge. The net result is 0 V_{DC}. DC voltage will increase in the negative direction as the gating angle (α) is increased from 90° to a maximum of 180°.

The above is displayed graphically in the following graph.

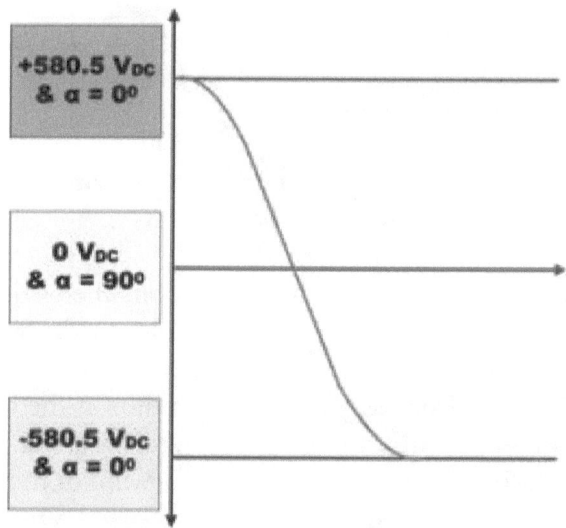

Figure 7.14 DC voltage as function of α.
Courtesy of Siemens AG

7.10 Basic Drive Operation

A thyristor bridge is a technique commonly used to control the speed of DC motor by varying the DC input voltage. It is important to note that the voltage applied to a DC motor can no longer be greater than the rated voltage specified by the manufacturer on the machine nameplate. This can be achieved by limiting the DC voltage to 0-430 V_{DC} by adjusting the control logic. Also, it is important to limit the shunt field voltage to a specified limit by the manufacturer.

The DC drive supplies voltage to the motor to operate at desired speed. The motor draws current from the power source in proportion to the torque or the load applied at the motor shaft.

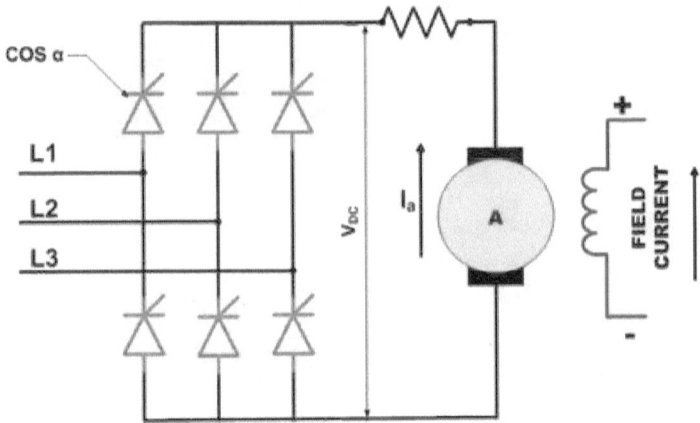

Figure 7.15 Oeration of DC motor
Courtesy of Siemens AG

If one needs motoring in only one direction, then single-quadrant drive with six thyristors is enough. Single-quadrant drives only operate in quadrant I. Motor torque (*M*) is developed in the forward or clockwise (CW) direction to drive the motor at the desired speed (*N*). It is analogous to driving a car forward on a flat surface from standstill to a desired speed. It takes more forward or motoring torque to accelerate the car from zero to the desired speed. Once the car is at desired speed, your foot can be let off the accelerator a little. When the car comes to an incline, a little more gas, controlled by the accelerator, maintains speed. To slow or stop a motor in single-quadrant operation, the drive lets the motor coast.

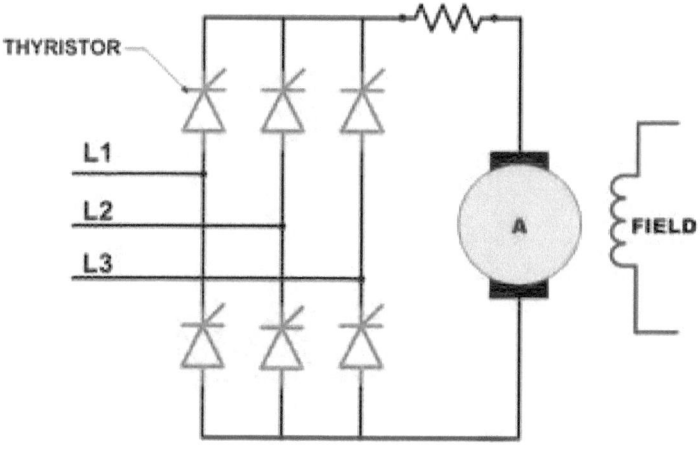

Figure 7.16 Single-quadrant drive.
Courtesy of Siemens AG

7.11 Changing the Direction of Rotation of DC Dynamometer

There are two ways to change the direction of a DC dynamometer/motor.

1. **Reverse Armature Polarity**

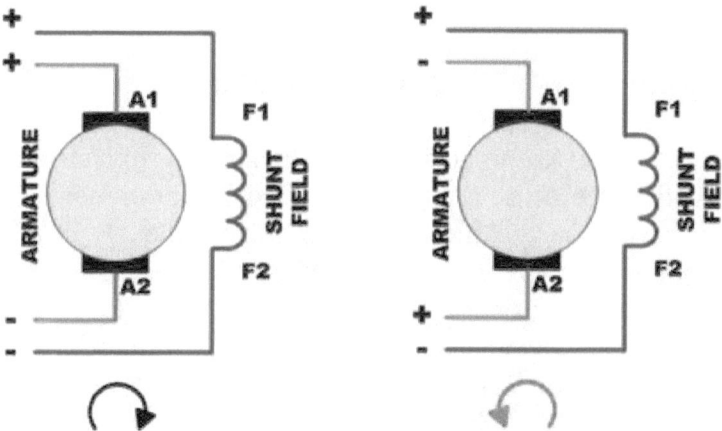

Figure 7.17 Change of direction—reverse armature polarity.
Courtesy of Siemens AG

2. Reverse Field Polarity

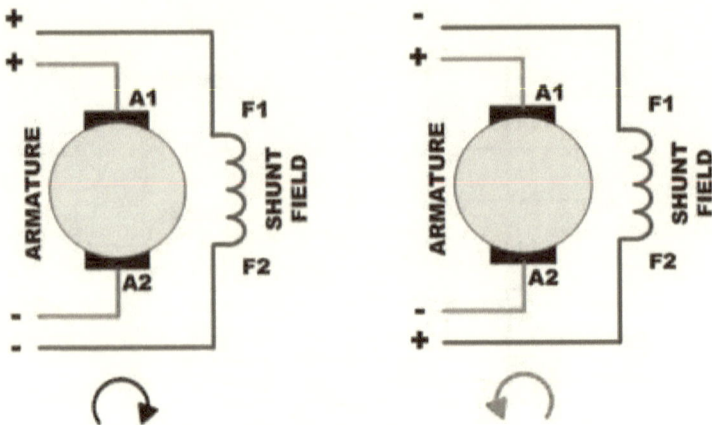

Figure 7.18 Change of direction—reverse field polarity.
Courtesy of Siemens AG

7.12 Stopping a Motor

Single-quadrant operation can be done by simply removing voltage to the motor and allowing the motor to coast to a stop. Alternatively, voltage can be reduced gradually until the motor is at a stop. The amount of time required to stop a motor depends on the inertia of the motor and the connected load. The more the inertia the longer is the time.

7.13 Regeneration—Four-quadrant Operation

In order for a drive to operate in all four quadrants, a means must exist to deal with the electrical energy returned by the motor. Electrical energy returned by the motor tends to drive the DC voltage up, resulting in excess voltage that can cause damage. One method of getting four-quadrant operation from a DC drive is to add a second bridge connected in reverse of the main bridge. The main bridge drives the motor. The second bridge returns excess energy from the motor to the AC line. This process is commonly referred to as regeneration. This configuration is also referred to as a four-quadrant design.

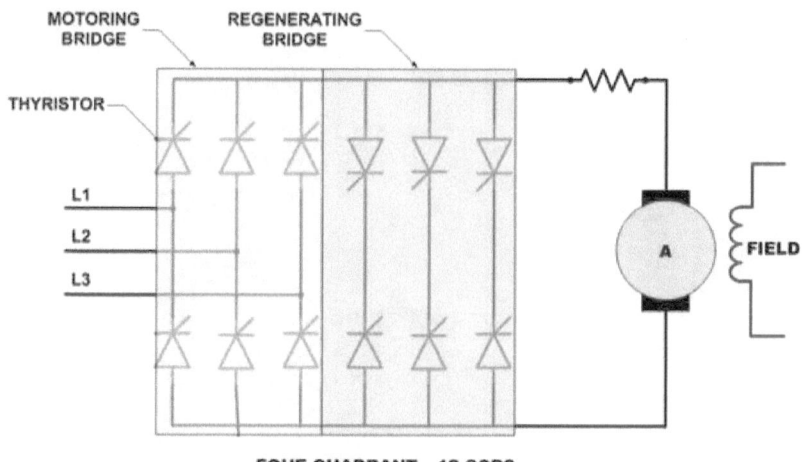

Figure 7.19 Four-quadrant twelve-thyristor bridge.
Courtesy of Siemens AG

7.14 Emergency Breaking

When the motor is required to stop quickly, the motoring bridge shuts off, and the regeneration bridge turns on. Due to the initial inertia of the connected load, the motor acts like a generator, converting mechanical power at the shaft into electrical power which is fed back to the AC line. The $I_a R_a$ voltage drop is of opposite polarity when the drive was supplying motoring power. Because the regeneration bridge is of opposite polarity, the voltage applied to the motor acts like an electrical brake for the connected load.

7.15 Speed Control Mode

When a DC machine is operating in speed control mode, that is, speed constant mode, it is usual to supply a current or speed feedback signal to the appropriate loop for more effective control of the drive. Current feedback sensors are built-in, whereas speed feedback is provided directly from the armature sensing circuit (default), or by taco generator, encoder to the relevant option board.

When in speed control mode, it is possible to modify the performance of the drive further by controlling the motor field, that is, field control. By weakening the field current, you can obtain an increase in motor speed beyond that normally achievable for the rated armature voltage of the DC motor.

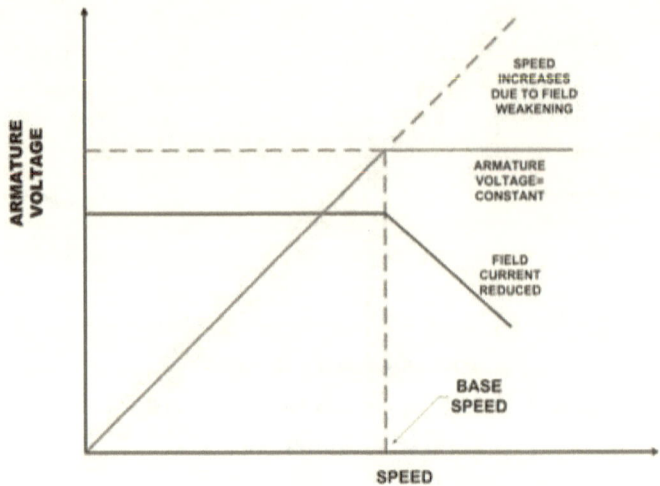

Figure 7.20 Effect of field weakening.
Courtesy of Siemens AG

7.16 Regeneration—Feedback to Mains

Line Reactor/Isolation Transformer

If no isolation transformer is used, it is recommended that user should install specified line reactor with the drive to provide a known supply impedance for effective operation of thyristor transient suppression circuits.

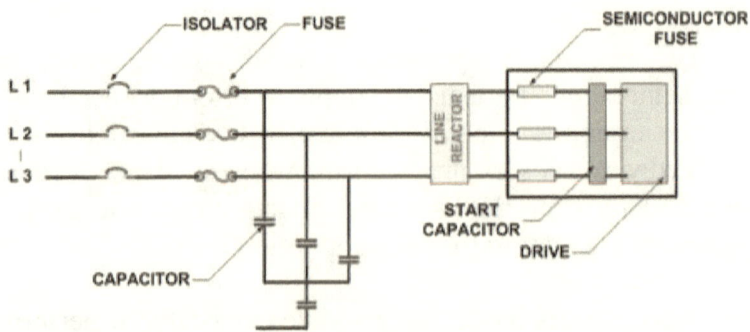

Figure 7.21 AC line reactor and capacitor fitted to drive.
Courtesy of Siemens AG

The DC drive panel system installation includes circuit breaker, choke, contactor, and control logic. Normally, the panels are offered as integrated fully built.

7.17 DC Dynamometer Application in Engine Testing

DC dynamometer is a regenerative dynamometer in which the prime mover under testing drives a DC motor which acts as a generator to create load which is fed AC power back into the commercial electrical power grid. The feeding back the power generated into the grid may generate some revenue for the dynamometer user.

DC dynamometer test cell will have additional components related to DC machine than the conventional test cell with eddy-current or hydraulic dynamometers. These components are mainly a DC drive, line reactors or isolation transformer, and a system to feed the electricity generated or the load bank to dissipate the energy.

Figure 7.22 DC dynamometer test bed.
Image courtesy of Dyne Systems, Inc. and Midwest & Dynamatic Dynamometers, USA. Copyright 2010 Dyne Systems, Inc.

Small DC dynamometer applications will use a load bank to dissipate electrical energy instead of feeding back to main grid. The load bank is a device, which is an electrical load, applies the load to an electrical power source which is a regenerative DC dynamometer in case of engine test cell with DC dynamometer. A load bank is intended to accurately simulate the operational or real life load which power source will encounter in actual application. However, unlike the "real" load, which is likely to be dispersed, unpredictable, and random in value, a load bank provides a contained, organized, and fully controllable load.

Load bank can be defined as self-contained, unitized, systematic device which includes both load elements with control and accessory devices required for operation.

7.18 Types of Load Banks

1. Resistive load bank.
2. Reactive load bank.
3. Capacitive load bank.

1. Resistive Load Bank

This is the most common type of load bank used. Resistive load bank proves equivalent loading of both generator and prime mover. The amount of load applied by load bank to the generator, the same amount of load is applied to the engine under test on AC/DC dynamometer.

The resistive load bank creates the load by the conversion of electrical energy to heat by means of resistors. This heat must be dissipated from the load bank, either by air or by water, by forced means or convection. Normally on small DC/AC dynamometers the load banks are dipped into the water tank. The variation in load is acheived by switching the number of resistances in the circuit.

2. Reactive Load Bank

A reactive load includes inductive (lagging power factor) and/or capacitive (leading power factor) loads.

Iron-core elements which act as inductive load when used in conjunction with resistive load bank create a lagging power factor.

Capacitive Load Bank

The capacitive load bank offers the leading power factor loads. These types of load banks are not very common in engine test bed.

Armature Choke for the DC Motor

A DC controller with six thyristor bridge configuration normally yields a waveform smooth enough to provide satisfactory operation DC motor without an armature choke.

But if excessive heating or excessive arcing and sparking at the brushes occur on start-up, an armature choke should be installed. The choke should be installed in series between the DC output of the controller and the motor.

Closure

The DC machine has capability of running in motoring or generating mode. The machine can be harness to be used as universal dynamometer with appropriate control system. The control system is most commonly called as DC drive unit. The advantage of using DC machine is while in generating mode, the power generated can be fed back to mains generating some revenue for the testing lab. The second advantage of DC machine (dynamometer) is it can be used to motor the engine to find frictional horsepower and also can be used for the starting of the engine. The DC machines are mostly air cooled, and the water piping is redundant in this case. The motoring and absorption ability of DC machine makes a good choice for a transient dynamometer and choice of loading unit in chassis dynamometer. Electrical machines such as DC and AC machines are the dynamometer of the future.

Bibliography:

1. Siemens drive manual.

Further reading on Drives and motors

1. Austin Hughes, *Electrical Motors and Drives*.
2. Theodore Wildi, *Electrical Machines, Drives and Power Systems*.

CHAPTER 8

Alternating Current (AC) Dynamometer

8.1 Introduction

We have seen in chapter 5 the aspects of DC dynamometer and its application in engine testing. AC dynamometer is much similar to it, except it uses alternating current motor commonly known as AC motor and powered by AC power source.

Alternation current (AC) dynamometer is a machine to test the mechanical power of the engine under test. A standard AC dynamometer system consists of an AC converter, a gearbox, speed sensor to measure RPM, torque transducer, and four quadrants of AC variable frequency and variable speed system.

8.2 AC Electrical Dynamometers

As we have seen in chapter 1, dynamometers use different methods to induce a braking force on the device under test. What exactly are the benefits of different types of dynamometers? In this chapter, we will study *AC dynamometers* in detail.

The modern invention of AC variable speed drives have made AC units acceptable for transient test cycles such as the ETC, the ELR, the ESC, and the FTP where motoring is required. AC machines are also used in chassis dynamometer application for testing of the vehicles.

AC dynamometers like DC dynamometers when used for engine testing are employed for

1. engine testing under load
2. motoring of the engine
3. engine starting.

The current trend is that AC machines are preferred over DC machines due to simplicity in construction of AC machine.

8.3 Principle of Operation

The working principle of the AC dynamometer is to convert the AC current from the generator into the direct current (DC) through ACS, and it will transform into the AC again for feeding to main grid line. AC variable frequency drive can adjust the current of the machine to control the rotation speed and the torque of the unit under testing (engine).

8.4 Rating of AC Dynamometer

The inertia of the AC dynamometer is lower than that of the DC dynamometer due to inherent difference in its construction for the same size of machine in terms of the power. This is due to less number of rotating parts. There is no commutator ring in AC machine which increases the inertia of DC machine substantially.

8.5 Construction of AC Dynamometer

The testing application that involves AC dynamometer will have two variants depending upon the mounting of AC motor/machine.

1. AC motor mounted in cradle using trunnion bearings and swinging arm with precision strain gauge load cell is used for measurement of torque.
2. AC motor mounted on its foot/base with in-line torque transducer is used for measurement of torque.

The following figure shows the cross-sectional view of AC dynamometer.

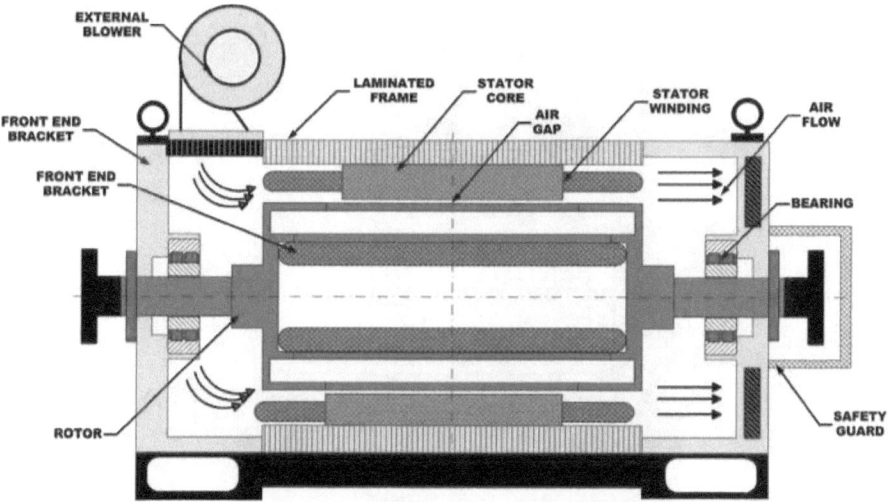

Figure 8.1 Cross section of AC machine.

8.5.1 Trunnion-mounted AC Dynamometer

The following figure shows the AC dynamometer mounted in trunnion bearings. The dynamometer carcass/motor frame is cradled in base frame using heavy-duty ball bearings. The swinging cradle is connected by using swinging arm, also known as lever arm, to a universal load cell. The other end of the load cell is connected to the baseplate.

The AC dynamometers are normally air cooled. The heat generated within the dynamometer is dissipated using the integral cooling fan, and it is known as forced air cooling.

Figure 8.2 Trunnion-mounted AC Dynamometer.
Image courtesy of Dyne Systems, Inc. and Midwest & Dynamatic Dynamometers, USA. Copyright 2010 Dyne Systems, Inc.

8.5.2 Foot-mounted

The foot-mounted AC dynamometer is easy to adapt for engine testing application. A foot-mounted AC machine is mounted on foundation baseplate. It is a robust construction.

Figure 8.3 AC dynamometer foot-mounted. Courtesy of Horiba Inc.

The foot-mounted AC dynamometer employs in-line torque transducer which is mounted between engine and dynamometer shaft. More commonly, a torque flange is used, which is a torque measuring device, that is bolted directly to the input flange of the AC dynamometer.

8.5.2.1 Advantage of Foot-mounted Machines

1. Heavy pedestals supporting the trunnion bearings are eliminated including expensive heavy-duty ball bearings.
2. Lever arm and load measuring chain mounted on dynamometer is also eliminated.
3. Standard foot-mounted AC motor can be adapted to work as dynamometer.
4. The weight of bare foot-mounted AC dynamometer is less than that of trunnion-mounted dynamometer.
5. Foot-mounted machines can be installed with in-line torque transducer or torque flange.

8.6 Liquid-cooled AC Dynamometer

The AC dynamometers are also available in liquid-cooled variant. The liquid-cooled machines are much more compact than the air-cooled machines of equivalent size. The heat dissipation in the water-cooled AC machines is more efficient that the heat recovery from air-cooled AC machine.

8.6.1 The Advantages of Liquid-cooled AC Dynamometer

1. Smaller dimensions for a given power capability.
2. Liquid cooling allows for a smaller rotor for a given torque capability.
3. Reduced noise, lower than standard air-cooled machine.
4. Higher torque ratings are possible.
5. A wider power range for a given dynamometer provides the capability of running a wider range of applications.

8.7 Mounting of In-line Torque Transducer

The AC dynamometer has two modes of operation:

1. constant torque mode
2. constant speed mode.

During the torque constant mode to maintain torque, it needs feedback on the real-time torque being exerted on the engine. The following figure shows the torque transducer mounting. The torque transducer is installed in line with the flange coupling of the AC dynamometer. It consists essentially of a flanged torque shaft fitted with strain gauges.

Figure 8.4 In-line torque flange mounting.
Courtesy of PowerLink—Power Testing Instrument Co., Ltd.

8.8 Comparison of AC and DC dynamometers (Bare Machines)

	AC dynamometer	DC dynamometer
1	AC machines can be classified as induction, wound rotor, and synchronous.	DC machines can be classified as self-excited, separately excited, permanent magnet (PM), and brushless or with brush
2	Induction motors can be further classified as three-phase and single-phase. A three-phase induction motor can be further classified as delta wound or star wound.	Self-excited machines can be further classified as shunt, series, and compound. Compound machines can be further classified as cumulative and differential.

3	Common terms you will hear discussed with AC machine include frequency, synchronous speed, and slip. An AC waveform is time varying or oscillatory. This means its amplitude starts at zero, rises to some maximum value, returns to zero, falls to some minimum value and then returns to zero. The number of times this occurs per unit of time is referred to as frequency	Common terms you will hear discussed with DC machine include commutator, brushes, counter electromotive force (EMF), torque, speed regulation, and speed-torque characteristic curve
4	AC machines are weightless than equivalent DC machine	DC machines are heavier for the same capacity AC machine
5	Power limitation is caused by the breakdown torque of AC motor decreasing as the square of speed ($1/n^2$)	Power limitation is caused by the commutation of DC motor

8.9 Power and Torque Envelopes

The following figure shows the typical power envelope in motoring and generating mode of an AC dynamometer. The constant torque line is corresponding to maximum current and excitation. The maximum performance limited by maximum permitted power output of the machine and mechanical design limits of the shaft and the machine structure.

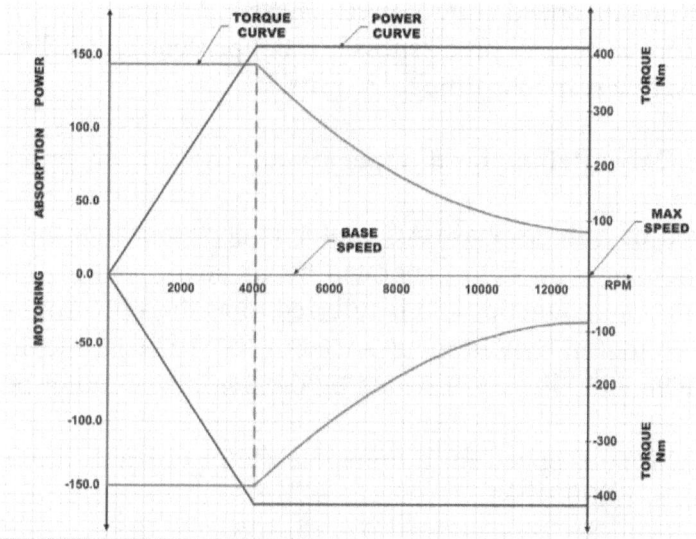

Figure 8.5 Typical power and torque envelope AC dynamometer.

Since these are four-quadrant machines, power absorbed can be reduced to zero, and there is no minimum curve.

In-line torque transducer is mounted between engine and dynamometer shaft. More commonly, a torque flange is used, which is a torque measuring device, that is bolted directly to the input flange of the brake and transmits data to a static stator.

8.10 Inertia of AC Dynamometer

The inertia of the AC dynamometer is normally less compared to the same size of DC machine. AC machine has less number of rotating parts. There is no commutator ring in AC machine. The response of the AC dynamometer is higher than the response of the DC dynamometer due to its low inertia.

8.11 Typical Accuracies of AC Machine

Typically, an AC machine can achieve the following accuracies:

1. Rotation speed testing preciseness : ±1 r/min.
2. Torque testing preciseness : ±0.1 percent FS.

3. Rotation speed control preciseness : ±1 rpm.
4. Torque control preciseness : ±0.2 percent FS (full scale).
5. Active response time : < 5 ms.

8.12 AC Drive Principles of Operation

The speed of a three-phase AC motor is dependent on the speed at which the magnetic field generated by its stator rotates. The speed of the rotating magnetic field is referred to as the synchronous speed (N_s) of the motor. Synchronous speed in RPM is equal to 120 times the frequency (F) in hertz divided by the number of motor poles (P).

The synchronous speed for a four-pole motor operated at 60 Hz, for example, is 1,800 RPM.

Synchronous speed decreases as the number of poles increases. The accompanying table shows the synchronous speed at 50 and 60 Hz for several different pole numbers.

For example,

$$N_S = \frac{120\,F}{P} \quad \ldots\ldots 8.1$$

$$N_S = \frac{120 * 60 H_z}{4}$$

$$N_S = 1800\ RPM,$$

where
N_S = synchronous speed
F = frequency
P = number of poles

An AC drive controls the speed of an AC motor in order to control the speed of the equipment mechanically connected to the motor. AC drives are most commonly used to control the speed of pumps and fans, but many other types of systems also use AC drives, i.e AC dynamometers used in engine and vehicle testing.

AC drive, inverter, and variable frequency drive are all terms that refer to equipment designed to control the speed of an AC dynamometer. However, the term inverter also refers to a part of an AC drive circuit.

The AC dynamometer with motoring and regeneration requires four-quadrant operation, where torque can be applied in either direction, regardless of the direction of AC machine rotation.

In quadrants II and IV, as shown in the accompanying speed—torque graph, the direction of torque is opposite to the direction of motion, so these are breaking quadrants.

Torque will always act to cause the rotor to run toward synchronous speed. If the synchronous speed is suddenly reduced, negative torque is developed in the motor. The motor acts like a generator by converting mechanical energy from the shaft into electrical energy which is returned to the AC drive. This is similar to driving a car downhill. The car's engine will act as a brake.

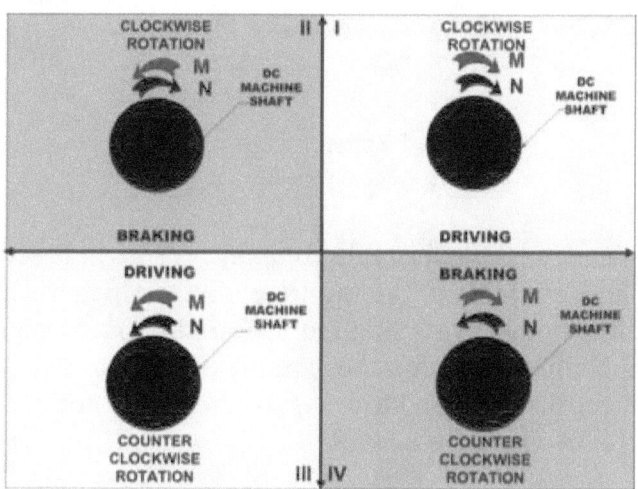

Figure 8.6 Four-quadrant operation of AC machine.
M, torque; N, speed.
Courtesy of Siemens AG

8.12.1 AC Drive Basics

Adjustable frequency AC motor drive controllers, frequently termed inverters, are typically more complex than DC controllers since they must perform two power section functions that of conversion of the AC line power source to DC and finally an inverter change from the DC to a coordinated adjustable frequency and voltage output to the AC motor. In other terms, AC drives function by converting a constant AC frequency and voltage into a variable frequency and voltage. The AC drive mainly consists of

1. incoming supply
2. converter
3. DC filter
4. inverter.

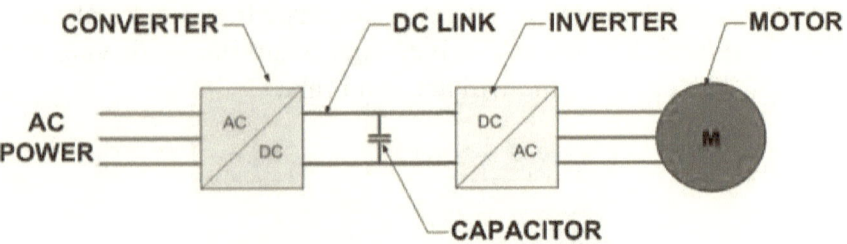

Figure 8.7 Schematic of AC drive.
Courtesy of Siemens AG

The acceptance of the adjustable frequency drive is based upon the simplicity and reliability of the AC drive motor, which has no brushes, commutator, or other parts that require routine maintenance, which more than compensates for the complexity of the AC controller. The robust construction and low cost of the AC motor makes it very desirable for a wide range of uses.

Also, the ability to make an existing standard constant speed AC motor an adjustable speed device simply by the addition of an adjustable frequency controller creates a very strong incentive for this type of drive.

8.12.2 Converter and DC Link

As the accompanying graphic shows, a converter section which is designed to convert an AC input to a DC output. In this example, as is often the case, the input is three-phase AC. In the absence of the DC link capacitor, the resultant DC has undesirable variations, referred to as ripple.

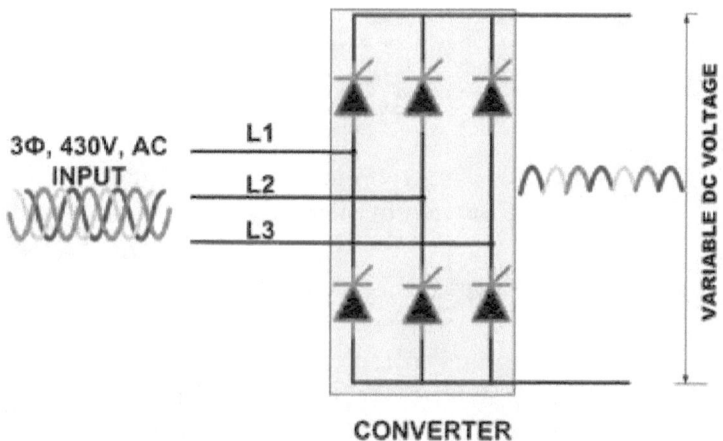

Figure 8.8 Converter output without capacitor.
Courtesy of Siemens AG

With the addition of the DC link capacitor, most of the ripple is removed. In essence, the DC link capacitor functions like a battery, providing power to the inverter section. Unlike a battery, however, the DC link capacitor is continually supplied with energy by the converter.

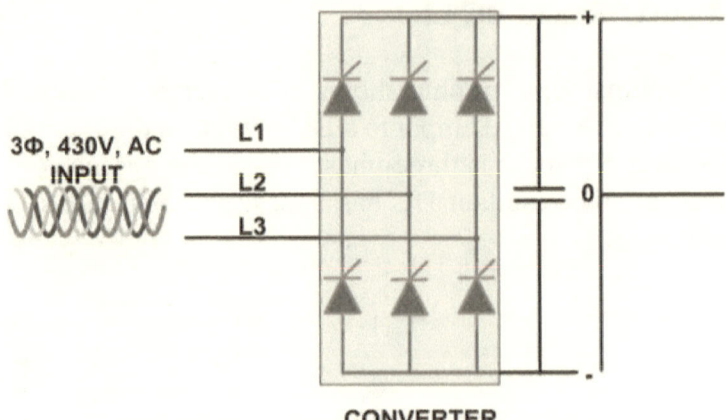

Figure 8.9 Converter output with capacitor.
Courtesy of Siemens AG

8.12.3 Insulated Gate Bipolar Transistor (IGBT)

AC drive inverter circuits often include insulated gate bipolar transistors (IGBTs). The IGBTs provide the high switching speed necessary for modern inverter operation. IGBTs are capable of switching on and off several thousand times a second. An IGBT can turn on in less than 400 nanoseconds and off in approximately 500 nanoseconds. An IGBT has three leads: a gate, a collector, and an emitter. When a positive voltage (typically +15-volt DC) is applied to the gate, the IGBT turns on. This is similar to closing a switch. When the IGBT is turned on, current flows between the collector and an emitter. An IGBT is turned off by removing the positive voltage from the gate. During the off state, the IGBT gate voltage is normally held at a small negative voltage (−15-volt DC) to prevent the device from turning on.

Typically, each switch element is made up of an IGBT and a diode, called a free-wheeling diode. The free-wheeling diode protects the IGBT by conducting current when the IGBT turns off, and the motor's collapsing magnetic field causes a current surge. This is also the path that is used when the motor is connected to an overhauling load. When a motor is connected to a load which is going faster than the speed set point, the drive will attempt to slow down the motor to regulate it to the set point.

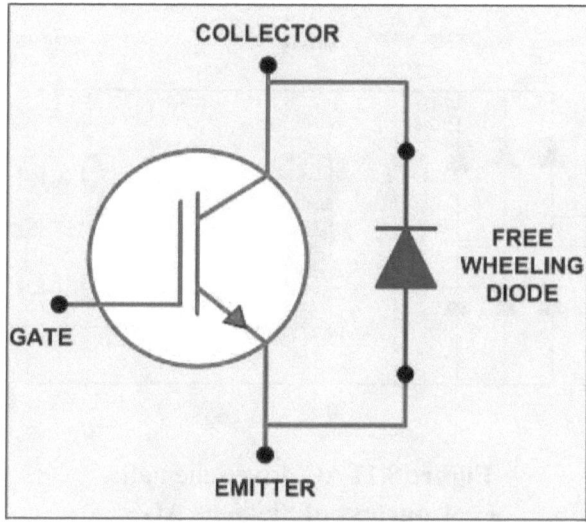

Figure 8.10 Schematic of IGBT.
Courtesy of Siemens AG

8.12.4 Converter and Control Logic

The control logic and inverter sections function to control the voltage and frequency applied to the AC motor. The accompanying graphic shows a basic AC drive with six IGBTs in the inverter section. The switching of each IGBT is controlled by the control logic. For the purpose of simplicity, the free-wheeling diodes are not shown.

There are a variety of ways to control the voltage and frequency applied to the motor. The most common approach is to switch the IGBTs in such a manner as to control the frequency and width of pulses applied to each motor phase. This is an approach referred to as pulse width modulation (PWM).

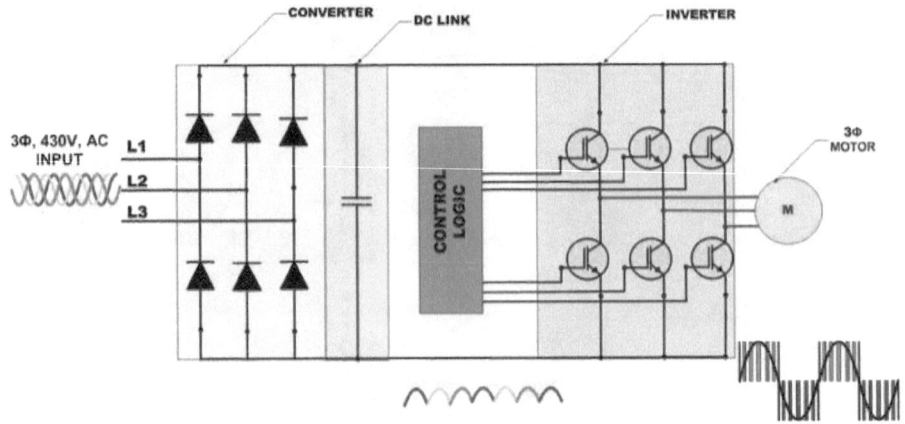

Figure 8.11 AC drive schematic.
Courtesy of Siemens AG

8.12.5 Pulse Width Modulation (PWM)

A pulse width modulation inverter controls the speed of an induction motor by varying the frequency and pulse widths of the voltage applied to the motor. This is done by controlling the IGBTs in the inverter.

Figure 8.12 shows the switching pattern which is known as pulse width modulation (PWM). If the length of time is increased for the IGBT to be on and then off, the motor responds to it as a sinusoidal waveform. The positive IGBT fires first in Figure 8.12 followed by its negative counterpart.

The voltage and frequency applied to the motor is the same for each phase, but each phase is shifted in time in the same manner as a typical three-phase voltage.

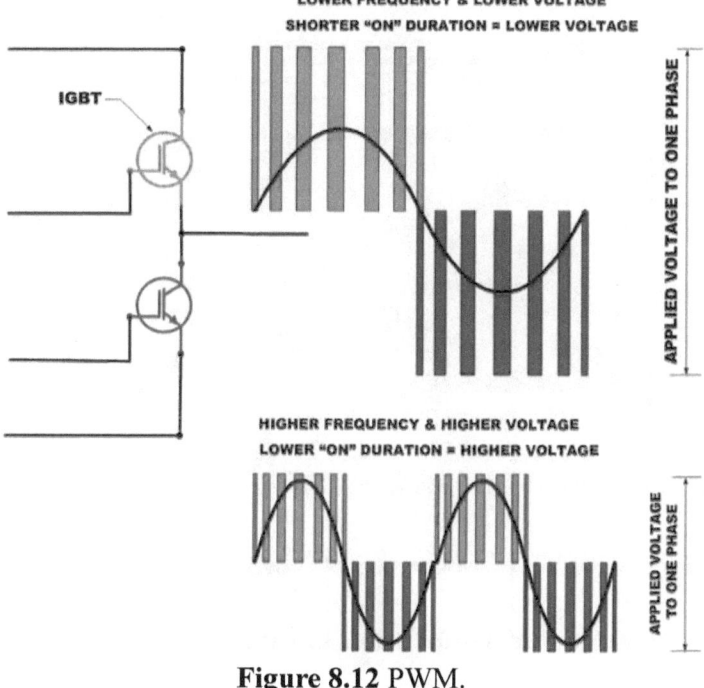

Figure 8.12 PWM.
Courtesy of Siemens AG

8.12.6 Control Panel

The enclosure is a steel cabinet which houses the incoming supply, circuit breaker, AC drive, and output lines to the AC machine. The cabinet structure is modular in nature. The AC drive enclosure is designed to meet different safety standards. The sizing of the enclosure depends upon

1. total number of drives housed in enclosure
2. total ventilation required
3. power output affected by derating due to altitude and ambient temperature.

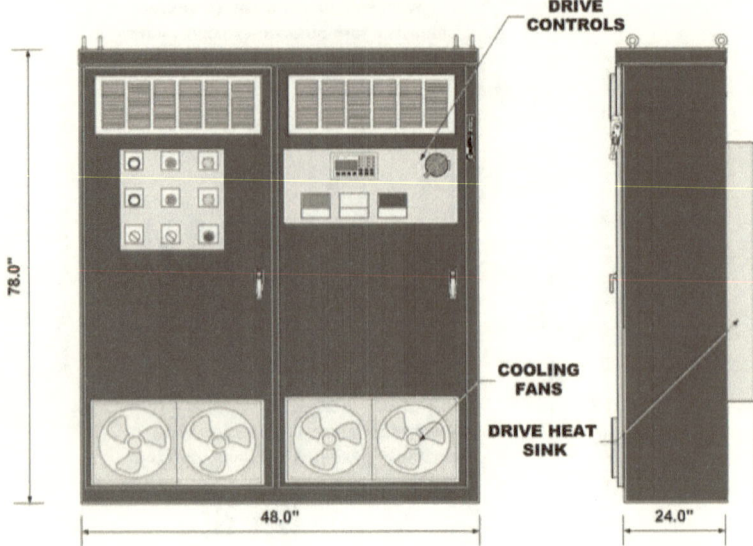

Figure 8.13 Typical layout of AC drive panel (enclosure).

8.12.7 Dynamometer Application in Engine Testing

The following figure shows the AC dynamometer test cell used for the engine testing purpose. The AC dynamometer is connected to a stand-alone computer that controls the speed and torque of the AC generator. The software that controls the dynamometer can reproduce any drive cycle to be simulated on engine.

Figure 8.14 AC machine—trunnion-mounted. Courtesy of Horiba.

Closure

Today, technological developments have made it feasible to replace hydraulic, eddy-current, and DC dynamometers, with AC dynamometers. Regenerative AC drive solutions provide energy recovery benefits from the device under test which can reduce the overall cost of testing. The AC dynamometer because of its capability to feed the power generated into mains is preferred solution. The AC dynamometer is similar to DC dynamometer with difference of power input. The input power to AC machines is alternating current. The AC machines have certain advantages such as machine is more compact and simpler than the DC dynamometer. The response of an AC machine is higher than the response of the DC machine due to low inertia of the AC machine. The inertia of the AC machine is lower than that of equivalent DC machine.

AC dynamometer with associated dynamometer controls and data acquisition system is a modern solution to test the engines and vehicles. AC dynamometer with four quadrants can function dynamically between quadrants. The AC dynamometer is an active dynamometer, specifically with the ability to transition dynamically between quadrants. Dynamic transition between quadrants means that the dynamometer system can simulate a vehicle load climbing up a hill (requiring generating torque or braking torque) and then going down the hill (requiring a motoring), thereby transitioning between two quadrants on the fly. The AC dynamometer system is capable of fully programmable, regenerative, and upgradeable.

Likewise DC machines, the AC machines are also used as transient dynamometers and find its application in chassis dynamometer. Regeneration also is an inherent part of test stands built with AC machines.

Bibliography:

1. In-line torque transducer. Picture form Powerlink catalogue.
2. *AC Drive training manual of Siemens.*

CHAPTER 9

Economics of Comparison

9.1 Introduction

In previous chapters, we have studied different types of dynamometers from simplest to most modern version like fan brake, hydraulic, eddy-current, DC and AC dynamometers, and magnetic powder. As designer of test cell question is raised, what dynamometer is suitable for application under consideration? Each type of dynamometer has its own advantages and disadvantages, and choosing the correct one will depend largely on the type of testing to be performed. In fact, beyond simple power and torque measurements, dynamometers can be used as part of a test bed for a variety of engine development activities such as the calibration of engine management controllers, detailed investigations into combustion behavior and tribology. (*Tribology* is the science and engineering of interacting surfaces in relative motion. It includes the study and application of the principles of friction, lubrication, and wear.)

9.2 Parameters Governing Selection of Dynamometer

The selection of dynamometer size or capacity depends upon the maximum torque, maximum speed, and maximum power of the engines to be tested.

1. Maximum Torque

It is important to consider all torque points that are to be tested, not only rated torque but also locked rotor and breakdown torque. Dynamometer selection should initially be based on the maximum torque requirement, subject to determining the maximum power requirements.

2. Maximum Speed

This rating is to be considered independent of torque and power requirements, and it is the maximum speed at which the dynamometer can be safely run under free run or lightly loaded conditions. It is not to be considered as the maximum speed at which full braking torque can be applied.

3. Maximum Power to be Absorbed

These ratings represent the maximum capability of the dynamometer braking system to absorb and dissipate heat generated when applying a braking load to the motor under test. The power absorbed and the heat generated by the dynamometer is a function of the torque (T) applied to the motor under test and the resulting speed (n) of the motor. This is expressed in these power (P) formulas.

9.3 Important Factors Influencing Selection of Dynamometer

1. Application of Dynamometer for Intended Testing

The type of testing is major governing factor which will decide the type of basic loading/braking device, that is, the bare dynamometer. The major types of testing are as follows:

1. production testing
2. quality testing/audit testing
3. R&D testing
4. emission testing
5. endurance testing
6. after-sales market—garages and service stations.

2. The Instrumentation Interfacing with Dynamometer

This is also an important factor as dynamometer should be capable of delivering proper output to be interfaced with modern data acquisition system.

3. Total Cost of Installation

The total cost of installation, that is, the complete test, should be evaluated with different types of dynamometer depending upon the application testing.

4. Motoring Requirement and Ability of Dynamometer to Perform Motoring

If this is a requirement from testing engineers, then this must be considered by design engineers to add this facility. It may change the total perspective of the test cell if one has to select the simple motoring attachment with power absorption unit (PAU), aka dynamometer, or one should select AC or DC dynamometer.

9.4 Comparison of Hydraulic and Eddy-current Dynamometers

When you think of hydraulic dynamometer, you have two types of hydraulic dynamometer for your consideration, that is,

1. Sluice gate type dynamometer.
2. Fast response servo-controlled dynamometers.

The comparison can be drawn between fast response servo-controlled hydraulic dynamometer with eddy-current dynamometers, as both these dynamometers are closed-loop-controlled machines. The following points are worth noting while you select a dynamometer for your test requirements.

	Hydraulic dynamometer	**Eddy-current dynamometer**
1	Weight of hydraulic dynamometer will be less for same capacity of power absorption	Weight of eddy-current dynamometer will be higher to that of same capacity hydraulic dynamometer
2	Torque and speed controls are affected by fluctuation in water pressure	No effect of water fluctuation on torque and speed controls
3	Normally, highly recommended for production tests	Highly recommended for R&D and quality audit-type test cells
4	Electrical power consumption is less. Only required for servo control	Electrical power is main source of eddy-current coils for load absorption
5	Water supply failure does not have instant effect on damage of dynamometer	Water supply failure is liable to major damage to power absorption unit
6	Servo-controlled hydraulic dynamometer is comparatively cheaper than same capacity eddy-current dynamometer	Eddy-current dynamometer will be little more expensive than equivalent capacity hydraulic dynamometer
7	Less maintenance required. Scale formation in stator and rotor compartments is very slow	More maintenance required. Frequency of scale formation in loss plate (water chamber) is more rapid

9.5 Water Supply System and Its Requirement

The dynamometer's ability to dissipate heat is a function of how long a load will be applied. Therefore, the maximum power ratings given are based on continuous operation under load, as well as a maximum of five minutes over load.

9.5.1 Influence of Water System on Working of the Hydraulic Dynamometer

The water supply system should supply the water under steady head, that is, constant pressure. The common header supplying the water to the test cells in test house should be designed in such way that it will maintain the required pressure even if all the test cells are operating those housing the hydraulic dynamometer. Let us say, two test cells are operating in a test house drawing water from common header which may maintain the pressure head. When all other remaining test cell starts operating, the header should be capable of supplying the required quantity of water without pressure drop.

The fluctuations in water pressure will create the load stabilization problem. Load variation is dependent on the pressure fluctuation in the water supply. The overhead tank is considered as one good source of constant water pressure supply to the dynamometer.

The fluctuation in water pressure is sometimes annoyance, and setting of proportional-integral-derivative (PID) becomes cumbersome task as pressure fluctuation affect dynamometer load. A water pressure regulator and/or an accumulator will correct most problems with inconsistent water pressure.

A water accumulator located near the test cell dampens the pressure fluctuations of the water and eliminates the pressure spikes associated with water system. However, this may add to the cost of the installation.

A dynamometer which is starving for the water quantity may tend to increase the water temperature and water vaporization inside the dynamometer and this will cause dynamometer control problems. It is a normal practice not to exceed the dynamometer water outlet temperature to 60°C.

Water pressure fluctuation does not pose a problem if the power absorption unit is eddy-current dynamometer and needs to have pressure regulator, and the accumulator is eliminated. However, the water supply system remains the same if the eddy-current dynamometer is of the same size to that of hydraulic dynamometer.

9.6 AC and DC Dynamometers

The test cell designer or the user of a dynamometer can decide whether they need the motoring facility or recover the energy/power generated while testing the prime mover. If these factors are important, then one may decide to go for motoring or regeneration type dynamometer. The ultimate choice of having a particular dynamometer is governed by a few factors listed below:

1. **Response of Machine/Dynamometer**

    The response time of lad, speed measurement, and control the governing factors while running a test sequence in steady-state testing or transient testing. Transient testing will need a type of drive technology and a number of quadrants the machines is required to operate considering motoring and absorption in both directions.

2. **Varying Load Cycle**

    If the load cycle is having frequent and rapid changes in load, then the AC or DC is preferred over the eddy-current dynamometer. The eddy-current dynamometer will be prone to loss plate damages due to the frequent and rapid changes in load.

3. **Minimum Load Capacity of Dynamometer**

    This is dependent on the minimum capacity curve of the dynamometer which influences the size of the smallest engine to be tested on the dynamometer. AC or DC machine will have the lowest minimum curve compared to eddy-current or hydraulic dynamometer.

4. Duty Cycle

The duty cycle of the testing also influences the selection of dynamometer, for example, overload testing of the engine. The dynamometer needs to be selected considering this factor.

5. Wide Range of Engines

The AC or DC can be selected to accommodate the wide range of dynamometers while maintaining the control accuracy. AC or DC machines can be operated from near-zero region to full power range of the power envelope.

6. Starting of Engine

If engine starting is employed using conventional dynamometers like eddy current or hydraulic, then it is necessary to install starting arrangement on the dynamometer. It can be in-line or piggyback motor mounted on the dynamometer to do initial cranking of the engine. This calls for additional cost and maintenance. However, the AC or DC can be used in motoring mode to start the engine.

7. Cooling Arrangement

The AC or DC motor will have their own cooling fan/blower to cool the windings. Installation with motoring/absorption dynamometer does not need water cooling system, thereby reducing the total cost of the installation.

Although the water-cooled versions of DC and AC machines are available, air-cooled machines are preferred due to simple installation. The only inherent advantage of adopting AC or DC water-cooled machine is the reduction in overall size of machine and longer life due to efficient cooling. However, it may add to installation and operating cost.

Advantages of Water-cooled AC or DC Machine

1. Water-cooled induction motors are designed to deliver a high power to weight ratio.
2. They serve in a variety of applications and may be good alternatives to ordinary air-cooled motors, especially where the temperature is high.
3. They frequently make sense in situations that call for minimal size and weight.
4. Water-cooled machines are also less noisy than fan-cooled motors and are highly reliable.

Disadvantages of Water-cooled AC or DC Machine

1. The installation cost is higher due to necessary water piping required to be run from the source to the installation point along with pumping machinery and piping accessories.
2. Moreover, the addition in operating cost is due to additional energy source for pumping the water supply to AC/DC dynamometer.
3. Water-cooled machines are custom-built and hence expensive. It is not easy to find the replacement for water-cooled machine.

After discussing above factors, question still remains unanswered that whether AC or DC dynamometer is economical. Let us consider the basic comparison between AC and DC before we could conclude.

9.7 Comparison of AC and DC Dynamometers

Factors	AC dynamometer	DC dynamometer
System price	It is more cheaper in some power ranges	Economical at very low power and very high power
Drive price	AC drive with IGBT is expensive	Uses well-proven SCR technology which is cost-effective

Motor price	Less expensive and more reliable	More costly due to commutator and brush technology
Installation cost	Several subtle issues may drive cost up	Usually less expensive and simple
Efficiency	Two power conversion stages	One stage conversion
Power factor	Constant and better	Good at high speed
Harmonics	6-, 12-, 18-, 24-pulse converters available at cost	6—or 12-pulse converter
Maintenance	Low maintenance or practically nil	Higher maintenance and need to monitor brush wear

9.7.1 System Price

It is the combined price of motor and drive. Cost for DC drive and motor is lower for small HP, whereas AC wins from 5 HP to about 100 HP. Then DC wins again as power increases. If DC motor is existing, then DC overall cost will be less. The system price is based on various cost factors.

9.7.2 Drive Price and Maintenance Costs

The main controller cost will be the same for AC and DC. Power and firing electronics dictate the total drive cost. DC drives use SCRs, whereas AC drives use IGBT as switching elements. IGBTs are more costly than SCRs, especially at higher power levels. AC drives comprises AC to DC conversion section. This uses diodes and bus capacitors that are not necessary in DC drive. These capacitors are expensive and pose maintenance problem at latter stage.

9.7.3 Motor Price

AC motors are usually less expensive than DC motors because they are simpler. AC motors don't have brushes like DC motors. Brush maintenance cost is usually sited in any cost comparison. The only good thing about DC brush maintenance is that it can be scheduled and usually doesn't result in unexpected downtime.

9.7.4 Installation Cost

In case of multiple drives, that is, multiple test cells common bus, AC drives can save money on system installation since only one mains connection is required versus individual DC drive mains connections. AC motors require only three power conductors, whereas DC motors require four. However, most AC drive suppliers recommend special shielded power cables to minimize electrical interference from the AC drive. The motor power lead filter is required if the cables to AC motor are over 100 feet long. Also, specially designed EMI filters may be required to be installed on the input side of the AC drive when it is installed in close proximity to EMI sensitive instrumentation devices. Add up all the extras when comparing installation costs.

The resistors used on most AC drives, with their additional stages and higher switching frequencies, generate more heat than DC and are less efficient.

9.7.5 Efficiency

DC drives are more energy efficient. AC drives have two stages of power conversion (AC to DC then DC to AC). DC drives have one stage (AC to DC). Each stage has energy losses in the form of heat generated in the drive. More heat is generated during switching (switching losses). The higher is the switching frequency, the more are the losses. If braking is required, the regeneration available in most DC drives will increase efficiency further beyond the DB resistors used on most AC drives. With their additional stages and higher switching frequencies, AC drives generate more heat than DC and are less efficient.

9.7.6 Power Factor

AC drives have a better power factor in most applications. DC drives approach the same power factor as AC drives only if you operate the DC drive at or near maximum speed.

9.7.7 Harmonics

AC and DC drives often produce similar harmonic problems on the AC mains. In either case, harmonics can be reduced by going to a 12-pulse (six phase) rectification scheme. This requires a phase-shifting transformer. The cost added for a 12-pulse AC drive is less than that for the 12-pulse DC drive. AC drives can also be purchased in 18—or 24-pulse configurations. An active front-end AC drive which utilizes an inverter for the DC bus supply, though costly, has the lowest harmonic content.

9.7.8 Performance

Higher switching frequency in the AC drive generally results in higher transient response capability than possible in a DC drive. AC motors permit to higher speeds. The AC motor may also have lower inertia. If the application requires "servo-like" performance or operation at high speed, then AC is usually a better choice. AC and DC can both be operated without an encoder. Performance suffers in either case. Low and zero speed performance is most severely affected.

9.7.9 Degree of Protection for Motors

The variable speed DC motors are customarily used with internal forced ventilation, whereas variable speed AC drives, asynchronous standard motors have predominantly been used for many decades, are with surface ventilations. The fact that AC motors with ratings of up to approximately 1,400 kW are supplied in degree of protection IP 54 as standard is a tribute to their simple and sturdy construction. DC motors have usually degree of protection IP 23.

9.7.10 Modernizing Existing DC Drives [1]

If old DC drive exists and needs to be modernized, then brainstorming session will help. The test cell designer needs to evaluate various possibilities or alternatives available to his disposal. The following alternatives if evaluated may help the test cell designer to arrive at a conclusion.

1. Replace the entire DC drive including converter and DC machine by a new DC drive.
2. Replace only the converter cubicle if the motor is still in good condition.
3. Replace the converter module by a modern digital unit.
4. Replace the old analog drive electronics by new digital electronics while continuing to use the power section.
5. Replace the entire drive system with a new AC drive, which is a new trend in industry.

The final solution of test cell designer will depend on the solutions to the following situations:

- The likely possibility of change of the requirements for the testing of engines in future, such as load requirements, regulatory changes by law which will force to change the capacity of the drive?
- System reliability of modernized drive.

Second choice will be replacing the existing DC drive with latest technology. AC drive will depend upon the following criteria's.

1. Will the new AC machine will fit on existing foundation, if not what will be the additional cost of new foundation?
2. Space availability for new AC motor as well as AC drive.
3. Outlay for new power cabling.
4. Interfacing of new AC drive with dynamometer control logic.
5. Is cooling and ventilation sufficient as per the requirement of new AC drive?

6. What is time duration of modernization with AC machine?
7. Whether the new AC machine with AC drive is commercially viable project?

A comparison of the two drive systems in this short overview shows that the question of whether the DC drive or the AC drive is the right choice for any particular user is entirely dependent on the individual engine/vehicle testing needs of the testing engineer.

9.8 Control Accuracy and Response

Engine testing dynamometers employ calibration routines that check the dynamometer's functionality and calibration. However, the standard calibrations carried out by test engineer do not ensure that the transient or steady-state loads applied by the dynamometer during a test meet response time or the loading accuracy requirements. The routing calibration does not account for checking response of the dynamometer. If the dynamometer has passed the routing calibration checks performed by quality assurance and quality audit, however, response time of the dynamometer is poor then there is possibility of error in loading during transient testing. The total response time is also a function of inertia of the dynamometer under consideration and response time of control system.

The measurement and control accuracy requirement offered by each type of dynamometer needs to be studied before any conclusion is drawn for the final selection of the dynamometer.

Closure

The overall economics of selection of dynamometer depends upon the user requirement of accuracy and response time and type of testing such as steady-state or transient testing. If motoring requirement is present, then the economics totally changes as it will demand a hybrid dynamometer, that is, combination of hydraulic/eddy current with motoring arrangement or AC or DC dynamometer system. However, factors such as test cell space, cooling arrangement, installation cost, and time frame for the test cell to be operative from inception to commission of the equipment.

Biblography:

1. A guide for users of variable-speed drives—ABB.
   Copyright 2012, material provided courtesy of ABB Ltd; reprint not allowed

CHAPTER 10

Control Modes—What Do We Control?

10.1 Introduction

The testing of engine is an interesting process, and it involves application of dynamometer in various modes. The characteristic parameters of an engine are speed, torque, and power output. The power output is calculated from the measured parameters of the prime movers. The dynamometer can be employed for engine testing in various modes, such as

1. open-loop mode
2. speed constant mode
3. trque constant mode
4. power law mode $M \alpha N^2$.

Similarly, when an engine throttle controller is employed to control the throttle lever of the engine, then the engine too can be operated in different modes, such as

1. open-loop or position mode
2. speed constant mode
3. torque constant mode.

It is irrelevant to the test engineer what amount of excitation current is supplied to eddy-current dynamometer or what amount of water pressure is inside of the dynamometer? It is significant to him to set the constant speed or constant torque mode which is useful for stabilized readings.

Let us first consider the dynamometer operation in different modes.

10.2 Open-loop mode

The dynamometer is allowed to run and load the engine by changing the outlet valve position. The operation of controlling the outlet valve can be remote through electrical controls. In this mode, merely a position of the outlet (usually it is butterfly valve) valve is changed, thereby changing the quantity of water flowing through dynamometer which will vary the load on the engine. Hence, more often, this mode is also called as position mode. In case of eddy-current dynamometer, it is achieved by changing the current flowing through the coils. In case of eddy-current dynamometer, open-loop mode is also called as constant current mode.

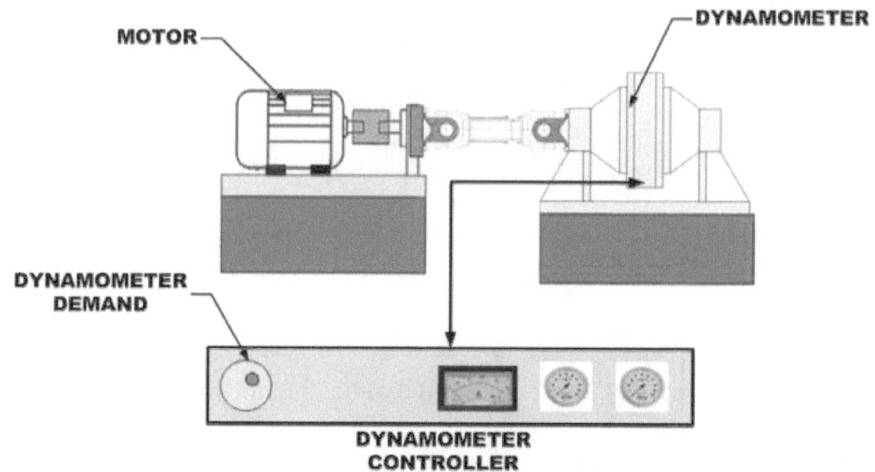

Figure 10.1 Open-loop control.

A typical open-loop system will consists of a dynamometer and open-loop controller. An open-loop system is often used for quick pass/fail testing on the production line or at incoming inspection. Predetermined current is set on dynamometer controller, and engine coupled is allowed to accelerate under pre-set load, and observation is made for the torque developed and compared with standard torque output. This is also called as "go/no-go" testing or comparative testing. In open-loop mode, there is no feedback of any of the parameters like "speed" or "torque." In case of eddy-current dynamometer, the current is increased as you increase the dynamometer demand control, thereby increasing the load. In case of hydraulic dynamometer, the position of outlet butterfly valve is adjusted depending on the dynamometer demand.

10.2.1 Open-loop Characteristics—Hydraulic Dynamometer

The following figure shows the open-loop characteristics of the hydraulic dynamometer with butterfly valve outlet-controlled. Three curves represent the different position of outlet valve.

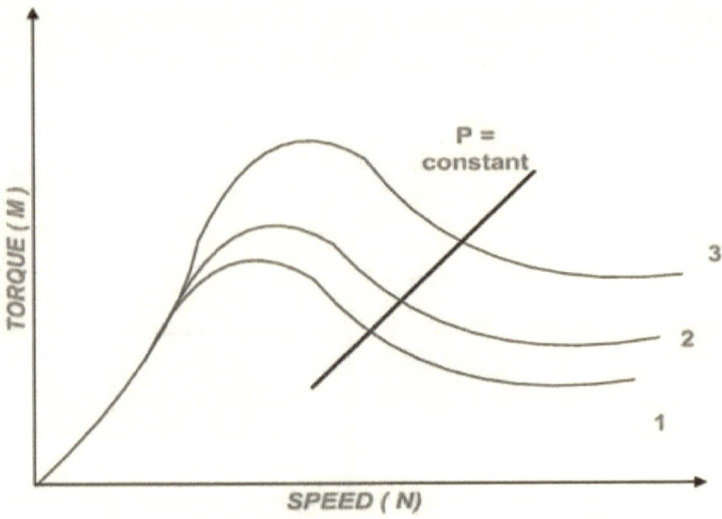

Figure 10.2 Open-loop hydraulic dynamometer.

10.2.2 Open-loop Characteristics—Eddy Dynamometer

As discussed above, the open-loop curves in the following figures are generated with different current settings.

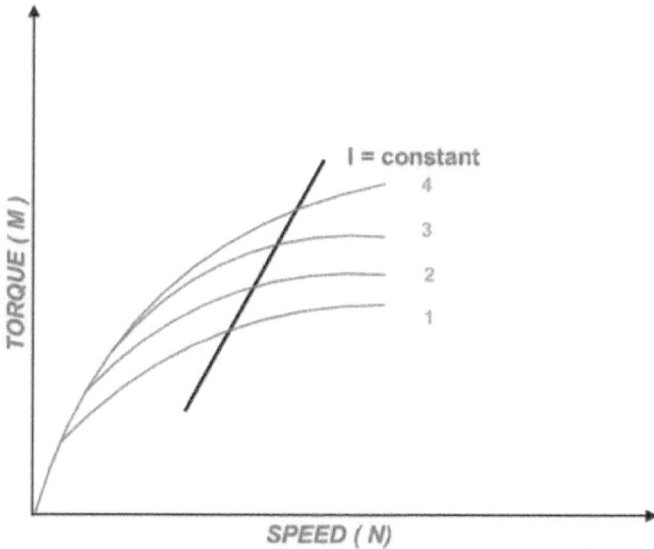

Figure 10.3 Open-loop characteristic for eddy-current dynamometer.

10.3 Closed Loop

What Exactly Closed Loop Means?

A closed-loop system uses the feedback information from the sensors, such as speed sensor or load cell. A closed-loop system is also called a servo mechanism or simply servo-controlled. In case of dynamometer, these sensors which give feedback information are usually for speed sensing a magnetic pulse pickup or encoder and for load sensing it is a universal load cell or in-line torque transducer.

10.3.1 Speed Constant Mode

In speed control mode, the actual speed is compared with the demanded speed from the control panel. The error (difference) signal is fed to the speed control circuit to increase or decrease the load. If the measured speed at dynamometer shaft is less than that of the set speed, the load is decreased. If the measured speed of the shaft is greater than that of the set speed, then the load is increased.

When a dynamometer is operating in "speed" constant mode, the feedback from speed sensor such as magnetic pickup or encoder is constantly compared with set torque.

10.3.2 Torque Constant Mode

If dynamometer is employed in "torque" control mode, engine torque is held constant at the cost of speed variation. When set torque and actual torque are different, then the difference in torque is determined and the error signal is fed to the torque controller. If the set torque is less than the measured torque, then the load is increased till the set torque is achieved, assuming that the engine has sufficient torque capability to attain the set torque. This will maintain the torque constant, while the speed will vary.

When a dynamometer is operating in torque constant mode, the feedback from load cell is constantly compared with set torque.

10.3.3 Power Law Mode

The power law mode is suitable for the variable speed gasoline or diesel engines used for the marine propeller drive. Propeller characteristic can be simulated approximately.

$$\text{Torque} = K(N^2) \quad \quad \quad 10.1$$

The factor K can be adjusted by the controller. This creates a duty cycle that is proportional to the square of the dynamometer speed. If the dynamometer has linear load characteristics, then the actual load will be proportional to the square of the speed.

10.4 Application of Modes for Engines

Operating mode		Type of the engine		
Dynamometer	Engine	Otto (gasoline)	Diesel automotive	Diesel genset (constant speed)
N	P	Full throttle performance possible	Full throttle performance possible	Full throttle performance possible
M	P	Engine can be tested till peak torque	Engine can be tested till peak torque	Engine can be tested for full range of torque
N	M	Speed control is better by dynamometer for full range	Speed control is better by dynamometer for full range	Not much suited as dynamometer speed control conflicts with engine governor
M	N	Full throttle performance possible	Full throttle performance possible	Speed and torque control are better
$M(N^2)$	P	Torque characteristic can be tested as predefined	Torque characteristic can be tested as predefined	n/a

N, speed constant; M, torque constant; P, throttle position for engine and butterfly valve position for dynamometer.

10.5 Dynamometer Modes with Relation to Torque Envelope

Constant speed or constant torque modes are achieved with the help of dynamometer and throttle controllers having PIDs.

10.5.1 Speed Constant Mode (N = Constant)

The speed controller PID maintains the engine speed constant by corresponding loading.

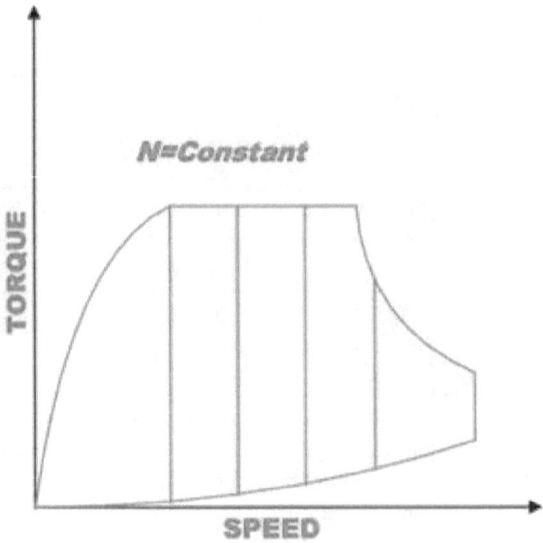

Figure 10.4 Constant speed mode for eddy-current dynamometer.

10.5.2 Torque Constant Mode (*M* = Constant)

The torque of the engine is maintained constant by changing the load and varying the speed.

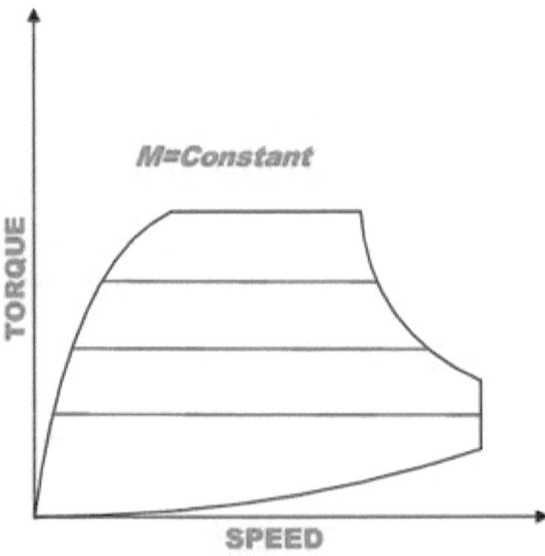

Figure 10.5 Constant torque mode for eddy-current dynamometer.

10.5.3 Power Law Mode $M(N^2)$

The torque of the engine is proportional to the engine square of the engine speed. In this case, the signal proportional to actual speed is set as command value to a torque controller as squared analog signal.

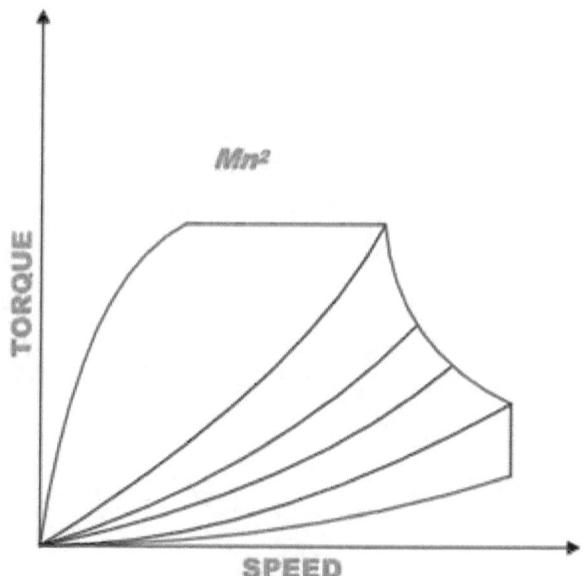

Figure 10.6 Power law mode for eddy-current dynamometer.

Torque of dynamometer is proportional and governed by the following equation:

$$T_{DYNO} \propto (RPM^n) \quad \text{............10.2}$$

$$T_{DYNO} = K.(RPM^n) \quad \text{............10.3}$$

The value of n can vary from 1 to 3.

10.6 Selection of Right Dynamometer Mode

The loading of the engine in steady-state condition by dynamometer should be such that a stable operating point is achieved. The operating point is arrived by selecting a point at the intersection of the dynamometer speed verses torque curve and engine speed verses torque curve. Therefore, it is necessary to select the dynamometer operation mode in such a way that it should satisfy the following two conditions:

1. The torque/speed curve of the dynamometer and torque/speed curve of the engine should intersect very clearly.
2. The slope of dynamometer speed/torque curve should be higher than the slope of the engine speed/torque curve.

Let us consider a variable speed gasoline engine used for automotive application. The following figure shows that dynamometer constant torque curve intersecting engine natural characteristics at point "A" and point "B."

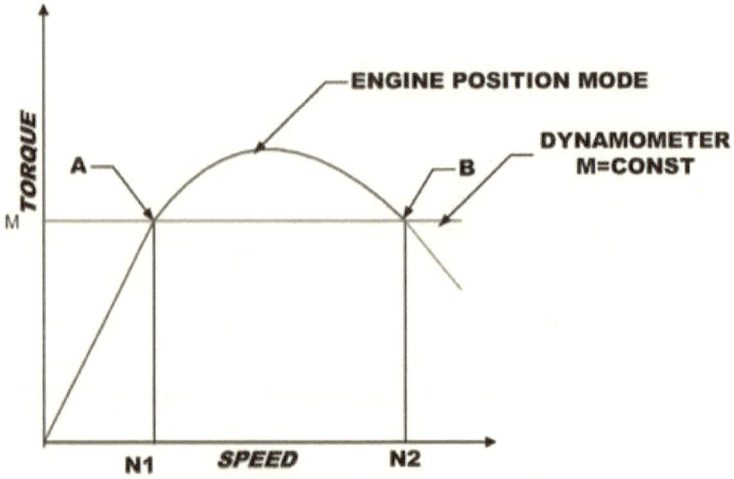

Figure 10.7 Incorrect mode selection.

The point "A" meets condition 1 mentioned above, and slope of engine curve is more than slope of dynamometer curve, which should be correct in other way. Hence, point "A" is unstable.

Now, consider the same engine in position mode and dynamometer in speed constant mode. The both operating points "A" and "B" satisfy both the conditions mentioned above, that is, points "A" and "B" are distinct intersection points, and the slope of dynamometer curve is greater than the slope of engine curve. Therefore, the speed constant mode selected for dynamometer is correct.

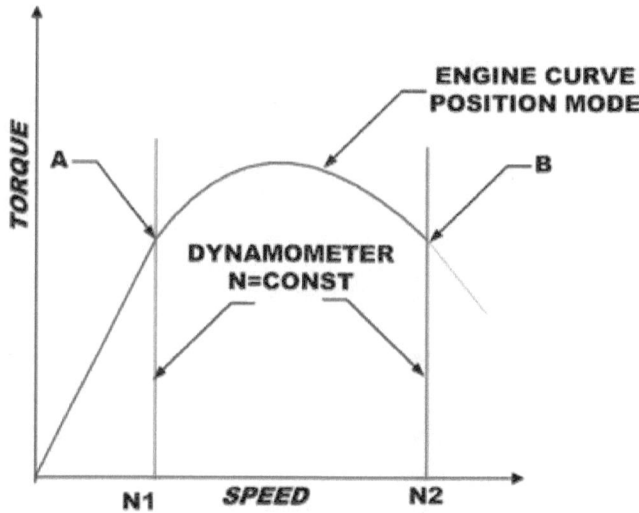

Figure 10.8 Right mode selection.

10.7 Manifold Vacuum/MAP Option Control Modes

This mode is provided for testing of the engines with fuel injection system. The manifold absolute pressure sensor (MAP sensor) is a pressure transducer used in an internal combustion engine's electronic control system. Engines that use a MAP sensor are typically fuel injected. The MAP sensor provides instantaneous manifold pressure information to the engine's electronic control unit (ECU). The data is used to calculate air density and determine the engine's air mass flow rate, which in turn determines the required fuel metering for optimum combustion. MAP sensors measure absolute pressure. The feedback from MAP sensor is used to control the throttle position of engine to maintain desired MAP.

Vacuum is the difference between the absolute pressures of the intake manifold and atmosphere. Vacuum is a "gauge" pressure, since gauges by nature measure a pressure difference, not an absolute pressure.

10.8 Bump-less Mode Transfer

The dynamometer controller is a dynamic system. It is necessary to make sure that the state of the system is correct when switching the controller between manual and automatic mode or between the different modes, such as speed mode to torque mode. When the system is in manual mode, the control algorithm produces a control signal that may be different from the manually generated control signal. It is necessary to make sure that the two outputs coincide at the time of switching. This is called *bump-less transfer*.

The system also provides a bump-less transfer capability between control modes so that you can easily change while running your test.

When controller is switched from manual mode to automatic (from open loop to closed loop), the results are expected to be uneventful. That is, we do not want the switchover to cause abrupt control actions that impact or disrupt our process.

The desired outcome is achieved at switchover by initializing the controller integral sum of error to zero. Also, the set point and controller bias value are initialized by setting:

1. set point value (SP) equal to the current measured value (MV)
2. controller bias (null value) value (CO_{bias}) equal to the current controller output value (CO).

With the integral sum of error set to zero, there is nothing to add or subtract from CO_{bias} that would cause a sudden change in the current controller output. With the set point equal to the measured process variable, there is no error to drive a change in our CO. And with the controller bias set to our current CO value, we are prepared by default to maintain current operation.

Thus, when we switch from manual mode to automatic, we have "bump-less transfer" with no surprises. This is what is expected by the test engineer.

10.9 Role of PID Controller

A proportional-integral-derivative controller (PID controller) is a generic control loop feedback controller widely used in industrial control systems. A PID is the most commonly used feedback controller.

The PID controller involves three separate constant parameters, and hence, it is sometimes called *three-term control*: the proportional, the integral, and derivative values denoted by P, I, and D, respectively.

PID is used to control the response of any closed-loop system. In the dynamometer test setup, it is used to control speed and torque or to that matter any physical parameter which is of interest of the test engineer. The dynamometer control system involving the control of speed and torque controllers, PID is used to provide zero steady-state error between set point and feedback, along with good transient performance.

Proportional Gain

This is used to adjust the basic response of the closed-loop control system. The PI error is multiplied by the proportional gain to produce an output.

Integral

The integral term is used to reduce steady-state error between the set point and the feedback values of the PI. If the integral is set to zero, then in most systems there willl always be a steady-state error.

Derivative

This is used to correct for certain types of control loop instability and, therefore, improve response. The derivative term has an associated filter to suppress high frequency signals.

The following figure shows the schematics of the PID controller.

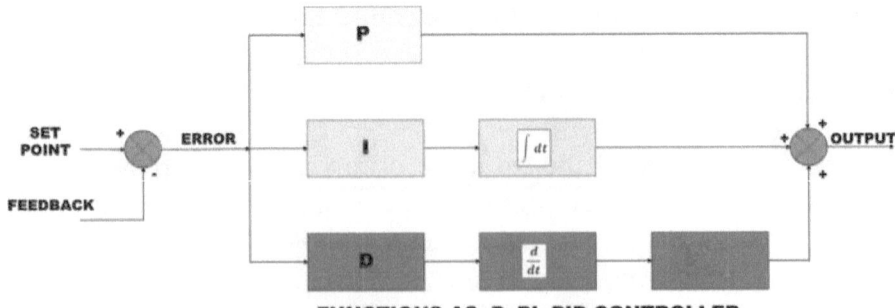

Figure 10.9 Schematics of PID.

10.10 The Methodology to Set Up a PI Controller

The gains should be setup so that a critically damped response is achieved for a step change in set point. When too much gain is set, then system will be an under-damped or oscillatory system. If the gain is too little, then system will behave like over-damped system.

To set up the proportional gain (P), set the integral gain ("I") to zero. Apply a step change in set point that is typical for the control system and observe the response. Increase the gain and repeat the test until the system becomes oscillatory. At this point, reduce the P gain until the oscillations disappear. This is the maximum value of P gain achievable.

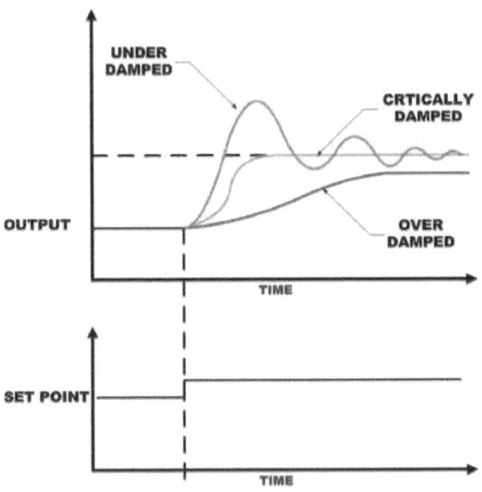

Figure 10.10 Response of PID.

If a steady-state error is present, that is, the feedback never reaches the set point value, the integral gain ("*I*") needs to be increased. As before, increase the integral gain ("*I*") and apply the step change. Monitor the output. If the output becomes oscillatory, reduce the *P* gain slightly. This should reduce the steady-state error. Increasing the integral gain ("*I*") further may reduce the time to achieve zero steady-state error. These values of proportional gain (*P*) and integral gain ("*I*") can now be adjusted to provide the exact response required for this step change.

In the interest of achieving a gradual convergence at the desired set point, the controller may wish to damp the anticipated future oscillations, and in order to compensate these oscillations, the derivative gain (*D*) controller may be adjusted.

Closure

The dynamometer control system and throttle control systems have control modes to facilitate the testing of engine in different modes. Test engineer can wisely use a combination of dynamometer and engine control mode to suit his application. Moreover, to simulate the power law, that is, propeller characteristics power law mode or $T \alpha N^2$ is used. The test engines can be constant speed or variable speed engines, and accordingly, modes can be selected. If engine is constant speed, then dynamometer would be selected for "torque = constant" mode and throttle would be selected for "speed = constant" mode. On the contrary, if the engine under test is automotive engine, that is, variable speed engine, then dynamometer would be selected for "speed = constant" mode and throttle would be selected for "torque = constant" mode. Some controllers provide the custom-built mode called as vacuum/MAP mode.

CHAPTER 11

Dynamometer: Torque and Power Measurement Accuracy

11.1 Introduction

Accuracy of power measurement is generally limited by the torque measurement accuracy of the system since the speed measurement can be achieved with any desired accuracy. Torque errors can arise from the application of extraneous (i.e., not indicated) torques from miss-calibration of the load cell and windage of external parts. In the process of torque measurement, the role of trunnion bearings is very important since they have a very little movement when a brinelling effect is generated. This sometimes leads to stickiness of the torque measuring chain. This problem can be alleviated by periodic turning of the trunnion bearing's outer races.

11.2 Measurement Characteristics

In the process of engine testing, the test engineer seeks to obtain numerical values for certain physical variables such as speed, torque, and engine exhaust temperature. These unknown variables are known as measurands. Engine testing system of dynamometer and measuring instruments consists of various types of displays, recorders, and indicators, such as digital and analog indicators.

The measurement system can be viewed as consisting of four subsystems as shown in Figure 11.1.

1. Sensing element.
2. Signal conditioning system.

3. Process and control.
4. Recording or indicating device.

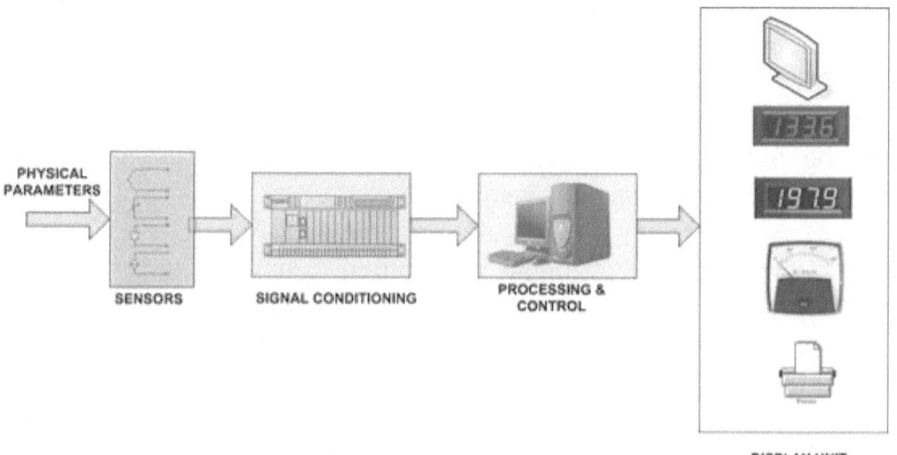

Figure 11.1 Measurement schematics.

The sensor sensing the parameter to be measured has a significant physical characteristic that changes in response to changes in the measurand. The signal conditioning system changes the output of the sensor so that it can be used conveniently for the processing and indicating purpose. The user is not concerned about the way the signal is processed. However, his main concern is how accurately it is produced and presented to him through indicators and recorders.

1. Accuracy

Accuracy, defined as the closeness of agreement between a measured value and true value, is a common term used to specify uncertainty. Dynamometer manufacturers frequently state a value for accuracy as part of the device specifications. There are two types of statements of accuracy:

I. An accuracy stated as a percent of reading.
II. An accuracy stated as a percent of full scale.

What Is the Difference between "Percent Reading" and "Percent Full-scale Reading?"

The measurement uncertainties associated with specifications for force-measuring devices are often expressed as a *percentage of full-scale reading*. This is not always the case, however, and sometimes, *percentage of* reading is used instead and the differences can be very significant, particularly when measuring forces that are quite small for a particular instrument.

For instance, if an instrument is specified to have an uncertainty of ±0.25 percent of full-scale output and its maximum force capacity is 2,000 N, then the user can reasonably hope that if the instrument is used correctly, the value of force it indicates will be correct, give or take 5 N. But if the force being measured is, say, 100 N, the uncertainty, expressed as a percentage of the force, will be ten times higher, that is, +/-2.5 percent. Similarly at 10 N, the force-proportional uncertainty would be one hundred times greater or ±25 percent.

The table below shows the uncertainties in the measurement of force, first given as 1 percent of reading and second expressed as 1 percent of full-scale reading to illustrate the difference; in the region marked with arrows, the device performing to 1 percent of full-scale reading is unlikely to make a meaningful measurement.

2. Precision

The precision of a measurement system, also called reproducibility or repeatability, is the degree to which repeated measurements under unchanged conditions show the same results.

3. Repeatability

Repeatability is the ability of a device to produce the same output reading when the same measurand is applied using the same procedure. Inability to repeat a measurement exactly, a random error known as repeatability error, is usually a component of the manufacturers' specification of instrument accuracy.

In the context of the dynamometer, the repeatability plays an important role. Results for torque and speed measurement are expected to be the same or repeatable within the system tolerance level. If the test engineer tests the same engine on the same dynamometer again and again and gets variation, then this variation is termed as repeatability. The variation in the results could be the inherent variations in the following parameters:

1. Dynamometer water pressure variation
2. Dynamometer PID tuning
3. Engine throttle setting and its PID tuning
4. Engine operating temperature
5. Engine water temperature
6. Test cell temperature and humidity in the test cell
7. Engine fuel temperature
8. Engine oil temperature.

If the system accuracy is ±0.2 percent FSD, then the readings taken in each test setup should be correct within ±0.2 percent FSD. It is possible to maintain engine operating temperature, fuel inlet temperature, and oil temperature at constant value within the specified tolerance by using engine fuel temperature, water temperature, and oil temperature using modern accurate controllers. However, it is rather difficult to predict the repeatability and blame on any one of the above-mentioned specific parameters.

The present day fully equipped test cell with combustion air controlling unit, engine fuel temperature, oil temperature, and water temperature controller the repeatability of the results is in close tolerance.

4. Reproducibility

Reproducibility is the degree of closeness with which a given value may be repeatedly measured. Reproducibility is also defined as the variability of the measurement system caused by differences in operator behavior, testing location, or variation in test equipment instruments. Mathematically, it is the variability of the average values obtained by several operators while measuring the same item.

In the context of engine testing, reproducibility is defined as variation in test results obtained when the same engine is tested on different dynamometer or by different operators or at different location. The reproducibility is largely dependent on the following factors:

1. Instrument and dynamometer accuracy at different location.
2. Operator skill level at various laboratories.
3. atmospheric conditions within test cell and control employed to maintain air pressure, temperature, and humidity.

To establish, repeatability and reproducibility is a complex method. The range and average method is most commonly used method to establish repeatability and reproducibility. Analysis of variance, commonly known as ANOVA method, is used to establish repeatability and reproducibility.

5. Sensitivity

This is ratio of the magnitude of output signal or response to input signals being measured. The sensitivity of the engine dynamometer and the measuring sensor, that is, load cell, are two different aspects of the test bed.

The sensitivity for dynamometer with trunnion-mounted can be defined as minimum torque required for affecting a recognizable deflection of the stator which can be transferred to measuring device or sensing device such as load cell. If minimum or the smallest weight is applied at the calibration pan and if the measuring system reflects the change, then dynamometer is said to be sensitive to the smallest grams applied.

The sensitivity is greatly influenced by the following factors:

1. Mounting of dynamometer in trunnion bearings and its alignment. This is greatly affected by the manufacturing accuracy of line boring of two pedestals housing trunnion bearings.
2. Lubrication of trunnion bearings.

3. Load cell mounting arrangement such as self-aligning bearing and its lubrication.
4. Prolonged use of trunnion bearings creates the brinelling effect, thereby increasing the sensitivity.

It can be inferred from the definition of sensitivity that measurement smaller than the sensitivity of dynamometer is not possible by measuring system. Therefore, the sensitivity should be smaller than the specified accuracy.

If the dynamometer is foot-mounted or fixed on bedplate, then the sensitivity is attributable only to the sensitivity of torque transducer. Also, the sensitivity of dynamometer mounted on flexure supports is dependent on the resilience of the mounting rubber bushes.

6. Resolution

If a measuring parameter is varied continuously, many measurement devices will show an output to those changes in discrete steps. This inability of the measurement system to follow changes in the measuring parameter exactly results in a *resolution error*. Resolution is usually treated as a random error. The sensing element itself may not produce a continuous output with a smoothly varying parameter under measurement. A wire wound potentiometer is used as position sensing device, which have a step type of output versus carbon film potentiometer.

If, for example, the digital torque indicator has a reading of 2.567, the reading resolution is simply a value of 1 in the last (least significant) digit. If the device rounds off values, the resolution uncertainty will be ±0.5 in the least significant digit. In instruments such as "analog indicators" in which the output is read by comparing a pointer to a scale, the ability to resolve a value of the measurand is limited by a characteristic called the scale *readability*.

7. Measurement Error

The *error* of a measurement is defined as the difference between the measured value and the true value of the parameter under measurement.

Error = Measured value – True value.

The following type of errors is important from point of view of engine testing and measurement of torque and speed.

    I. Random error

Random Error = readding – Average of readings.

    II. Systematic error (E_s)

E_s = *Average of Readings – Ture value.*

11.3 Calibration of Measured Parameters

Calibration is a comparison between measurements—one of known magnitude or correctness made or set with one device and another measurement made in as similar a way as possible with a second device. Calibration, in its purest sense, is the comparison of an instrument to a known standard.

Definition

Calibration is defined as a set of operations that establish, under specified conditions, the relationship between the values of quantities indicated by a measuring instrument or measuring system and the corresponding values realized by standards.

The word *calibration* is misused as an adjustment. However, adjustment is defined as the operation of bringing a measuring instrument into a state of performance suitable for its use.

The first step in the calibration process is to take a set of data consisting of measurement system output as a function of the measured parameter such as torque. A graphical presentation of these data is known as the *calibration curve.*

Calibration Frequency

The calibration frequency really depends upon how extensively dynamometer is used. The dynamometer consists of sensors, signal conditioning circuits, and displays. Dynamometer calibration frequency should be decided by the test engineer. For this purpose, test engineer needs to collect and monitor the instruments and establish the history trend, following this criteria will help decide the calibration frequency.

When Calibration Frequency Can Be Decreased

- If the instrument has performed to specification and the drift has been insignificant compared to its specified tolerance.
- If the instrument is deemed to be noncritical or in a low priority location.

When Calibration Frequency Should Be Increased

- If the sensor has drifted outside of its specified tolerances during a given time period.
- If the sensor is located in a critical process or area of the plant and has drifted significantly compared to its specified tolerance over a given time period.
- When measuring a sensor that is located in an area of the plant that has high economic importance for the plant.
- Where costly production downtime may occur as a result of a "faulty" sensor.
- Where a false measurement from a sensor could lead to inferior quality batches or a safety issue.

Standards for Calibration

Standards of measurement have been important to commerce for a very long time since it is important to the purchaser to know that the weight or length of a purchase is accurate and traceable to know standards.

The standard for mass is the International Prototype Kilogram, which is platinum-iridium cylinder kept at the International Bureau of Weights and Measures in France.

11.4 What Do We Understand by Traceability?

Traceability is documentation, essentially a pedigree showing a direct link to the official US 1 kg weight standards housed at the National Institute of Standards and Technology (NIST) in Gaithersburg, MD, if you are a user in United States. These NIST standards are in turn calibrated to the international 1 kg standard maintained at the Bureau of Weights and Measures in France.

Traceability not only means that a weight or mass standard has links to the national 1 kg standard, but also means that the measurements were appropriate for the accuracy class required for the application. Traceability also requires proof that all environmental factors affecting accuracy were considered at the lab doing the measurement. There are two types of traceability: direct and indirect.

1. **Direct Traceability**

Direct traceability means a weight or mass standard has been tested by national institute of the national level. This national institute then issues a report number to the organization for which they have performed measurements. The local commercial calibration labs get their weights calibrated by national institute and get a certificate. This certificate provides direct traceability for the lab's mass standards, which can then be used in calibrating weights for clients.

2. Indirect Traceability

Indirect traceability exists when a client's weight or mass standard is tested by a metrology lab that has direct traceability and has the necessary measurement control program in place.

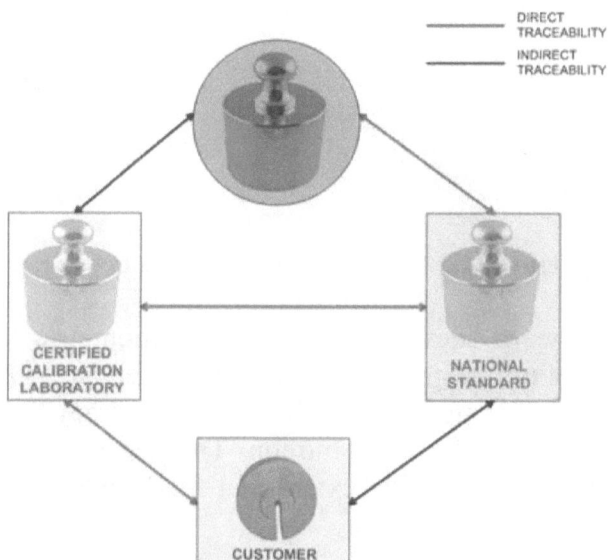

Figure 11.2 Traceability of calibration.

11.5 Torque Measurement

Torque is not an absolute quantity. Torque is a derived quantity and product of the force and the total length of the calibration arm. The following figure shows the arrangement of the lever arm (L_a), calibration lever arm (L_c), and weight pan.

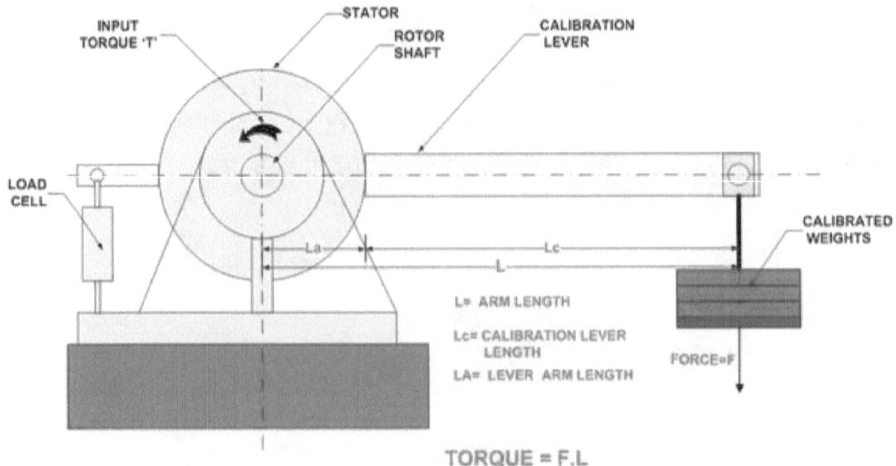

Figure 11.3 Principle of torque measurement.

The credibility of development of dynamometric method of measurement of torque goes to the hydraulic scientist William Froude who developed this method in 1877. The accuracy is guaranteed by the Newton's third law of motion. The accuracy of the torque measurements assured to the extent that the Newton's third law of motion cannot be broken.

The torque measurement is independent on the type of power absorption method or the method of the cooling adopted in the dynamometer. The rotor rotates in bearings supported by stator. The stator is mounted and cradled on large ball/roller bearings called as "trunnion bearings," allowing only oscillatory movement. The "trunnion bearings" are supported in pedestal.

When a dynamometer is in action, a torque "T" is applied at the rotor shaft by the prime mover. It rotates and tries to carry the stator along with it in the same direction. The movement of the stator is dynamometrically restricted from rotation by the force "F" acting on the lever arm.

The force "F" invariably acts vertically up or down. The perpendicular distance between the center of the dynamometer and the line of action of force is called "calibration arm length," which is shown as "L" in Figure 11.3.

$$Torque = F.L \quad \quad \quad \quad \quad \quad \quad \quad \quad \quad \quad \quad \quad \quad \quad (11.1)$$

It is the torque applied to prevent the stator from the rotation while the input torque at the rotor shaft tries to rotate it. As per the Newton's third law of motion, the "action" must be equal to "reaction." The impressed torque must be exactly equal to opposing torque. The above equation cannot be violated by external factors such as working conditions or the type of absorption method.

11.6 Torque Transmission Process

The torque generated at the flywheel of the engine is different than the torque at the dynamometer input coupling. The torque input fed at the dynamometer coupling is actually measured. This is T_{Dyno} as shown in Figure 11.4.

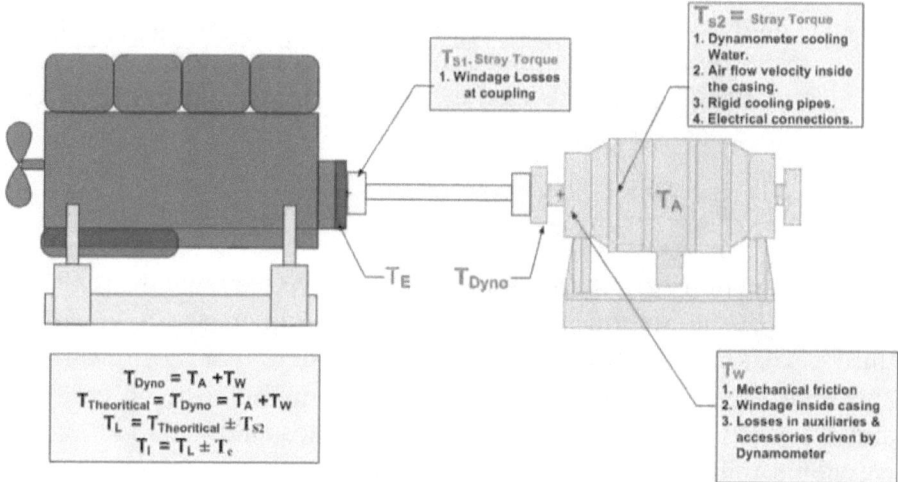

Figure 11.4 Account of accuracy.

A. The difference between T_E and T_{Dyno} is one source of error, and let us denote this as T_{S1}, called as stray torque.

$$T_{Dyno} = T_E - T_{S1} \quad \quad \quad \quad \quad \quad \quad \quad \quad \quad \quad \quad (11.2)$$

The source of stray torque T_{s1} is as mentioned below:

1. Windage losses at coupling and driveshaft between dynamometer and engine. This is practically negligible and hence neglected. However, this is significant for large shafts and flexible couplings used on flywheel side which can be estimated if the need arises.
2. Accessories driven by dynamometer such as servo oil pump, taco generator which is not cradled. The power absorbed by such devices should be very low so that it does not affect the accuracy of the dynamometer. If it is to be driven by dynamometer, it should be cradled to dynamometer so torque reaction is transferred to dynamometer.

B. The substantial part of dynamometer input torque T_{Dyno} is absorbed into the power absorption mechanism inside the casing denoted as T_A, and other small part of the power is absorbed in mechanical friction in shaft bearings, seals, and windage inside the casing. Let us denote this part of the torque as T_W.

$$T_{Dyno} = T_A + T_W \quad \quad \quad \quad \quad \quad \quad \quad (11.3)$$

From the section Torque Measurement, we have seen that the reaction generated by the friction is accounted in measurement as per Newton's third law of motion. It is worth to point out here that those bearings and seals are in the dynamometer and dynamometer being cradled, and these losses are accounted for in the measurement. Therefore, Newton's third law ensures that dynamometer input torque T_{Dyno} is exactly equal to reaction torque. This is also known as theoretical torque T_{TH}.

$$T_{Dyno} = T_{TH} \quad \quad \quad \quad \quad \quad \quad \quad (11.4)$$

D. Torque reaction sensed by the measuring device T_L will be different than the theoretical torque T_{TH} due to presence of stray torque T_{s2}.

$$T_L = T_{TH} \pm T_{S2} \quad \ldots \ldots \ldots \ldots \ldots \ldots \ldots \ldots \ldots \ldots \ldots \ldots (11.5)$$

E. Indicated torque T_I: The indicated torque will be different than the sensed or theoretical torque (T_L) due to systematic or accidental errors which can be classified under measurement errors (T_e).

$$T_I = T_L \pm T_e \quad \ldots \ldots \ldots \ldots \ldots \ldots \ldots \ldots \ldots \ldots \ldots \ldots (11.6)$$

Summary of the Errors

Now, let us analyze the equations. From equations (11.2) and (11.3), we get

$$T_E - T_{S1} = T_A + T_W \quad \ldots \ldots \ldots \ldots \ldots \ldots \ldots \ldots \ldots \ldots (11.7)$$

Therefore,

$$T_E = T_A + T_W + T_{S1} \quad \ldots \ldots \ldots \ldots \ldots \ldots \ldots \ldots \ldots \ldots (11.8)$$

From equations (11.3), (11.4), (11.5) and (11.6) torque indicated by the dynamometer is given by fowling equation:

$$T_I = T_A + T_W \pm T_{S2} \pm T_e \quad \ldots \ldots \ldots \ldots \ldots \ldots \ldots \ldots \ldots (11.9)$$

Now substitute for $(T_A + T_W)$ from equation (11.8) into equation (11.9)

$$T_I = T_E - T_{S1} \pm T_{S2} \pm T_e.$$

Therefore, the total error is difference between the indicated torque and measured torque.

$$\pm | T_I - T_E | = T_{S1} \pm T_{S2} \pm T_e \quad \ldots \ldots \ldots \ldots \ldots \ldots \ldots (11.10)$$

It is worth noting here that the error is independent of T_A and T_W. It does not matter how torque is absorbed, i.e whether hydraulically or electromagnetically in dynamometer. It also does not depend on mechanical friction.

Conclusion

The error depends only on stray torques T_{s1}, T_{s2} and measurement errors (T_e).

The error is normally expressed as percentage of the torque being measured.

$$e = \frac{\pm |T_I - T_E|}{T_E} 100\% \quad \ldots \ldots (11.11)$$

The numerator of equation is known from the calibration curve. However, the value of T_E is not known. *Actually, the whole purpose of testing the engine is to find T_E*

It can be predicted that T_E must lie between the two values $T_I + T_e$ or $T_I - T_e$.

Since T_I is the mean value of these two limits and is most likely the value of T_E.

Therefore, equation (11.11) can be modified by substituting T_I for T_E.

$$e = \frac{\pm |T_I - T_E|}{T_I} 100\% = \frac{T_{s1} \pm T_{s2} \pm T_e}{T_I} 100\% \quad \ldots (11.12)$$

The following figure summarizes the errors associated in torque transmission process from engine flywheel to dynamometer.

DYNAMOMETER

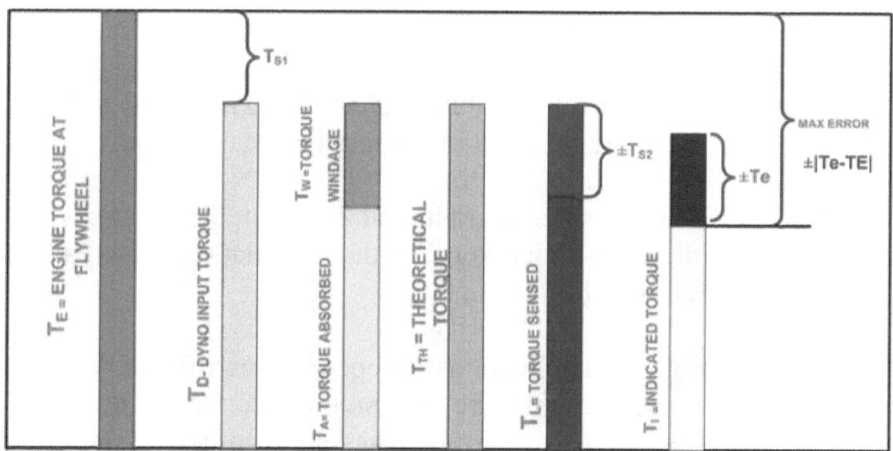

Figure 11.5 Errors in torque transmission process.

T_E	ENGINE TORQUE AT FLYWHEEL
T_{S1}	STRAY TORQUE
T_{DYNO}	DYNAMOMETER INPUT TORQUE
T_A	TORQUE ABSORBED IN CASING
T_W	WINDAGE INSIDE CASING, MECHANICAL FRICTION, LOSSES IN FRICTION, AUXILIARIES
$T_{THEORETICAL}$	THEORETICAL REACTION TORQUE
T_{S1}	STRAY TORQUE
T_L	TORQUE SENSED BY SENSOR
T_e	MEASUREMENT ERROR
T_I	INDICATED TORQUE

11.7 Errors Associated with Torque Measurement

As we have got enough background about measurement and its characteristics, let us now discuss the possible errors in the measurement of torque.

1. Calibrated Deadweights

The deadweights used for calibration should be calibrated to know standards, and it should be traceable to international standard.

The standard for mass is the International Prototype Kilogram, which is platinum-iridium cylinder kept at the International Bureau of Weights and Measures in France.

The correct weight for applications is dependent on the use. Weights should be more accurate than the precision of the weighing device, and it is recommended that a weight has an accuracy of one-third (1/3) of the weighing device readability.

Let us assume that the weight used for the calibration is 100 gm and actually weighs 99 gm. As we have seen in earlier sections, the torque is a derived quantity, and it is a product of the length of the lever arm and force (weight applied).

Assuming lever arm length of one meter (m) and absolute accurate, the actual torque will be

$$Torque\ (T) = F.L$$

$$Torque\ (T) = \left(\frac{99}{1000}\right) kg.\, 1m = 0.099\ kgm.$$

Whereas theoretical torque will be

$$Torque\ (T) = \left(\frac{100}{1000}\right) kg.\, 1m = 0.1\ kgm.$$

Therefore, the % $Error = \frac{Theoritical\ Reading - Actual\ reading}{Theoritical\ reading}.\,100$

$$\%\ Error = \frac{0.1 - 0.099}{0.1}.\,100$$

$$\%\ Error = 1\ \text{percent}.$$

It is up to the test engineer's discretion and evaluation what accuracies should be allowed for the deadweights calibration.

2. Gravitational Effect

The gravity acting on a mass varies depending on the latitude and height of the location where measurement is being conducted. For this reason, the same mass may show different weight indications from one place to another. Therefore, make it a rule to calibrate the balance every time it is relocated. The weights used for dynamometer calibration should be recalibrated accordingly.

Weight, defined as a force and often denoted by a bold letter **W**, is a vector. Its magnitude (i.e., as a purely scalar quantity), often denoted by an italic letter W, is the product of the mass m of the object and the magnitude of the local gravitational acceleration g:

Thus,

$W = mg$.

In the International System of Units (SI), the unit of measurement for weight is that of force, the Newton.

$$F = m.g,$$

where
F = force in (N)
m = mass (kg)
g = local gravitational acceleration (m/s²)

Local gravitation is function of geographical location where measurement is being made. The geographical location is defined by its latitude and altitude. The equation for local gravitation is

$g_0 = 9.780327\,(1 + 0.0053024\,sin^2 2\emptyset - 0.0000058 sin^2 2\emptyset) - 3.155 \times 10^{-7} h \left[\frac{m}{s^2}\right]$ ... 11.13

True weight can be calculated by the following formula:

$$W = \frac{g_o}{g} W_A, \quad \ldots\ldots\ldots\ldots\ldots\ldots\ldots\ldots 11.14$$

where

W = true weight

g_o = local acceleration due to gravitation

g = standard acceleration due to gravity

3. Calibration Lever Arm Length

A. Manufacturing Accuracy

The length of the calibration lever is manufactured to a great precision. The part of the arm length is built into dynamometer during the manufacturing and utmost care is taken to maintain the accuracy of measuring. If the typical manufacturing tolerance is ±0.1 mm and arm length is 500 mm, then the error (e) is

$$e = \frac{\pm 0.1}{500}(100)\%$$

$$e = 0.02\%.$$

B. Position of Calibration Arm

Calibration lever (L_c) should be perfectly horizontal during calibration. However, while adding or removing the weight on the calibration arm, certain amount of vibrations (oscillations) will be crated. These vibration play an important role when a dynamometer is using dial-type self-indicating weight bridge. In case of load cell arrangement, the vibrations are much less as movement is restricted to small amount. If dynamometer turns by say and angle of ±a, then the effective arm "L_e" decreases to $L \cos(\pm a)$ and the error (e) is

$$e = \{1 - \cos(\pm\alpha)\}(100)\% \quad \ldots\ldots\ldots\ldots\ldots 11.15$$

DYNAMOMETER

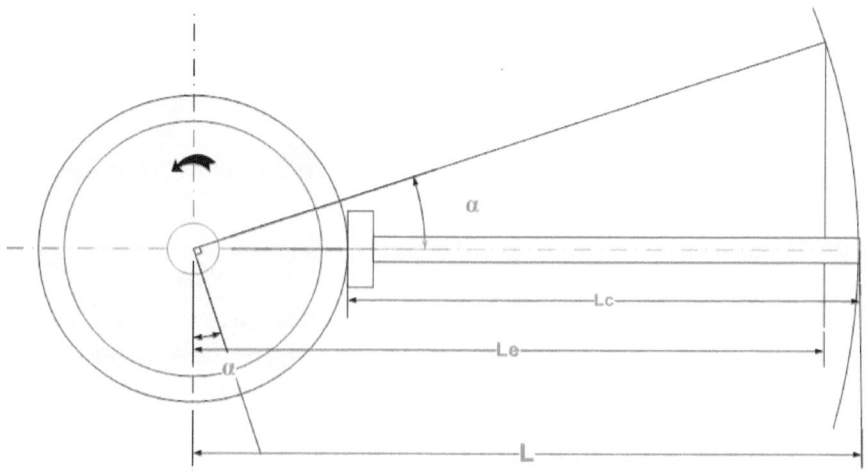

Figure 11.6 Effective lever arm.

C. Deflection of Lever Arm

The lever arm is very long in case of very large dynamometer and when large amount of weight is applied. Due to such large weight and cantilever arm effect, the arm may deflect. This deflection will not keep the lever arm horizontal, and effective length will decrease as discussed above.

D. Temperature Effect

Normally, calibration is done at room temperature. However, dynamometer is in the engine room, and the room temperature might increase to say 45 to 55°C. The lever arm (L_a) being built in dynamometer will also be at this temperature, and there will be an increase in the torque arm length due to expansion. The calibration lever (L_c) is attached to this lever arm which is at different temperature. The total length of the calibration lever (L) will change by (Δl) as effect of thermal expansion and the change in temperature (Δt).

$$\Delta l = \sum \alpha . l . \Delta t \quad \dots\dots\dots\dots\dots\dots\dots\dots 11.16$$

And the error (*e*) is given by the following formula:

$$e = \frac{\sum \alpha . l . \Delta t}{L} (100)\% \quad \ldots \ldots 11.17$$

The total error will be the sum of all the errors listed above.

Closure

The torque measurement is dependent on the accuracy of the calibration weight and the calibration arm length. The entire measuring chain is responsible for the final torque accuracy. The measuring chain which consists of load cell, signal processing unit, and the final display unit. Each will have its own accuracy limit. The total overall accuracy should not be more than the accuracy specified by the manufacturer. Torque is a derived quantity, so is the power. The torque measurement accuracy will therefore affect the power measuring accuracy.

Bibliography:

1. *Rice Lakes Reference Manual for Dead Weights.* Courtesy of Rice Lake Weighing systems
2. Richard Boynton, *Precise Measurements of Mass* (Technical Paper).
3. Beamex, How often should instruments be calibrated? (White Paper).

CHAPTER 12

Correction Factor and Horsepower

12.1 Introduction

The engines are tested daily by Development team, production team and quality control team using the dynamometer and associated control system and Instrumentation. Each of these teams wants to be sure that every time they test the engine the test set up produces the correct results. If the dynamometer produces the consistent results then one can trust that result set is outcome of the modification or development work done on engine. The dynamometer results are affected by various parameters such as

1. Test cell environment (weather conditions)
2. Engine water temperature variation,
3. Engine lubricating oil temperature
4. Engine air intake conditions

The engine oil temperature alone can have significant influence on engine power as engine power lost in friction is function of engine oil temperature. The weather conditions play important role in engine power output. It practically means that it is directly proportional to the amount of oxygen present in air which will help in burning the fuel efficiently. The power output is sensitive to the air intake conditions. When barometric pressure goes up, the air density goes up. We also know that when air temperature goes up air expands and its density goes down. Similarly when Humidity increase the water vapor displaces the oxygen content in the air. This is complicated phenomena has lead the scientist and engineers develop formulae which will apply the correction in engine power measurement. These formulae are commonly known as correction factors.

12.2 How correction factor and Horse power is related?

The performance of an engine is influenced by the ambient conditions. The most important ambient conditions influencing gasoline engine performance are:

1. Humidity of the air.
2. Inlet air temperature.
3. Altitude of the test cell where engine is being tested.

Ambient conditions depend on the geographical location and may vary with time of the year. Engine controls may compensate for some effects of these variations, but engine performance deviations are to be expected. The given performance data of an engine refers to the specified ambient conditions. If the site ambient conditions deviate from these, actual engine efficiency and emissions will differ from the specified values as well.

The horsepower and torque produced by a naturally aspirated internal combustion (IC) engines are dependent upon the density of the air supplied to the combustion chamber. The higher density means more oxygen molecules and more power and lower density means less oxygen and less power.

The observed horsepower, on the dynamometer test bed and the correction factor, allow mathematical calculation of the affects of air density and the humidity of the air on the wide-open-throttle (WOT) horsepower and torque. The dynamometer correction factor is simply the mathematical reciprocal of the relative horsepower value.

The typical correction factor (CF) is calculated based on the absolute barometric pressure, air temperature and water content of the air used for combustion by the engine under test.

The correction factor attempts to predict the horsepower that would be developed if the engine were tested at sea level under standard pressure and temperature conditions.

12.3 The STP and NTP conditions

12.3.1 STP—Standard Temperature and Pressure

STP is commonly used to define Standard conditions for Temperature and Pressure which is important for the measurements and documentation of chemical and physical processes:

- STP—Standard Temperature and Pressure—is defined by IUPAC (International Union of Pure and Applied Chemistry)

  1. Air having properties

Temperature:—0 °C (273.15 °K, 32 °F)

Pressure:—10^5 Pascal (1 Pa = 10^{-6} N/mm^2 = 10^{-5} bar = 0.1012 kp/m^2 = 1.02 x 10^{-4} m H$_2$O = 9.869 x 10^{-6} atm = 1.45 x 10^{-4} psi (lbf/in^2)

  2. STP—commonly used in the Imperial and USA system of units—as air at 60 °F (512 °R) and 14.696 psia (15.6 °C, 1 atm)

12.3.2 NTP—Normal Temperature and Pressure

NTP is commonly used as a standard condition for testing and documentation of fan capacities:

- NTP—Normal Temperature and Pressure—

  1. *Temperature of air at 12 °C (293.15 °K, 68 °F)*
  2. *Pressure of Air:—1 atm (101.325 kN / m², 101.325 kPa, 14.7 psia, 0 psig, 29.92 in Hg, 760 torr).*
  3. *Density:-1.124 kg/m³ (0.075 pounds per cubic foot)*

12.4 Use of correction factors

1. The most important benefit of the using the correction factor is to standardize the horsepower and torque readings, so that the effects of the ambient temperature and pressure are removed from the readings.
2. By using the correction factor, power and torque readings can be directly compared to the readings taken on some other day, or even taken at some other altitude.

From above it can be concluded that corrected readings are the same as the result that you would get by taking the engine under test to a certain temperature, humidity, pressure controlled dynamometer test bed /cell to measure "standard" power.

If you test your engine on the dynamometer on a cold day at low altitude, it will make a lot of power. And if you take exactly the same engine to the same dynamometer on a hot day, it will make less power. This is mainly due to density of air and in turn density of oxygen in air will be high at cold temperature than at hot temperature. On the other hand if you test the exact same Engine to the same dynamometer where the temperature, humidity and pressure are all carefully controlled on those different days, it will always make exactly the same power. There can be some difference due to personnel errors taking readings. But in this chapter we are mainly concerned about the effects of atmospheric temperature, pressure and humidity.

It is also important to know how much power your engine is really making on specific day due to the temperature, humidity and pressure of that day. In such case you should look at the uncorrected power readings. However in case of refurbished engines you may want to see how much more power you are getting solely due to the new valves, pistons and new piston rings then corrected power is more useful, since it removes the effects of the temperature, humidity and atmospheric pressure and just indicate how much more (or less) power you have than in your previous tests taken before the refurbishment of the engine.

There is no "right" answer. it's simply a matter of how you want to use the information.

12.5 Horsepower and Torque:

We are well aware of that by definition "power is the rate at which work is done". Power, in mechanical terms, is the ability to accomplish a specified amount of work in a given amount of time. By definition, one horsepower (HP) is equal to applying a 550 pound force through a distance of one foot in one second. In real terms, it would take one HP to raise a 550 pound weight up one foot in one second. As discussed in previous chapters horsepower may be expressed as:—

$$HP = \frac{2.\pi.(RPM).T}{33000} \quad \ldots \ldots \ldots \ldots \ldots \ldots \ldots \ldots \ldots \ldots 12.1$$

This is further simplified as:—

$$HP = \frac{(RPM).T}{5252} \quad \ldots \ldots \ldots \ldots \ldots \ldots \ldots \ldots \ldots \ldots 12.2$$

Where: HP = horsepower, hp
T = torque, ft-lbs
RPM = engine speed—revolutions per minute

This is a very important formula. The mere observation of the formula will tell us that speed and torque are inversely proportional. In very simple words it informs you that if you can keep the same amount of torque, then the more rpm you can turn the more horsepower you get or keeping the horse power same if you increase the RPM you will get the less torque.

12.6 Effect of altitude on the power

Absolute barometric pressure is a measure of how hard the air molecules are being pushed closer to one another. The unit of measurement is typically inches of mercury (inches Hg) in *FPS* system and in kPa in **S.I.** system. The more pressure, the more molecules there are in a liter of air and the more air the engine takes in during the intake stroke (increasing horsepower potential). Absolute barometric pressure is equal to Relative barometric pressure only at sea level. Relative barometric pressure is reported at airports and by weather barometers.

A good approximation for converting relative barometric pressure to absolute barometric pressure is:

$$H_{ABS} = H_{Rel} - \left(\frac{Elevation}{1000}\right) \quad \ldots\ldots\ldots\ldots\ldots\ldots 12.3$$

Where: H_{ABS} = Absolute barometric pressure.
H_{Rel} = Relative barometric pressure.
Elevation = Test cell elevation in feet above sea level.

Let us imagine that you are testing the engine in dynamometer test cell located at the height of 4500 feet above the sea level. The most of the engines will struggle to produce designed power at such a high elevation, due to the lack of air density, as we know that air is thinner at higher altitude. As a result, power output falls drastically compared to identical/similar engines that operate at or near sea level. For this reason, just about every engine tested applies a altitude correction roughly in the range of 10 to of 12%. This implies that an engine that produced 100 hp "corrected" up to 112 hp.

On the other hand if engine is "Turbocharged" then power output rarely falls as drastically in response to air density reduction. This is due to their turbo control systems that combat air density reductions by allowing for higher boost pressures. These increased boost pressures can almost completely offset the ambient pressure reduction and make the "altitude correction" almost completely unnecessary.

12.7 Effect of Humidity on the power.

The humidity is again correlated to density of the air in the atmosphere where engine is being tested. **The idea behind this is that air density**

reduces as moisture content increases. In other words, the more humid the day is, the less power the engine will make.

Water content in air is calculated from the ambient wet and dry bulb temperatures. Dry bulb temperature is normal room temperature. Wet bulb temperature is always less than or equal to dry bulb temperature. As air is blown over the wet bulb thermometer the water evaporates and cools the thermometer. The dryer the air, the cooler the wet thermometer indicates. If the ambient air is saturated i.e. humidity equal to 100%, very little water evaporates and the wet bulb temperature is equal to the dry bulb temperature. These measurements are then converted to partial pressure in inches of mercury and used in the correction formula. Water vapor displaces oxygen and reduces the amount of combustion air supplied during the intake stroke while decreasing horsepower potential.

The relative humidity can be established by wet and dry bulb temperatures using psychometric charts.

12.8 Effect of Temperature on the Power

As with altitude, increases in ambient air temperatures almost always yield reductions in engine output. Conversely, reductions in ambient temperature just about always yield increases in engine output.

Normally the refurbishment workshops may use the trick of strategic placement of air temperature sensor in dynamometer test cell, which is being used for correction factor calculation. When need for lower-than-normal dynamometer result arises, it's easy to place the air temperature sensor in a slightly colder environment i.e. colder spot in the test cell. Similarly, when a higher-than-normal result is needed, all one has to do is to place the sensor in a hot environment e.g. near the exhaust header in the test cell where the engine is being tested.

12.9 Power Correction Factors

There is more than one correction factor. All are a simplistic compromise but they are widely used.

- DIN 70012 Also popular, and used by people who like bigger numbers! In a lot of conditions it gives figures a few percent higher than the rest.
- SAE-J1349 is commonly used by lots of dynamometer companies as default.
- JIS D 1001 Method
- EEC 80/1269
- ISO 1585

12.10 Society of Automotive Engineers

The Society of Automotive Engineers (SAE) created the SAE J1349, June, 1990 standard method for correcting horsepower and torque readings so that they will seem as if the readings had all been taken at the same "standard" test cell where the air pressure, humidity and air temperature are held constant. Furthermore, the SAE J1349 (June, 1990) standard includes an assumed mechanical efficiency of 85% in order to provide an estimate of the true engine horsepower.

The equation for the dynamometer correction factor given in SAE J1349 JUN 1990, converted to use pressure in millibar is:

$$CF_{SAE} = 1.180 \left[\left(\frac{990}{P_D}\right) \cdot \left(\frac{T_C + 273}{T_{REF} + 273}\right)^{0.5} \right] - 0.180 \quad \ldots \ldots \ldots \quad 12.4$$

Where CF = The dynamometer correction factor
P_d = The pressure of the dry air in millibar
T_c = Ambient temperature, °C

The obtain the pressure of the dry air Pd, one needs to subtract the vapor pressure P_v from the actual air pressure.

12.10.1 SAE J1349 Update:

In August 1204 the SAE released J 1349 Revised which specifies that the preferred method of determining the friction power used by the motor accessories is actual measurement, and that the assumption of 85% mechanical efficiency (as formerly used in SAE J 1349 Revision

JUN 1990) should only be used when actual friction data are not available.

The equation for computing brake horsepower, assuming 85% mechanical efficiency is:

$$CF_{SAE} = 1.176 \cdot \left[\left(\frac{990}{P_d}\right)\left(\frac{T_c + 273}{T_{REF} + 273}\right)^{0.5}\right] - 0.176 \quad \ldots\ldots\ldots 12.5$$

The above formula is also represented in the following format:

$$CF_{SAE} = \frac{P}{P_0} = \left[\frac{(p_0 - p_{V0})}{P - p_V}\right] \cdot \left(\frac{T}{T_0}\right)^{0.5} \quad \ldots\ldots\ldots\ldots\ldots\ldots 12.6$$

Where:
P = Output power (W)
P_0 = Reference power output (W)
p = Atmospheric Pressure (bar)
p_0 = Reference atmospheric pressure (bar)
p_v = Partial pressure of water vapor (bar)
P_{vo} = Partial pressure of reference water (bar)
T = Ambient air temperature (K)
T_0 = Reference ambient air temperature (K)
Reference atmospheric Pressure = P_0 = 1.11325 bar
Reference atmospheric temperature T_0 = 293 K
Reference atmospheric pressure p_0 = 0.99 bar or 990 mbar
Reference ambient air Temperature T_0 = 302.4 K
Reference partial pressure of water vapor P_{vo} = 0.013 bar

Section 5.1 of the SAE J 1349 AUG, 1204 revision also makes it clear that this correction factor is not intended to provide accurate corrections over an extremely wide range, but rather that the intended range of air temperatures is 15 to 35 deg C, and the intended range of dry air pressures is 900 to 1050 millibar

12.10.2 Derivation of SAE Correction Factor Formula

The mass air flow rate of one dimensional steady state compressible fluid flow through orifice is given by the equation:

$$\dot{m} = \frac{A_{EFF} P_0}{\sqrt{R_S \cdot T_0}} \left\{ \frac{2 \cdot \gamma}{\gamma - 1} \left[\left(\frac{P}{P_0} \right)^{2/\gamma} - \left(\frac{P}{P_0} \right)^{(\gamma+1)/\gamma} \right] \right\}^{\frac{1}{2}} \quad \dots \dots \dots \dots \dots 12.7$$

Where A_{EFF} = effective area of the orifice

γ = ratio of the specific heat of air = $\dfrac{C_p}{C_v}$

P = Pressure at orifice

P_0 = upstream absolute pressure °R

T_0 = upstream absolute temperature °R

If we assume $\left(\dfrac{P}{P_0} \right)$ constant for a given engine, the mass flow rate will be as:

$$\dot{m} \, \alpha \, \frac{P_o}{\sqrt{T_0}} \quad \dots\dots\dots\dots 12.8$$

Since the indicated horse power for a given air fuel ration is directly proportional to mass of dry air injected. It follows from equation 12.8 that indicated power can be corrected by the following scaling factor,

$$\alpha = \frac{P_{REF}}{P_D} \sqrt{\frac{273 + T_F}{273 + T_{REF}}} \quad \dots\dots\dots\dots 12.9$$

Where P_D = dry air pressure = $P_o - P_{VAP}$

P_{REF} = Reference dry air pressure

T_{REF} = Reference temperature °F

T_F = Measured ambient temperature °F

In accordance with the standard, only the indicated power (IHP) is adjusted by the correction factor. The frictional horsepower (FHP) is not considered in the correction factor.

Please note that indicated horse power is the work done by combustion of fuel on the piston. The BHP, by definition, is the difference between IHP and FHP.

Therefore,

$$BHP = IHP - FHP$$

The corrected brake horsepower is given by:

$$BHP_{Corrected} = \alpha.IHP - FHP \quad \dots\dots\dots\dots\dots\dots\dots\dots\dots\dots\dots. \quad 12.10$$

Now by definition correction factor is:

$$CF(\alpha) = \frac{BHP_{Corrected}}{BHP_{Measured}}$$

$$CF = \frac{\alpha.IHP - FHP}{IHP - FHP} \quad \ldots\ldots\ldots\ldots\ldots\ldots\ldots\ldots\ldots\ldots 12.11$$

$$CF = \frac{\alpha. - \frac{FHP}{IHP}}{1 - \frac{FHP}{IHP}} \quad \ldots\ldots\ldots\ldots\ldots\ldots\ldots\ldots\ldots\ldots 12.12$$

By definition, the mechanical efficiency is given as η_m

$$\eta_m = \frac{BHP}{IHP} = 1 - \frac{FHP}{IHP} \quad \ldots\ldots\ldots\ldots\ldots\ldots\ldots\ldots 12.13$$

Therefore, by substitution equation 12.13 into equation 12.12 we get,

$$CF = \frac{\alpha. - \frac{FHP}{IHP}}{\eta_m}$$

Further simplifying we get,

$$CF = \frac{1}{\eta_m}\alpha - \frac{1}{\eta_m}\frac{FHP}{IHP} \quad \ldots\ldots\ldots\ldots\ldots\ldots\ldots\ldots 12.14$$

However, from equation 12.13 we get

$$1 - \eta_m = \frac{FHP}{IHP}$$

Therefore, the equation 12.14 is further simplified. To

$$CF = \frac{1}{\eta_m}\alpha - \left(\frac{1}{\eta_m} - 1\right) \quad\ldots\ldots\ldots\ldots\ldots\ldots\ldots\ldots\ldots 12.15$$

By substituting equation 12.9 into equation 12.15 we get

$$CF = \frac{1}{\eta_m}\frac{P_{REF}}{P_D}\sqrt{\frac{273 + T_F}{273 + T_{REF}}} - \left(\frac{1}{\eta_m} - 1\right) \ldots\ldots\ldots\ldots\ldots .12.16$$

Now let us substitute mechanical efficiency 0.85(85%) and SAE reference values for pressure and temperature. After simplifying we get equation 12.5 which is rewritten below

$$CF_{SAE} = 1.176.\left[\left(\frac{990}{P_d}\right)\left(\frac{T_c + 273}{T_{REF} + 273}\right)^{0.5}\right] - 0.176$$

12.11 DIN—70012 Method.

The "Deustches Institut für Normung, German Institute for Standardization" commonly know as DIN has provided the correction factor.

The DIN 70012 is a standard from German DIN regarding road vehicles. Because the German word for horsepower is Pferdestärke, in Germany it is commonly abbreviated to PS. DIN hp is measured at the engine's output shaft, and is usually expressed in metric (Pferdestärke) rather than mechanical horsepower.

This method for power correction, recommended by DIN-70012 standard (DIN, 1986), does not account for changes in air humidity. If changes in the atmospheric conditions are small with respect to a standard condition, then the engine overall efficiency, fuel specific heat and air/fuel ratio can all be considered constants. Taking the engine volumetric efficiency varying proportionally to the square root of the temperature, the following correlation is written:

$$CF_{DIN} = \frac{P}{P_0} = \left(\frac{p_0}{p}\right) \cdot \left(\frac{T}{T_0}\right)^{0.5} \quad \ldots\ldots\ldots\ldots\ldots\ldots 12.17$$

Where:
P = Output power (W)
P_0 = Reference power output (W)
p = Atmospheric Pressure (bar)
p_0 = Reference atmospheric pressure (bar)
T = Ambient air temperature (K)
T_0 = Reference ambient air temperature (K)
Reference atmospheric Pressure = P_0 = 1.11325 bar
Reference atmospheric temperature T_0 = 293 K

12.12 JIS D 1001 Method

Engine power test codes determine how the power and torque of an automobile engine is measured and corrected. Correction factors are used to adjust power and torque measurements to standard atmospheric conditions to provide a more accurate comparison between engines as they are affected by the pressure, humidity, and temperature of ambient air

The JIS D 1001 standard (JIS 1993) recommends the following correction factor (CF):

$$CF_{JIS} = \frac{P}{P_0} = \left[\frac{(P_0 - P_{v0})}{(P - P_v)}\right] \cdot \left(\frac{T}{T_0}\right)^{0.75} \quad \ldots\ldots\ldots\ldots\ldots\ldots 12.18$$

Where:
P = Output power (W)
P_0 = Reference power output (W)
p = Atmospheric Pressure (bar)
p_o = Reference atmospheric pressure (bar)
p_v = Partial pressure of water vapor (bar)
P_{vo} = Partial pressure of reference water (bar)
T = Ambient air temperature (K)
T_0 = Reference ambient air temperature (K)
Reference atmospheric Pressure = P_0 = 1.01325 bar
Reference atmospheric temperature T_0 = 293 K
Reference partial pressure of water vapor P_{vo} = 0.013 bar

12.13 ISO 1585 Method

SO 1585 is an engine net power test code intended for road vehicles. This method is given by ISO 1585 standard (ISO, 1992). The suggested power correction factor (*CF*) is:

$$CF = \frac{P}{P_0} = \left[\frac{(P_0 - P_{v0})}{(p - P_v)}\right]^{1.2} \cdot \left(\frac{T}{T_0}\right)^{0.6} \quad \ldots\ldots\ldots\ldots\ldots\ldots 12.19$$

Where:
P = Output power (W)
P_0 = Reference power output (W)
p = Atmospheric Pressure (bar)
p_o = Reference atmospheric pressure (bar)
p_v = Partial pressure of water vapor (bar)
P_{vo} = Partial pressure of reference water (bar)

T = Ambient air temperature (K)
T_0 = Reference ambient air temperature (K)
Reference atmospheric Pressure = P_0 = 1.0 bar
Reference atmospheric temperature T_0 = 298 K
Reference partial pressure of water vapor P_{vo} = 0.013 bar

12.14 Correct sensing of parameters—Ambient Air temperature, Pressure and RH in Test cell.

After having discussed the various correction factors provided by different standards, we are sure that we need to measure Air temperature, Air Humidity and Atmospheric pressure in test cell.

This pressure and temperature data has to be measured, and entered as figures, or sensed automatically depending on which dynamometer system we are using directly just before each "run" is made. If this is not done the data or graph will not be accurate!

The most important point to remember while carrying out the engine test is that any carbon monoxide in the room will affect the engine power. So exhaust leaks and poor air circulation can lead to substantial power variations, which are not the result of engine differences. If you can smell exhaust gases in the room, when the engine is running, the engine test data will be affected.

The mounting of the relative humidity sensors in the test cell plays important role as they are exposed to hydrocarbon vapors from gasoline or oil. Their calibration drifts rapidly, because they are just as susceptible to gasoline vapors as they are to moisture in the air. For this reason, many engine manufacturers will have the air ducts brought from outside of test cell to the carburetor. The relative humidity sensor to be mounted near the duct entrance outside of the room to protect the sensor from gasoline vapors. Then the relative humidity must be converted to absolute humidity for power correction

Similarly ambient pressure transducers mounted inside the test cell need to be protected from moisture and hydrocarbon vapors.

Air intake temperature sensor equally plays an important role and care should be taken for its proper mounting. It should not be mounted near exhaust extraction pipes as radiation heat of exhaust pipe will cause Air intake temperature sensor to provide false reading.

Closure:

Power correction standards try to estimate what engine power would be under reference conditions. The correction factors cannot actually calculate exactly what power output would be. The greater the difference between the ambient conditions during the test and the reference conditions, the greater the error in the estimate. Most power correction standards include limits on their applicability. This limit is typically +/−7%. This means if the correction factor is greater than 1.07 or less than 0.93, the corrected power numbers are not officially considered to be acceptable, and the test should be performed again under conditions which are closer to the reference conditions.

Power corrections are only valid for Wide Open Throttle (WOT) / full throttle performance tests. You should disregard corrected power numbers for any test performed under partial throttle conditions.

Ref:

1. ISO, 1992, "Road Vehicles-Engine Test Code-Net Power", ISO 1585, International Organization for standardization.
2. DIN, 1986, "Automotive Engineering; Maximum Speed; Acceleration and Other Terms; Definitions and Tests", DIN 70012, Deustches Institut für Normung.
3. JIS, 1993, "Road Vehicles—Engine Power Test Code", JIS D 1001, Japanese Industrial Standard.
4. SAE, 1995, "Engine Power Test Code—Spark Ignition and Compression Ignition—Net Power Rating", SAE J 1349, Society of Automotive Engineers.
5. Measurement of combustion engine power characteristics correction factors by Radim ČECH, Petr TOMČÍK.
6. Comparison of Engine Power Correction Factors for Varying Atmospheric Conditions by J. R. Sodré, S. M. C. Soares

PART II

DATA ACQUISITION AND CONTROL SYSTEM

CHAPTER 13

The Sensors and Transducers

13.1 Introduction

In the field of engine testing, sensors and transducers play a major role. Modern engine is monitored for various parameters while testing and sensors and transducers play a major role in acquiring data. In our daily life of humans, we sense outer world parameters like taste, touch, vision, smell, and hearing through our sensors such as tongue, hands, eyes, nose, and ears. Our senses need a certain amount of energy to work properly. For example, eyes need certain amount of brightness to see, and ears need certain loudness of sound to be able to hear. Similarly pressure on the skin must be great enough to feel the touch. Some of the sensors used in the physical world to measure the physical parameters also need certain amount of energy, sometimes called as excitation voltage. The modern inventions or the devices can extend the human physical senses of touch (pressure, temperature) and hearing (sound pressure).

Various kinds of sensors and transducers are employed to measure a wide range of parameters. The modern sensors are highly complex in nature. When you study sensors, two questions are raised:

1. What they do? This gives information about the function of sensor.
2. How they do? This gives information about their structure.

A transducer is a device which transforms the input signal of one energy form into an output of another energy form, doing so in such a way that a prescribed relationship is maintained between the input and output signal.

Sensors are devices that are used to measure physical variables like temperature, pH, velocity, rotational rate, flow rate, pressure, and many others. Some of the sensors are mentioned in the table below.

The role of environmental sensors can be appreciated by considering them as an extension of human senses. Sensors sense the same phenomena as human senses; however,

- Physical sensors are available 24 hours a day and round the year.
- Their measurements are precise.
- Their measurements are reproducible.

Table 13.1 Sensors and transducers

Input	Sensor/transducer	Output	Excitation
Temperature	Thermocouple RTD Thermistor	Voltage Resistance Resistance	Current, voltage
Pressure	Strain gauge Piezoelectric	Strain gauge Piezoelectric	Voltage
Force and torque	Strain gauge Piezoelectric	Voltage Voltage	Voltage
Acceleration/ Vibration	Strain gauge Piezoelectric Variable capacitance	Voltage Charge Voltage	Voltage Voltage
Position/ Displacement	LVDT and RVDT Potentiometer	AC Voltage Voltage	Voltage Voltage
Light intensity	Photodiode	Current	
pH	Electrode	Voltage	
Sound	Strain gauge Piezoelectric	Strain gauge Piezoelectric	Voltage

| Flow rate | Coriolis
Vortex shedding
Turbine | Frequency
Pulse/
Frequency
Pulse/
Frequency | Voltage |

13.2 Thermocouples

13.2.1 Theory of Operation

A *thermocouple* is a junction between two different metals that produces a voltage related to a temperature difference. Thermocouples are a widely used type of temperature sensor for measurement and control and can also be used to convert heat into electric power. When two wires composed of dissimilar metals are joined at both ends and one of the ends is heated, there is a continuous current which flows in the thermoelectric circuit. This is now known as the thermoelectric effect or Seebeck effect. *Thomas Johann Seebeck* made this discovery in 1821.

Thermocouples come with different pairings of materials allowing for a very wide range of applications. The different compositions are standardized into thermocouple types. The different types are given letter names which are standardized across the industry. This helps user purchase a "J" type thermocouple temperature indicator from one manufacturer and "J" type thermocouples from a different manufacturer and be able to put together a system that will work. The manufacturer of thermocouples will provide tables indicating the best thermocouple type for a particular temperature range. The manufacturer will also provide data regarding the temperature and voltage relationship for each type of thermocouple.

The voltage relationship with the temperature is not always a simple linear relationship. Therefore, a "Look up" table may be needed to convert the voltage readings into temperature readings. This information is available from the thermocouple manufacturer. Most of the manufacturers of the thermocouple also make a digital indicator that converts the voltages into "engineering" units which makes the life easy for the user. Simplicity of the device "Thermocouple" may be deceiving though.

13.2.2 Factors Affecting the Accuracy of the Thermocouple

1. Closeness of the junction to measurement point

The voltage at the thermocouple terminals is proportional to the temperature of the junction. Thus when using a thermocouple, it is very important that the junction of the thermocouple be in very close contact with the object that is being measured.

2. Compensating cables

The thermocouple is an electrically conducting wire; care must be taken so that there is no possibility of coming in contact with other exposed electrical conductors.

3. Time duration

The insulation between the two wires of the thermocouple can breakdown over a considerable long time of use and cause errors in the temperature reading.

Different thermocouple types (e.g., J, K, T, E, etc) use different mixtures of metals in the wire.

13.2.3 Identification for Insulated Thermocouple Wire

The insulation on thermocouple wire is color coded for identification. Common guidelines include that the negative lead in insulated thermocouple wire is red. The positive lead has the color of the thermocouple as well as the overall color of insulated extension grade wire. The outer jacket of thermocouple grade wire is typically brown. For high temperature wire, it is common to have a color-coded tracer thread in the white material.

13.2.4 Thermocouple Wire

The thermocouple wire is one that is used in a thermocouple from the point of sensing to the point of cold junction compensation where the signal is measured.

13.2.5 Accuracy of Thermocouple

Accuracy of thermocouples varies with thermocouple types, for example, for the lower temperature ranges type T composed of copper wire in the positive lead and constantan (copper-nickel mixture) for the negative has good specifications for accuracy.

13.2.6 Thermocouple Grade and Extension Grade Wire

The wire that is used to make the sensing point (probe part) of the thermocouple is the thermocouple grade wire. Extension grade wire is only used to extend a thermocouple signal from a probe back to the instrument reading the signal.

13.2.7 Types of Thermocouples Commonly Used

Type K: Junction Material—Chrome l/Alumel

Type K is the "general purpose" thermocouple. It is low cost, and, owing to its popularity, it is available in a wide variety of probes. Thermocouples are available in the range from –200°C to +1300°C. Sensitivity is approximately 41 µV/°C. Use type K unless you have a good reason not to.

Type E: Junction Material—Chromel/Constantan

Type E has a high output (68 µV/°C) which makes it well suited to use at low temperature (cryogenic). Another property is that it is nonmagnetic.

Type J: Junction Material—Iron/Constantan

Limited range (–40°C to +750°C) makes type J less popular than type K. The main application is with old equipment that cannot accept "modern" thermocouples. J types should not be used above 760°C as an abrupt magnetic transformation will cause permanent decalibration.

Type N: Junction Material—Nicrosil/Nisil

High stability and resistance to high temperature oxidation makes type N suitable for high temperature measurements without the cost of platinum.

Type B: Junction Material—Platinum/Rhodium

It is suited for high temperature measurements up to 1800°C. Unusually type B thermocouples (due to the shape of their temperature/voltage curve) give the same output at 0°C and 42°C. This makes them useless below 50°C.

Type R: Junction Material—Platinum/Rhodium

It is suited for high temperature measurements up to 1600°C. Low sensitivity (10 µV/°C) and high cost make them unsuitable for general purpose use.

Type S: Junction Material—Platinum/Rhodium

It is suited for high temperature measurements up to 1600°C. Low sensitivity (10 µV/°C) and high cost makes them unsuitable for general purpose use. Due to its high stability, type S is used as the standard of calibration for the melting point of gold (1064.43°C).

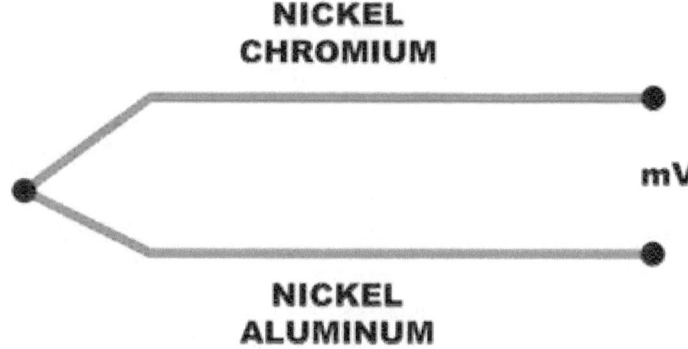

Figure 13.1 Typical thermocouple junction.

13.2.8 Cold Junction Compensation (CJC)

It is not possible to simply connect up a voltmeter to the thermocouple to measure this voltage, because the connection of the voltmeter leads will make a second, undesired thermocouple junction. To make accurate measurements, this must be compensated for by using a technique known as cold junction compensation (CJC). In case you are wondering why connecting a voltmeter to a thermocouple does not make several additional thermocouple junctions (leads connecting to the thermocouple, leads to the meter, inside the meter, etc), the law of intermediate metals states that a third metal inserted between the two dissimilar metals of a thermocouple junction will have no effect provided that the two junctions are at the same temperature. This law is also important in the construction of thermocouple junctions. It is acceptable to make a thermocouple junction by soldering the two metals together as the solder will not affect the reading. In practice, thermocouple junctions are made by welding the two metals together (usually by capacitive discharge). This ensures that the performance is not limited by the melting point of solder.

All standard thermocouple tables allow for this second thermocouple junction by assuming that it is kept at exactly zero degrees centigrade. Traditionally this was done with a carefully constructed ice bath (hence, the term "cold" junction compensation). Maintaining an ice bath is not practical for most measurement applications. Instead the actual temperature at the point of connection of the thermocouple wires to the measuring instrument is recorded.

Typically cold junction temperature is sensed by a precision thermistor in good thermal contact with the input connectors of the measuring instrument. This second temperature reading, along with the reading from the thermocouple itself, is used by the measuring instrument to calculate the true temperature at the thermocouple tip. For less critical applications, the CJC is performed by a semiconductor temperature sensor. By combining the signal from this semiconductor with the signal from the thermocouple, the correct reading can be obtained without the need or expense to record two temperatures. Understanding of cold junction compensation is important; any error in the measurement of cold junction temperature will lead to the same error in the measured temperature from the thermocouple tip.

13.2.9 Precautions and Considerations for Using Thermocouples

Most measurement problems and errors with thermocouples are due to a lack of understanding of how thermocouples work. Thermocouples can suffer from aging, and accuracy may vary consequently especially after prolonged exposure to temperatures at the extremities of their useful operating range. Listed below are some of the more common problems and pitfalls to be aware of.

13.2.10 Connection Problems

Many measurement errors are caused by unintentional thermocouple junctions. Remember that any junction of two different metals will cause a junction. If you need to increase the length of the leads from your thermocouple, you must use the correct type of thermocouple extension wire (e.g., type K for type K thermocouples). Using any other type of wire will introduce a thermocouple junction. Any connectors used must be made of the correct thermocouple material, and correct polarity must be observed.

13.2.11 Lead Resistance

To minimize thermal shunting and improve response times, thermocouples are made of thin wire. (In the case of platinum types, cost is also a consideration.) This can cause the thermocouple to have a high resistance which can make it sensitive to noise and can also cause errors due to the input impedance of the measuring instrument. If thermocouples with thin leads or long cables are needed, it is worth keeping the thermocouple leads short and then using thermocouple extension wire (which is much thicker, so has a lower resistance) to run between the thermocouple and measuring instrument. It is always a good precaution to measure the resistance of your thermocouple before use.

13.2.12 Decalibration

Decalibration is the process of unintentionally altering the makeup of thermocouple wire. The usual cause is the diffusion of atmospheric particles into the metal at the extremes of operating temperature. Another cause is impurities and chemicals from the insulation diffusing into the thermocouple wire. If operating at high temperatures, check the specifications of the probe insulation.

13.2.13 Noise

The output from a thermocouple is a small signal, so it is prone to electrical noise pickup. Most measuring instruments reject any common mode noise (signals that are the same on both wires) so noise can be minimized by twisting the cable together to help ensure both wires pick up the same noise signal. Additionally, an integrating analog to digital converter can be used to help average out any remaining noise.

13.2.14 Common Mode Voltage

Although thermocouple signal is very small, much larger voltages often exist at the input to the measuring instrument. These voltages can be caused either by inductive pickup (a problem when testing the temperature of motor windings and transformers) or by "earthed" junctions. A typical example of an "earthed" junction would be measuring the temperature of a hot water pipe with a noninsulated thermocouple. If there are any poor earth connections, a few volts may exist between the pipe and the earth of the measuring instrument. These signals are again common mode (the same in both thermocouple wires) so will not cause a problem with most instruments provided they are not too large.

13.2.15 Thermal Shunting

All thermocouples have some mass. Heating this mass takes energy which will affect the temperature you are trying to measure. Consider for example measuring the temperature of liquid in a test tube. There are two potential problems. The first is that heat energy will travel up the thermocouple wire and dissipate to the atmosphere, therefore, reducing the temperature of the liquid around the wires. A similar problem can occur if the thermocouple is not sufficiently immersed in the liquid, due to the cooler ambient air temperature on the wires; thermal conduction may cause the thermocouple junction to be a different temperature to the liquid itself. In the above example, a thermocouple with thinner wires may help, as it will cause a steeper gradient of temperature along the thermocouple wire at the junction between the liquid and ambient air. If thermocouples with thin wires are used, consideration must be paid to lead resistance. The use of a thermocouple with thin wires connected to much thicker thermocouple extension wire often offers the best compromise.

13.3 Resistance Temperature Detector (RTD)

13.3.1 Theory of Operation

The use of resistance temperature detector (RTD) for measurement of temperature is very popular. Temperature measurements are perhaps the oldest known measurements. The resistivity of the material has dependence on the temperature. This was first discovered by Sir Humphrey Davy. The RTD's resistance verses temperature characteristics are stable and reproducible and have a near linear positive temperature coefficient from 200°C to 800°C. These characteristics establish RTDs as popular choice for the measurement of temperature. The platinum is used as the primary element in all high-accuracy resistance thermometers. RTD is most commonly known as PT-100 sensor.

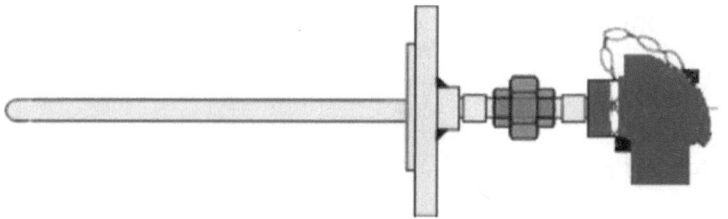

Figure 13.2 RTD—flange mounting.

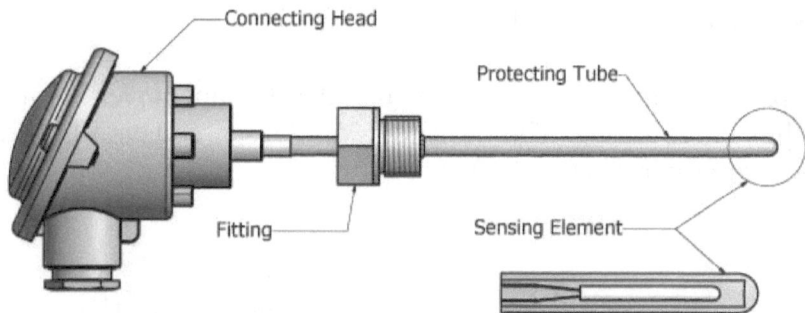

Figure 13.3 RTD—screw mounting.

The principle of operation is to measure the resistance of a platinum element. The most common type (PT-100) has a resistance as shown in the table below:

Table 13.2 Resistance of PT-100

Temperature	Resistance
0°C	100 Ohm
100°C	138.4 Ohm

All metals produce a positive change in resistance for a positive change in temperature. This, of course, is the main function of a RTD. System error is minimized when the nominal value of the RTD resistance is large. This implies a metal wire with a high resistivity. The lower the resistivity of the metal, the more material we will have to use.

The most common RTDs are made of

- Platinum
- Nickel
- Nickel alloys

For measurement integrity, platinum is the obvious choice.

13.3.2 Linearization of RTD

The RTD is a more linear device than the thermocouple, but it still requires curve-fitting. The Callendar-Van Dusen equation is widely used to approximate the RTD curve:

$$R_T = R_0 + R_0\alpha \left[T - \delta - \left(\frac{T}{100} - 1\right)\left(\frac{T}{100}\right) - \beta\left(\frac{T}{100} - 1\right)\left(\frac{T}{100}\right)^3\right] \quad \ldots\ldots 13.1$$

where
R_T = resistance at temperature T
R_0 = resistance at T = 0°C
α = temperature coefficient at T = 0°C
  (typically +0.00392 $\Omega/\Omega/°C$)
δ = 1.49 (typical value for .00392 platinum)
$\beta = 0$ T > 0
  = 0.11 T < 0

The exact values for coefficients α, β, and δ are determined by testing the RTD at four temperatures and solving the resultant equations. This familiar equation was replaced in 1968 by a 20th order polynomial in order to provide a more accurate curve fit. The plot of this equation shows the RTD to be a more linear device than the thermocouple.

The maximum temperature rating for RTDs is based on two different factors:

1. Element material. Platinum RTDs can be used as high as 650°C (1302°F).

2. Second determining factor for temperature rating is probe construction.

13.3.3 Advantages of RTD

1. RTD is cheaper than thermocouple and good solution for low-range measurement.
2. Since the RTD circuit is just a resistance circuit, no special extension lead wires or connectors are required, making this portion of the circuit less expensive than that of a thermocouple

13.2.4 Disadvantages of RTD

1. RTDs are fragile at temperatures above 320°C (600°F). The RTD sensor will not hold up well at these elevated temperatures if there is any vibration present.
2. The tolerance or accuracy of a RTD generally decreases as temperature increases.
3. RTDs are generally more expensive to manufacture or purchase than thermocouples.

13.3.5 Tolerance and Accuracy

The tolerance or accuracy for RTD sensors is stated at one point only, which is usually ± 0.10 percent @ 0°C (32°F). It is sometimes stated with respect to full scale of measurement such as 0.10 percent of 200 percent of full scale.

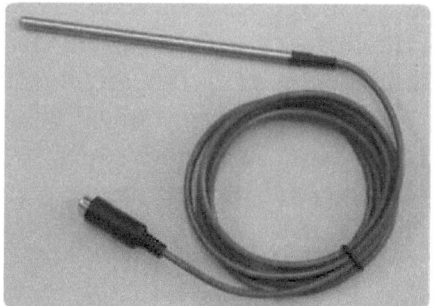

Figure 13.4 RTD without bulk head.

13.3.6 Comparison of Thermocouple and RTD

The engine testing using dynamometer is not complete without measurement of various temperatures of dynamometer and engine.

Following temperatures are measured on dynamometer:

1. Dynamometer water outlet temperature (range: 0-100°C)
2. Dynamometer bearing temperature (range: 0-100°C)
3. Eddy-current dynamometer loss plate temperature (range: 0-400°C)
4. Eddy-current dynamometer coil temperature (range: 0-400°C)

Following temperatures are measured on engine:

1. Engine water outlet temperature (range: 0-100°C)
2. Engine lubricating oil temperature (range: 0-200°C)
3. Engine air intake temperature (range: 0-200°C)
4. Engine exhaust manifold temperature (range: 0-1000°C)
5. Engine fuel inlet temperature (range: 0-200°C)
6. Engine fuel spill back temperature (range: 0-200°C)

The measurement of the above parameters can be done by selecting proper temperature sensor. When we start to cross the boundaries between choosing one type of sensor over another, the optimum choice between thermocouple and RTD can be difficult. There is a lot of overlap between these sensors at the more popular lower end of the operating temperature range. So for sensors that cover the same operating range, and applications where response time is not a driving issue, plus stability, accuracy, and sensitivity are acceptable, we really have to drill deeper and compare characteristics between sensors to find the best fit for a given application.

The following table shows 13.1 the detailed comparison of the thermocouple and the RTD:

Table 13.1 Comparison of thermocouple and RTD

Characteristic	Thermocouple (T/C)	Temperature Detector (RTD)
Measurement Range	Wide, −250°C to +2600°C	Narrower, −200°C to +850°C, often limited to a lower temperature by its insulation.
Output Signal	Voltage wrt difference in end-to-end temperature	Resistance change wrt actual temperature
Accuracy	Less accurate 2-4°C typical	More accurate up to 1°C typical
Long-term Stability	Fair limited to shorter periods good	Good, stable over long periods
Stability/Drift	Good, but more subject to drift	Excellent, better long-term stability
Sensitivity	Lower	Higher sensitivity
Interchangeability	Good	Excellent
Linearity	Fair linearity, special linearization generally required	Better linearity, special linearization still required, but to a lesser degree
Self-heating Error	No self-heating error	Some self-heating error, but low
Extension Cable	High effect, must match T/C type and is more expensive	Lower effect, can use different material, but ultimately limited by lead wire resistance
Response Time	Fast (≤ 0.1 seconds typical), but cold junction compensation has thermal lag	Slower (1-7 seconds typical)
Repeatability	Reasonable	Better and greater standardization
Hysteresis Excellent Good	Hysteresis excellent good	Hysteresis excellent good
Signal Strength Low	prone to EMI higher	more EMI resistant
Vibration/Shock Resistance	Good resistance	Less resistant than T/C
Robustness/ Ruggedness	Very good	Good
Sensor Dimensions	Very small to very large	Very small to very large

Measurement Area	Small, single point-of-contact	Larger, whole element must contact, 1" typical
Fine Wire Diameter	Small down to 0.25 mm diameter	Larger up to 3 mm diameter
Reference Junction	Required and a significant source of measurement error. Usually requires a stable ambient at cold junction	Not required and not a source of error
Excitation Required	Not required, self-powered	Yes, reference voltage or current source
Lead-Wire Resistance	High, but often mitigated by mating technology	Must be considered wrt maximum added resistance and potential resistive imbalance between leads
Cost	Less expensive	More expensive
Complexity	Very simple and less subject to mechanical stress	Physically larger and has a more complex construction making it more subject to mechanical stress
Calibration Ease	More difficult and adds CJC calibration	Less difficult, no CJC to contend with
Noise Immunity	Lower noise immunity but often mitigated by good wiring practice. Small signals and high impedance leads can easily pick up noise.	Better noise immunity than a T/C

In general, if your application requires the highest accuracy, cost is not a concern, and your operating ambient is less than 800°C, then the choice of a RTD over a thermocouple sensor is probably the right one. The RTD is more accurate, more stable, more repeatable, and offers a more robust output signal with better sensitivity and linearity than a thermocouple. However, the RTD does have a narrower operating range with a lower maximum operating temperature, it is generally more expensive, and it does require excitation which might drive the need for an external power source (e.g., a Wheatstone bridge). Please review Part 2 of this series for other differentiating features of RTD sensors.

The modern trend is to use the temperature transmitters wherein signal conditioning is done at the sensor level. The following figure shows the modern temperature transmitter.

Figure 13.5 The temperature transmitter. Courtesy of Acromag INC.

13.4 Strain Gauge

The strain gauges are widely used to measure the force and pressure in the application of engine testing.

13.4.1 Theory of Operation

A strain gauge is a sensor whose resistance varies with applied force. It converts force, pressure, tension, weight, etc., into a change in electrical resistance which can then be measured. The strain gauge is defined as a device whose electrical resistance varies in proportion to the amount of strain in the device.

Strain gauges are widely employed in sensors that detect and measure force and force-related parameters, such as torque, acceleration, pressure, and vibration. The strain gauge is the building block for strain sensors that often employ multiple strain gauges in their construction. A strain gauge will undergo a small mechanical deformation with an applied force that results in a small change in gauge resistance proportional to the applied force. Because this change in resistance with applied force is so small, strain gauges are commonly wired using a Wheatstone bridge.

The application of external force on a stationary, nonmoving body will result in producing the stress and strain in the stationary body. Strain is a measure of the deformation of a body when subject to an applied force. Specifically, strain (ε) is the fractional change in dimension (length, width, or height) of a body when subject to a force along that dimension. Mathematically, the strain is defined as the ratio of elongation to the gauge length of the test specimen or the change in length per unit of original length.

The strain (ε) is the ratio of the change in length, ΔL, of the sample to its original length L_0.

$$\varepsilon = \frac{L-L_0}{L} = \frac{\Delta L}{L_0}, \quad\quad\quad\quad\quad\quad\quad\quad\quad\quad\quad\quad (13.2)$$

where
ε = strain
L = final length
L_0 = original length

Strain can be positive (tensile) or negative (compressive). Although dimensionless, strain is sometimes expressed in units such as in/in. or mm/mm.

There are various methods to measure the strain. However, strain measurement using strain gauge is most popular method.

The strain of a body is always caused by an external influence or an internal effect. Strain might be caused by forces, pressures, moments, heat, structural changes of the material, and the like. The most widely used gauge is the bonded metallic strain gauge.

The proper mounting of a strain gauge is critical to its performance in ensuring that the applied strain of a material is accurately transferred through the adhesive and backing material, to the foil itself.

The relationship between the resultant fractional changes of gauge resistance to the applied strain (fractional change of length) is called the gauge factor (GF), or sensitivity to strain which is fundamental factor of strain gauge.

Specifically, the gauge factor is the ratio of the fractional change in resistance to the strain:

$$GF = \frac{\Delta R/R}{\Delta L/L} = \frac{\Delta R/R}{\varepsilon}, \quad \ldots\ldots(13.3)$$

13.4.2 Factors Influencing Selection of Strain Gauges

The strain gauge system consists of the following components:

1. Strain gauge
2. Bonding adhesive
3. Protective coating
4. Lead wires

Following factors are important while selecting the proper strain gauge:

1. Static Strain

Select a strain gauge and other components of installation that will provide the maximum electrical and dimensional stability and repeatability with minimum difficulty of installation.

2. Dynamic Strain

Select a strain gauge constructed for longer fatigue life. Note the maximum strain level in the dynamic load cycle.

3. **Current Carrying Capacity**

   Strain gauge size, grid design, and thermal conductivity of bonding material and adhesive as well as the "heat-sink" capability of element to which these are attached place a practical limit on current carrying capacity of the strain gauge.

4. **Operating Temperature Range**

   Using a strain gauge with temperature compensation can significantly reduce temperature sensitivity. Use of dummy gauge is explored when a strain gauge with preferred temperature compensation is not available.

5. **Strain Range**

   The strain of strain gauge is determined by degree of anneal of the foil, geometry of the strain sensing grid, and the elastic properties of the strain gauge backing and the adhesive used.

6. **Strain Sensing Materials**

   Copper-nickel alloy or constantan is the most frequently used material for strain gauges.

7. **Lead Wire Material**

   Select leads with materials that have low and stable resistivity.

13.4.3 How Strain Gauge Is Attached to Specimen?

Strain gauge is constructed by bonding a fine electric resistance wire or photographically etched metallic resistance foil to an electrical insulation base (backing), and attaching gauge leads. Strain gauge is used for strain measurement by bonding it on the surface of the specimen with specified adhesive.

Figure 13.6 Strain gauge.

The measurement of such small changes in resistance is really important and hence strain gauges are almost always used in a bridge configuration with a voltage excitation source. The general Wheatstone bridge, illustrated in Figure 13.7 below, consists of four resistive arms with an excitation voltage, V_{EX}, that is applied across the bridge.

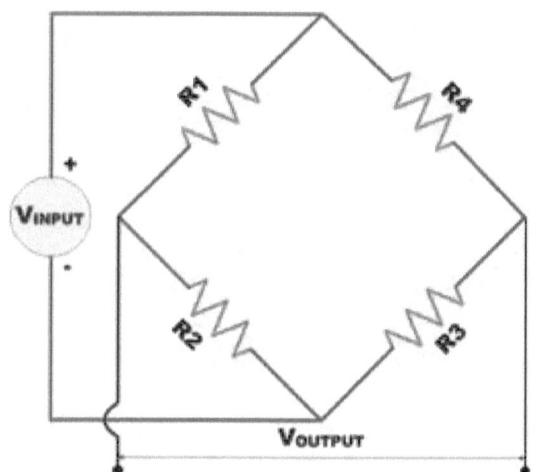

Figure 13.7 Wheatstone bridge.

The strain generated in the specimen is transmitted to the resistor through the gauge base where expansion or contraction occurs. As a result, the resistor experiences a variation in resistance. This variation is proportional to the strain as indicated in the following equation.

$$\varepsilon = \frac{\Delta L}{L} = \frac{\Delta R/R}{K}, \quad \text{.............................13.4}$$

where
ε = strain
R = gauge resistance
ΔR = resistance change due to strain
K = gauge factor as shown on the package

13.4.4 Strain Measurement Using a Wheatstone Bridge Circuit

Resistance of a strain gauge changes proportionally to the received strain. To measure strain is to measure this resistance change. Since this resistance change is very small in usual case, it requires a Wheatstone bridge circuit to convert the resistance change into voltage output.

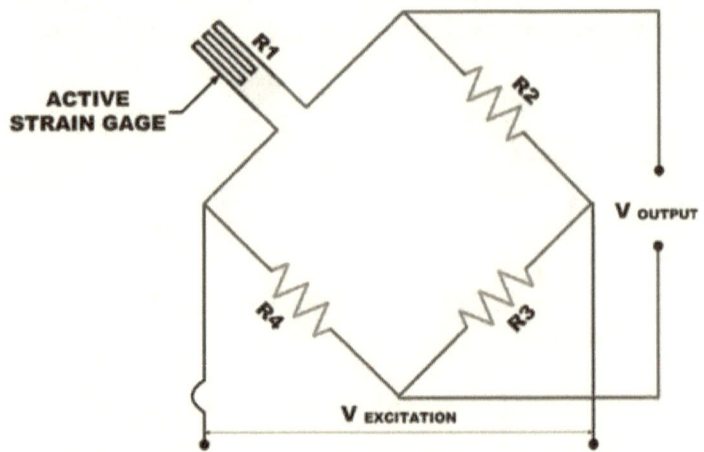

Figure 13.8 Strain gauge in bridge configuration.

The output voltage of a bridge circuit is given as follows.

$$V_{OUTPUT} = \frac{R_1 R_3 - R_2 R_4}{(R_1 + R_2)(R_3 + R_4)} E, \quad \text{.............13.5}$$

where
V_{OUTPUT} = output voltage
E = exciting voltage

R_1 = gauge resistance
R_2, R_3, R_4 = resistance of fixed resistors

13.3.5 Signal Conditioning for Strain Gauges

Strain gauge measurement involves sensing extremely small changes in resistance. Therefore, proper selection and use of the bridge, signal conditioning, wiring, and data acquisition components are required for reliable measurements. To ensure accurate strain measurements, it is important to consider the following:

- Bridge completion
- Excitation
- Remote sensing
- Amplification
- Filtering
- Offset
- Shunt calibration

The most important application of strain gauge is the load cell. A simpler form of the Wheatstone bridge is the load cell. The load cell is a device principally used in weighing systems that utilizes strain gauge technology internally. Unlike the strain gauge, the output of a load cell will be expressed in equivalent units of force. As a result, processing a load cell signal does not require intimate knowledge of its bridge type, gauge factor, or Poisson's ratio. Rather, the important considerations for a load cell are its rated output (mV/V), its excitation, and its rated capacity.

13.5 Load Cell

The load cell is commonly used transducer on dynamometer to measure the force/torque produced by the engine.

13.5.1 Theory of Operation

The load cell is a transducer that converts mechanical force into electrical signals. When a force is applied to it in a specific manner, a load cell produces an output signal that is proportional to the applied force. There are many different types of load cells that operate in different ways, but the most commonly used load cell today is the strain gauge (or strain gauge) load cell. The strain gauge load cells use an array of strain gauges to measure the deformation of a structural member and convert it into an electrical signal.

13.5.2 Load Cell Operating Principles

The load cells are classified based on the operating principle and the type of output signal generated.

1. Hydraulic Load Cell

When weight is applied, the hydraulic load cell changes the pressure of the load cell fluid in proportion to the load. As force increases, the pressure of the hydraulic fluid rises. This pressure can be locally indicated or transmitted for remote indication or control. The hydraulic load cell does not need the electrical signal; they are preferred in hazardous applications.

2. Pneumatic Load Cells

When weight is applied, the pneumatic load cell changes the pressure of the air in the load cell chamber in proportion to the load. As force increases, the pressure of the air rises proportionately. These types of load cells are proffered in the applications requiring explosion proof devices. The pneumatic load cells are however sensitive to temperatures. In the past, some of the hydraulic dynamometers were using pneumatic load cells along with analog dial indicator for the measurement of load.

3. Strain Gauge Load Cell

The strain gauge load cells convert the load acting on them into an electrical signal. The gauges themselves are bonded onto a beam or structural member that deforms when weight is applied. The strain gauges are arranged in the form of Wheatstone bridge. When weight is applied, the strain changes the electrical resistance of the gauges in proportion to the load.

The most widely used load cell is the strain gauge load cell. The strain gauge load cell plays a major role in the torque measurement by dynamometer. The strain gauge is bonded to an insulating backing material that in turn is bonded to the surface of the load cell. As the load cell is compressed or elongated due to the application of external load, the strain gauge also experiences this movement of load cell. The strain gauges are connected together to form a Wheatstone bridge as shown in Figure 13.7.

13.5.3 Load Cell Classification Based on Working Principle

1. Cantilever or bending beam
2. Compression type
3. Tension type
4. Universal
5. Shear

13.5.4 Types of Load Cells Classified Based on Construction

1. Beam-type load cell
2. Column load cell
3. S-type load cell

The following figure shows how load cells are classified:

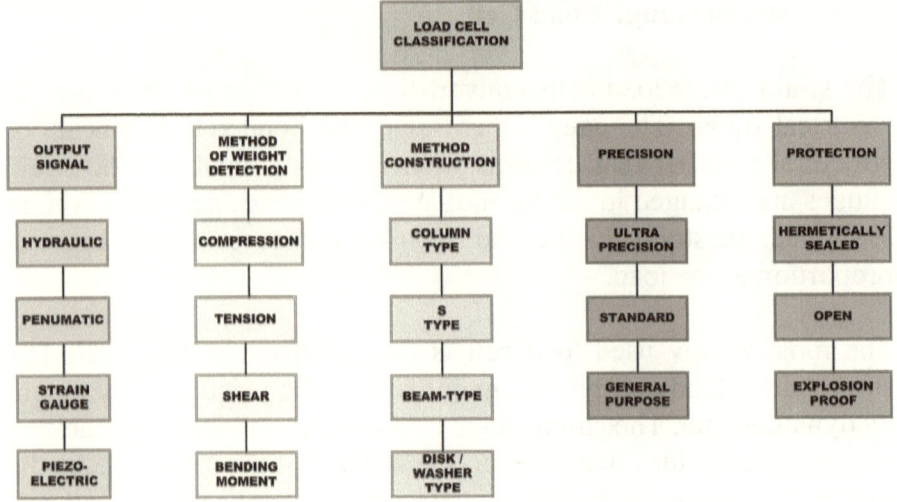

Figure 13.9 Load cell classification.

1. Beam-type Load Cells

The beam-type load cell is made of stainless steel to which strain gauges are applied. The strain gauges are normally arranged at 45° to the longitudinal axis on the side of the spring body and are therefore subject to shear forces. Under the influence of the load acting in the measuring direction, the load cell bodies and therefore the strain gauges are elastically deformed. This generates a measuring signal voltage that is proportional to the load.

The advantage of beam-type load cell is that it is easiest load cell to install.

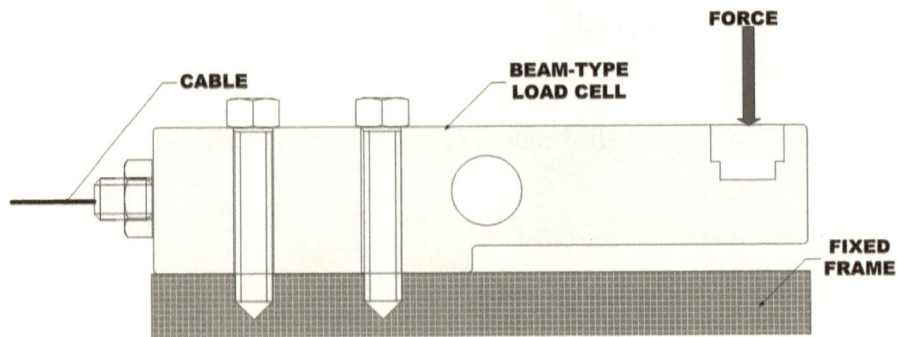

Figure 13.10 Beam-type load cell.

2. Column Load Cell

The column-type load cells are universal loading type load cells. It is designed for easy installation, usually between two flat faces bearing on its loading rings. Alternatively tensile load transfer can be achieved via a tie rod assembly through the center hole as shown in Figure 13.11:

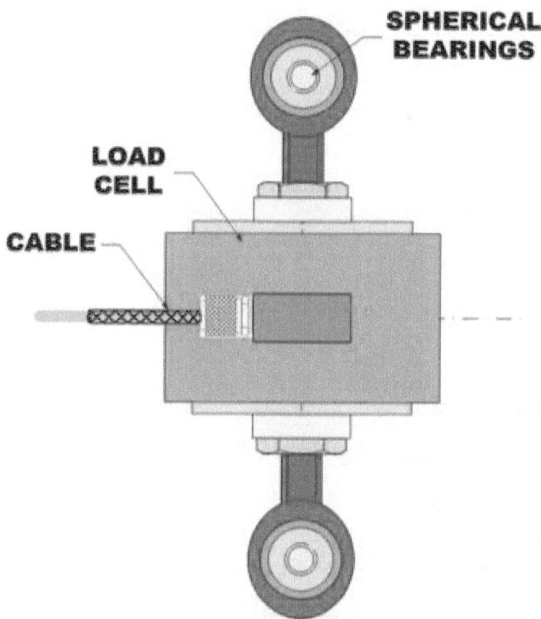

Figure 13.11 Column-type load cell.

3. S-type Load Cells

S-Beam type load cell get their name from their "s" or "z" shape. The S-Beam provides an output under tension as well as compression loading pull. One of the important applications for S-Beams load cell is dynamometer.

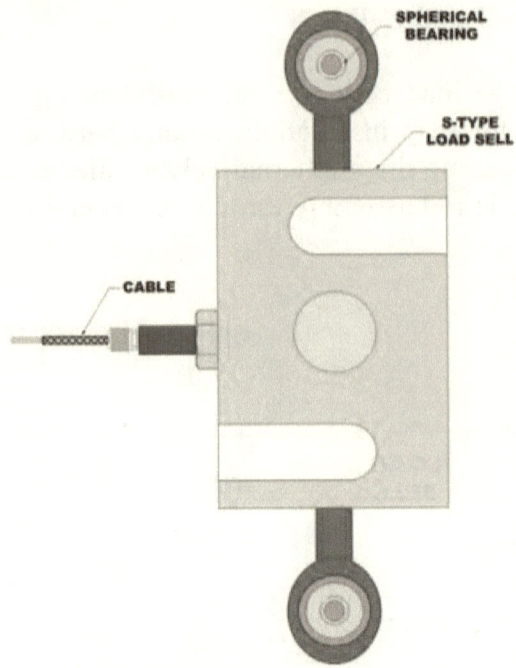

Figure 13.12 S-type load cell.

4. Diaphragm Load Cells

These load cells convert the deflection of diaphragms into a change in resistance, inductance, or capacitance. These are mainly used in the pressure, strain, and tension applications. The diaphragm-type load cell is designed for simple mounting on a smooth flat surface. The diaphragm-type load cells are widely used in force measurement in compression loading.

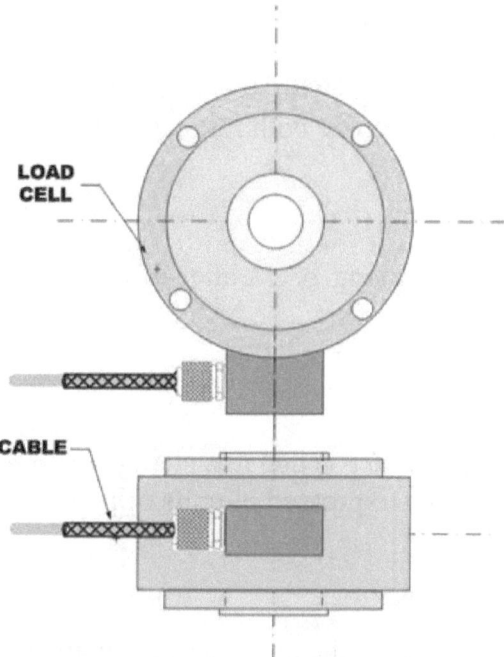

Figure 13.13 Diaphragm-type load cell.

13.5.5 The Important Terminology of the Load Cell

Ambient Conditions

These are the conditions (humidity, pressure, temperature, etc.) of the medium surrounding the load cell.

Ambient Temperature

This is the temperature of the medium surrounding the load cell.

Angular Load Eccentric

A load applied eccentric with the primary axis at the point of application and at some angle with respect to the primary axis.

Angular Load Concentric

A load applied concentric with the primary axis at the point of application and at some angle with respect to the primary axis.

Axial Load

It is the load applied along or parallel to and concentric with the primary axis.

Barometric Sensitivity

The change in *zero balance* is due to a change in ambient barometric pressure. It is normally expressed in units of percent RO/atm.

Calibration

The load cell calibration is the process of the comparison of load cell outputs against standard test loads.

Calibration Curve

The graph of the comparison of the load cell outputs against standard test loads.

Combined Error

Combine error includes nonlinearity and hysteresis. This is the maximum deviation from the straight line drawn between original no-load and rated load outputs expressed as percentage of the rated output and measured on both increasing and decreasing loads.

The following figure shows the calibration curve, load cell linearity, and the combined error.

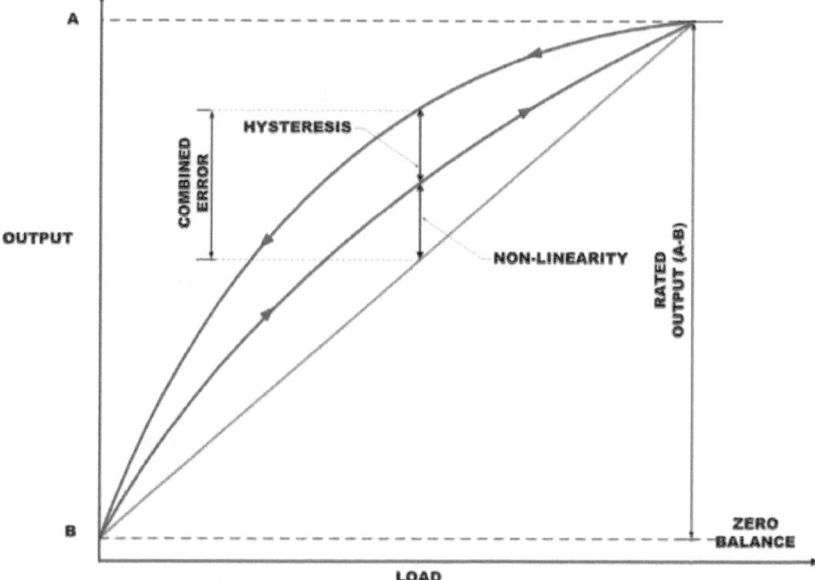

Figure 13.14 Combined error.

Compensation

It is defined as the utilization of supplementary devices, materials, or process to minimize known sources of error.

Creep

The creep is the change in load cell output occurring with time while under load and with all environmental conditions and other variables remaining constant.

Creep Recovery

This is a change in no-load output occurring with time after removal of a load which had been applied for a specific period of time. Usually measured over a specific time period immediately following removal of rated load and expressed as a percent of rated output over a specific period of time.

Deflection

It is the change of length along the primary axis of the load cell between no-load and rated load conditions.

Drift

Drift is a random change in output under constant load conditions.

Eccentric Load

Any load applied parallel but not concentric with the primary axis.

Error

Error is defined as the algebraic difference between the indicated and true value of the load being measured.

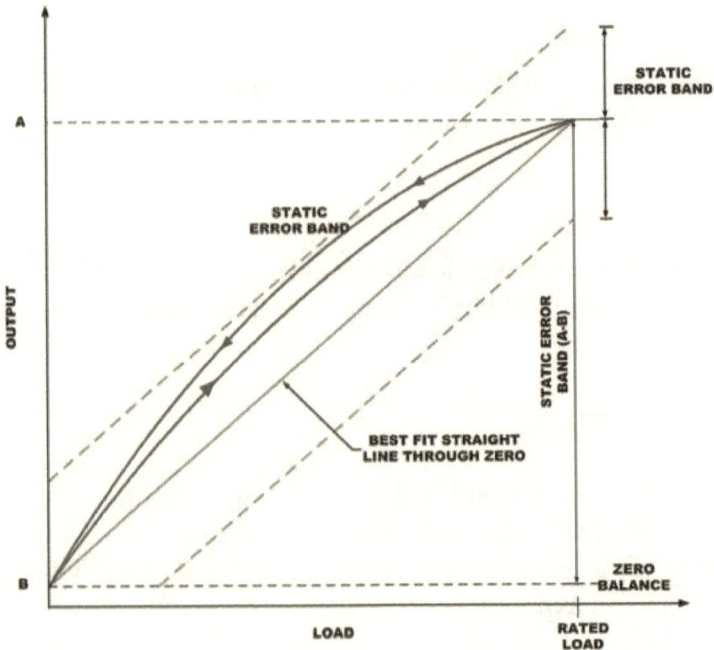

Figure 13.15 Static error band.

Excitation

The voltage or current applied to the input terminals of the load cell.

Frequency Response

The range of frequencies over which the load cell output will follow the sinusoidal varying mechanical input within specified limits.

Hysteresis

The maximum difference between load cell output readings for the same applied load; one reading obtained by increasing the load from zero and the other by decreasing the load from rated output.

Insulation Resistance

The dc resistance measured between the load cell circuit and the load cell structure. Normally measured at fifty volts and under standard test conditions.

Linearity

The maximum deviation of the calibration curve from a straight line between zero and full scale expressed as a percent of full scale output and measured on increasing measures only.

Load

The weight or force applied to the load cell.

Load Cell

The load cell is the device which produces an output signal proportional to the applied weight or force.

Natural Frequency

It is defined as the frequency of free oscillations under no-load load conditions.

Nonlinearity

The maximum deviation of the calibration curve from a straight line drawn between the no-load and rated outputs; expressed as a percentage of the rated output and measured on increasing load only.

Output

The signal (voltage, current, pressure, etc.) produced by the load cell. Where the output is directly proportional to excitation, the signal must be expressed in terms of volts per volt, per ampere, etc., of excitation.

Output, Rated

The algebraic difference between the outputs at no-load and at rated load.

Overload Rating, Safe

The maximum load in percent of rated capacity which can be applied without producing a permanent shift in performance characteristics beyond those specified.

Overload Rating, Ultimate

The overload rating is defined as the maximum load in percent of rated capacity which can be applied without producing a structural failure.

Primary Axis

The primary axis is the axis along which the load cell is designed to be loaded; normally its geometric centerline.

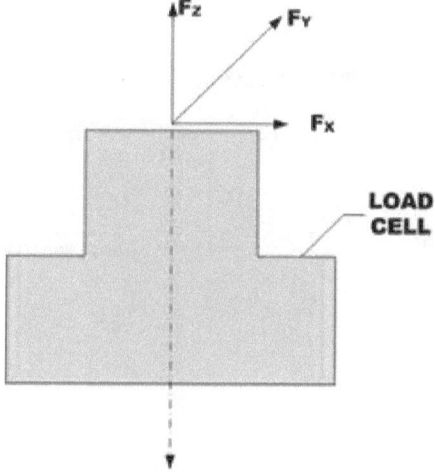

Figure 13.16 Load Primary axis drawing.

Rated Capacity (Rated Load):

The maximum axial load the load cell is designed to measure within its specifications.

Reference Standard

It is a force measuring device whose characteristics are precisely known in relation to a primary standard.

Repeatability

The maximum difference between load cell output readings for repeated loadings under identical loading and environmental conditions.

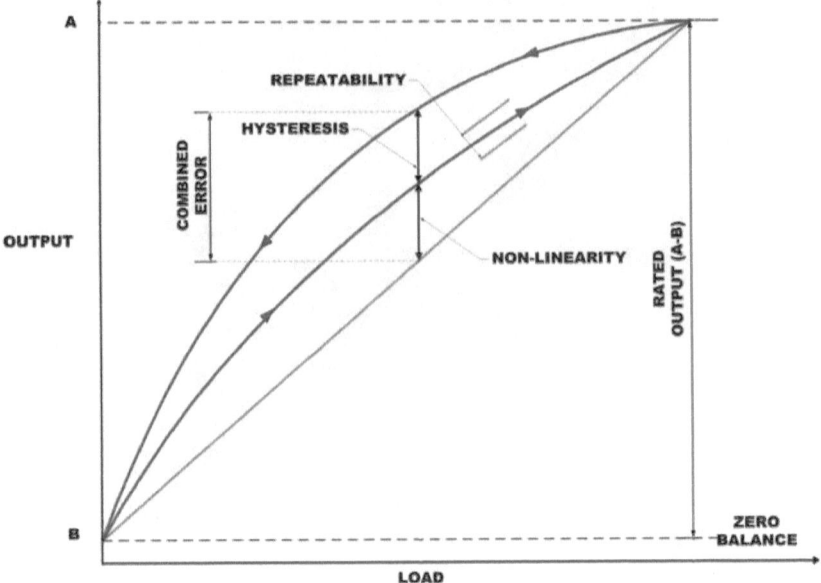

Figure 13.17 Repeatability.

Resolution

The resolution is defined as the smallest change in mechanical input which produces a change in the output signal.

Sensitivity

The sensitivity is the ratio of the change in output to the mechanical input.

Shunt Calibration

Electrical simulation of load cell output by insertion of known shunt resistors between appropriate points within the circuitry.

Shunt-To-Load Correlation

The difference in output readings obtained through electrically simulated and actual applied loads.

Side Load

Any load acting 90 degrees to the primary axis at the point of axial load application.

Stabilization Period

The time required to insure that any further change in the parameter being measured is tolerable.

Standard Test Conditions

These are the environmental conditions under which measurements should be made when measurements under any other condition may result in disagreement between various observers at different times and places. These conditions are as follows: temperature 23°C ± 2°C (73.4°F ± 3.6°F).

Temperature Effect on Rated Output

This is the change in rated output due to a change in ambient temperature.

Temperature Range Compensated

This is the range of temperature over which the load cell is compensated to maintain rated output and zero balance within specific limits.

Temperature Range Safe

The temperature rage is defined as the extremes of temperature within which the load cell will operate within permanent adverse change to any of its performance characteristics.

Terminal Resistance Corner To Corner

The resistance of the load cell circuit measured at specific adjacent bridge terminals at standard temperature, with no load applied, and with the excitation and output terminals open-circuited.

Terminal Resistance Input

The resistance of the load cell circuit measured at the excitation terminals at standard temperature, with no load applied and with the output terminals open-circuited.

Zero Balance/Zero Float

This is the signal of the load cell in the no load condition.

Zero Stability

The degree to which *zero balance* is maintained over a specified period of time with all environmental conditions, loading history, and other variables remaining constant.

13.6 Pressure Transducer

The pressure transducers are used to measure the pressure of various parameters of the engine. Pressure is defined as a force per unit area.

13.6.1 Theory of Operation

Pressure transducers operate under the same principle. Strain gauges, mounted on a diaphragm where the pressure is applied, measure the deformation of the diaphragm that is proportional to the pressure. The following sections describe the principle of operation of the strain gauge load cells and how to make a measurement from them, although the same applies for strain gauge pressure transducers.

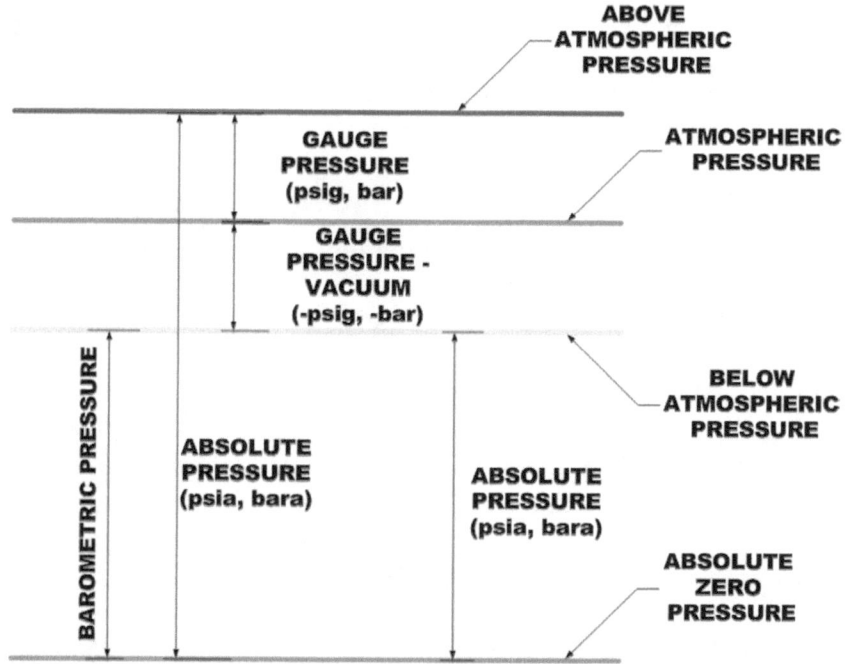

Figure 13.18 Understanding pressure.

13.6.2 Gauge Pressure

A **gauge** is often used to measure the pressure difference between a system and the surrounding atmosphere. This pressure is often called the **gauge pressure** and can be expressed as

$$Pg = Ps - Pa, \quad \quad \quad 13.6$$

where
Pg = gauge pressure
Ps = system pressure
Pa = atmospheric pressure

13.6.3 Atmospheric Pressure

Atmospheric pressure is pressure in the surrounding air at—or "close" to—the surface of the earth. The atmospheric pressure varies with temperature and altitude above sea level.

13.6.4 Absolute Pressure

Absolute Pressure is the sum of the available atmospheric pressure and the gauge pressure in the pumping system.

Absolute Pressure (PISA) = Gauge Pressure + Atmospheric Pressure..13.7

13.6.5 Standard Atmospheric Pressure

Standard Atmospheric Pressure (*atm*) is used as a reference for gas densities and volumes. The Standard Atmospheric Pressure is defined at sea level at *273°K (0 C)* and is **1.01325 bar** or *101325 Pa (absolute)*. The temperature of *293°K (20°C)* is also used.

In imperial units, the Standard Atmospheric Pressure is *14.696 psi*.

1 atm = 1.01325 bar = 101.3 kPa = 14.696 psi (lb/in^2) = 760 mmHg = 10.33 mH_2O = 760 torr = 29.92 inHg = 1013 mbar = 1.0332 kg/cm^2 = 33.90 ftH_2O

13.6.6 Pressure Measurement Devices/Instruments

There are various instruments available to measure the pressure. The traditional pressure measuring instruments such as manometers and analog pressure gauges are still in practice. However if the application demands more accuracy, then pressure transducer in conjunction with digital indicators are used.

13.6.6.1 Manometers

The manometer is an instrument for measuring the pressure acting on a column of fluid. Manometers are based on the principle of balancing the column of liquid whose pressure is to be found. They are of two types:

U-tube Manometers

These gauges use the hydrostatic pressure of a liquid column for pressure reading. Mostly mercury is used as the liquid because of its low vapor pressure and its cohesion.

The most basic version of U-tube manometer is U-shaped glass tube closed at the one end and connected at the other to the chamber whose pressure is to be measured. The tube is filled with mercury so that in the volume between the closed end and the level of the mercury column, there is only the vapor pressure of the mercury. This is called Torricelli's vacuum. The difference in heights of the mercury levels in the two legs of the U-shaped tube is proportional to the pressure.

$$P = \rho.g.h, \quad\quad\quad\quad\quad\quad\quad\quad\quad 13.8$$

where
ρ = density of the used liquid, usually mercury
g = gravitational acceleration
H = difference in heights of the levels of the two columns

Example—Differential Pressure Measurement in an Orifice

A water manometer connects the upstream and downstream of an orifice located in an air flow. The difference height of the water column is *10 mm*.

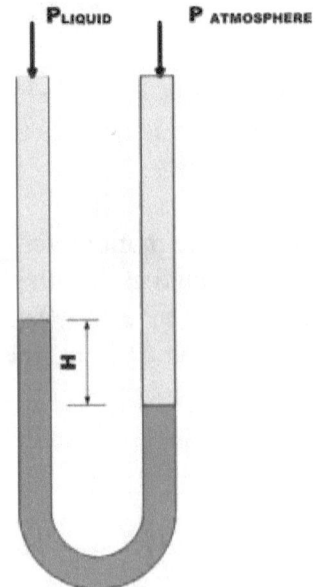

Figure 13.19 U-tube manometer.

The pressure difference head can then be expressed as:

$$p_d = (9.8 \text{ kN/m}^3)(10^3 \text{ N/kN})(10 \text{ mm})(10^{-3} \text{ m/mm})$$
$$= 98 \text{ N/m}^2 \text{ (Pa)},$$

Where 9.8 (kN/m^3) is the specific weight of water in SI units.

Inclined Tube Manometer

Inclining the tube manometer will increase the accuracy of the measurement. The sensitivity to pressure change can be increased further by a greater inclination of the manometer arm; alternatively the density of the manometric fluid may be changed.

The pressure difference in an inclined u-tube can be expressed as

$$p_d = \gamma \, h \, \sin(\theta), \quad \dots\dots\dots\dots\dots\dots\dots\dots 13.9$$

where

θ = *angle of column relative the horizontal plane*

Example—Differential pressure measurement with an inclined U-tube manometer.

We use the same data as in the example above, except that the U-tube is inclined to *30°*.

The pressure difference head can then be expressed as:

$$\mathbf{p}_d = (9.8 \text{ kN/m}^3)(10^3 \text{ N/kN})(10 \text{ mm})(10^{-3} \text{ m/mm}) \sin(30)$$
$$= 69.3 \text{ N/m}^2 \text{ (Pa)}$$

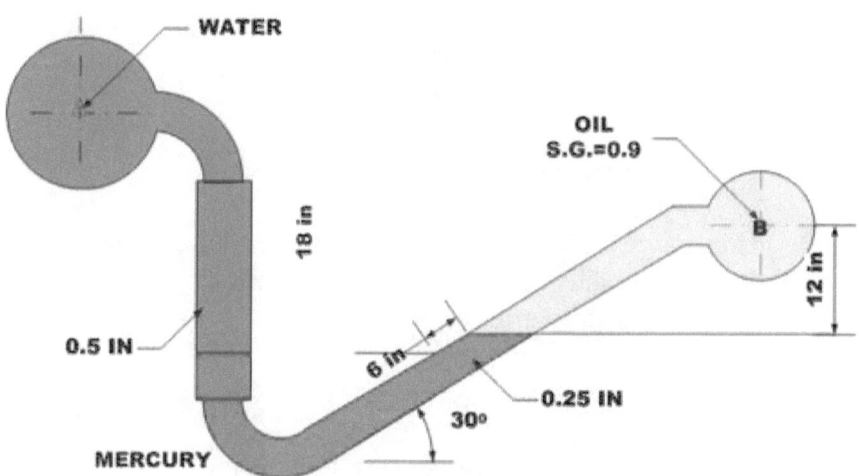

Figure 13.20 Inclined tube manometer.

13.6.6.2 Pressure Gauges

Bourdon Mechanical Gauge

Bourdon gauge consists of a thin-walled metal tube, somewhat flattened and having shape in the form of a C. The linkage attached to its free end magnifies any motion of the free end of the tube. As pressure increases within the system on which gauge is mounted, it travels through the tube. The metal tube begins to straighten as the pressure increases inside of it. This motion is transferred through a linkage to a gear train connected to an indicating needle which moves over a calibrated dial to show the pressure in either PSI or kg/cm^2.

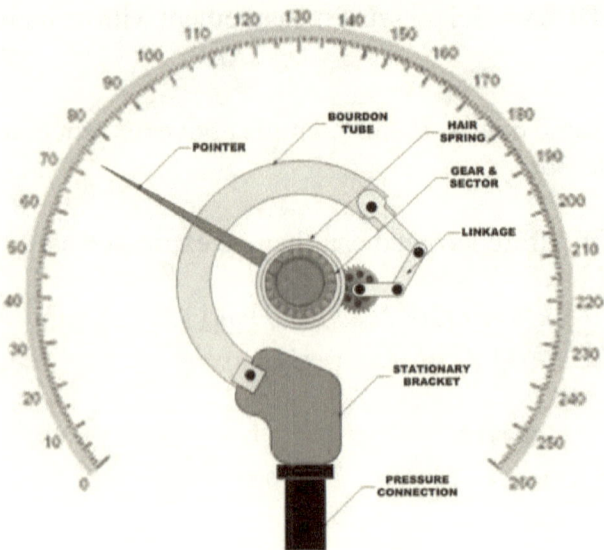

Figure 13.21 Bourdon mechanism.

The Bourdon gauge is a highly accurate but rather delicate instrument. It gets damaged easily due to pressure surges. In addition, it malfunctions if pressure varies rapidly.

Diaphragm Pressure Gauge

The **diaphragm pressure gauge** uses the elastic deformation of a diaphragm also knows as membrane instead of Bourdon tube to measure the pressure. If a diaphragm or a bellows separates two regions with different pressures (P_1, P_2,), the difference Dp ($D_p = P_1 - P_2$) of these two pressures causes a force that deforms the diaphragm or bellows. There are many possibilities for measuring this deformation, for example,

1. Mechanically by a lever and a pointer.
2. Optically by a mirror and a light pointer.
3. Electrically by changes of the capacity of a capacitor formed by the diaphragm and an additional electrode which is usually placed in a region of very low pressure.

For precision measurements, one side of the diaphragm is evacuated to very low pressure. This is called a reference vacuum. The other side is exposed to the pressure to be measured. The deformation of the diaphragm depends on the pressure difference. However, the deformation is not linear.

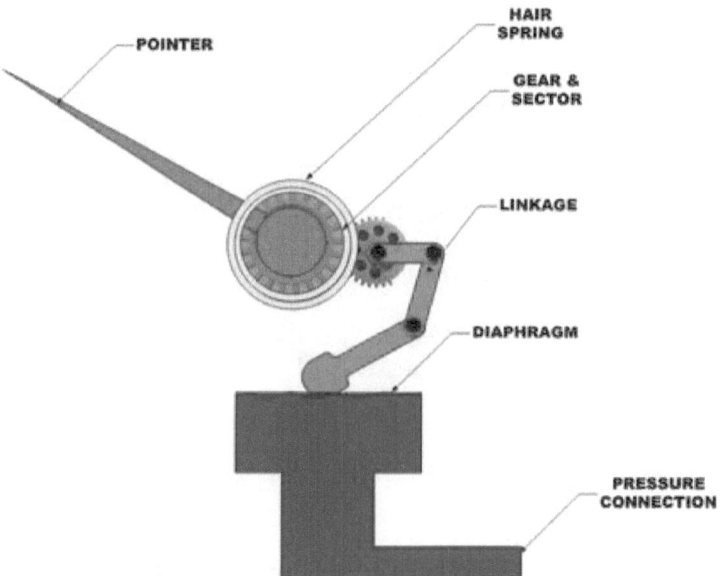

Figure 13.22 Diaphragm mechanical pressure gauge.

13.6.6.3 Pressure Transducer

The pressure transducers alongwith digital pressure indicators provide precise and accurate measurements of gauge pressure, absolute pressure and differential pressure. The digital pressure transducer is made by diffusing a fully active four arm strain gauge bridge into the surface of single crystal silicon diaphragm. The pressure acting on the diaphragm changes the resistance of the strain gauge and in turn changes the output signal. The signal is conditioned, calibrated, and fed to digital indicator which displays the measured parameter in engineering units. The following figure shows the cross section of a typical pressure transducer.

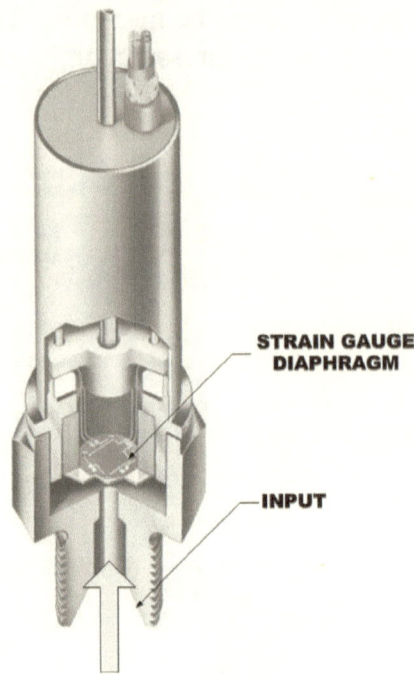

Figure 13.23 Cross section of pressure transducer.
Courtesy of Formerly known Druck Transducers.

13.7 Rotary Torque Transducer

Torque is rotational force. The traditional way to measure the torque is to mount the load cell on the dynamometer with Trunnion bearing arrangement. The torque transducer eliminates the dynamometer mounting with trunnion bearings. In case of DC or AC dynamometer, they can be adopted as received from the manufactures and can be used as foot-mounted machines.

13.7.1 Theory of Operation

Torque transducer senses the shear stresses in a torsion bar as consequence of applied torque. Torque applied to an object with angular acceleration is called dynamic torque. And torque transducer is an ideal sensor to measure the dynamic torque.

Figure 13.24 Typical torque transducer. Courtesy of Magtrol, Inc. USA.

The amount of torque developed is proportional to the force applied and to how far from the center of the rotation the force is applied.

13.7.2 Mechanical Design of Torque Transducer

The following illustration shows the typical mechanical construction of the rotary torque transducer. The mechanical torque is converted into strain on the shaft surface.

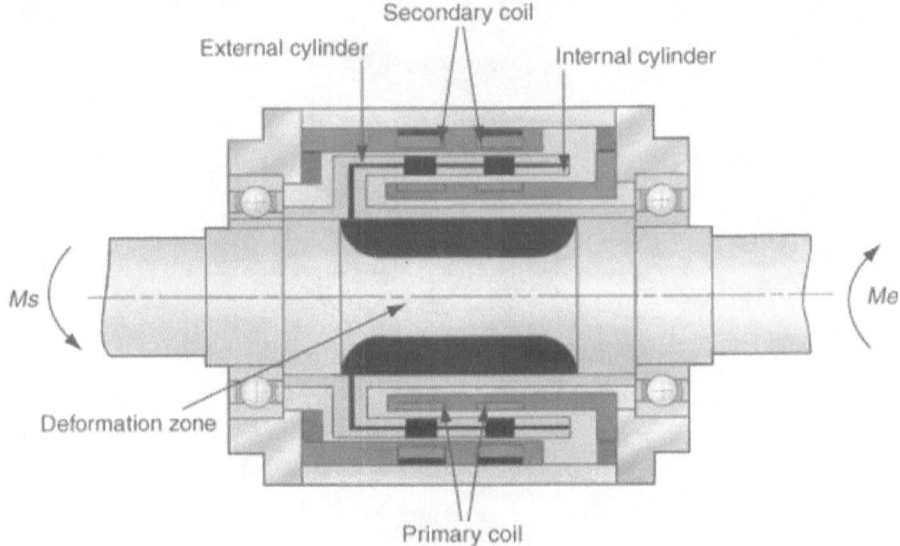

Figure 13.25 Construction of torque transducer.
Courtesy of Magtrol Inc. USA.

The transducer consists of shaft mounted in bearing which is supported in stator. The part of the shaft length is utilized to convert the torque into a proportional tensional angle. This torsional angle or angle of twist is measured between the two ends of this shaft length by an inductive angular position measurement system. The inductive angular position measurement system converts the angle of twist into a proportional electrical signal.

13.7.3 Electrical Design

The measuring system, based on the principle of a variable, torque-proportional transformer coupling, consists of two concentric cylinders shrunk on the shaft on each side of the shaft's deformation zone and two concentric coils attached to the housing.

Both cylinders have a circularly disposed coinciding row of slots and rotate with the shaft inside the coils. An alternating current with the frequency of 20 kHz flows through the primary coil. When no torque is applied, the slots on the two cylinders fail to overlap. When torque is applied, the deformation zone undergoes an angular deformation and the slots begin to overlap. Thus a torque-proportional voltage is on the secondary coil. The conditioning electronic circuit incorporated in the transducer converts the voltage to a nominal torque signal of 0 to ± 5 V_{DC}.

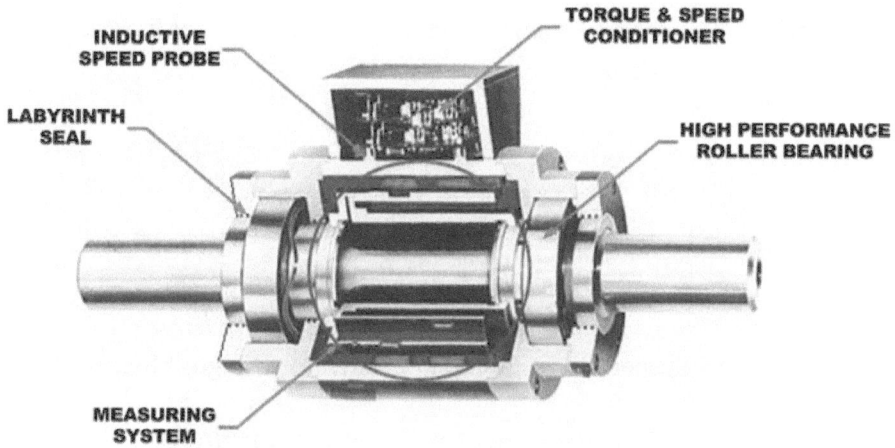

Figure 13.26 Cross section of torque transducer.
Courtesy of Magtrol Inc. USA.

An optical sensor reads the speed on a toothed path machined directly on the measuring system. The electronic conditioner outputs a frequency signal proportional to the shaft rotational speed.

The built-in electronics comprise the following functional groups:

- Oscillator for generation of the AC input voltage
- Preamplifier for the output signal
- Phase-sensitive rectifier for conversion of the AC output signal into a DC voltage
- Output amplifier

13.7.4 Installation Schematics

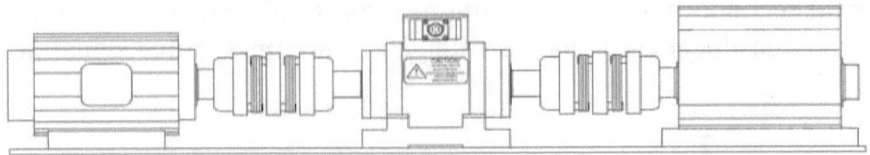

Figure 13.27 Supported installation. Courtesy of Magtrol Inc. USA.

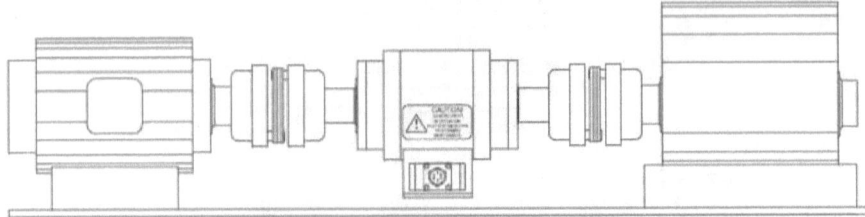

Figure 13.28 Suspended installation. Courtesy of Magtrol Inc. USA.

13.7.5 Coupling Requirements for Torque Transducer

The criteria for selecting appropriate couplings for torque measurement are as follows:

1. High torsional spring rate: ensures a high torsional stiffness and angular precision (should be greater than three times the torque transducer stiffness)
2. Clamping quality (should be self-centering and of adequate strength)
3. Speed range
4. Balancing quality (according to speed range)
5. Alignment capability

The higher the speed of the application, the more care is required in selecting the coupling and assembling (alignment and balancing) the drive train configuration.

13.7.6 Mechanical Calibration

To perform this, a calibration setup with lever arm and weights is needed to create a torque.

Steps for Calibration

1. Apply the rated torque to the transducer and then remove the load again.
2. Precisely adjust the transducer to the zero point.
3. Apply a known torque to the transducer.
4. Set the indicator to the proper value.

A Simple Calibrating Setup

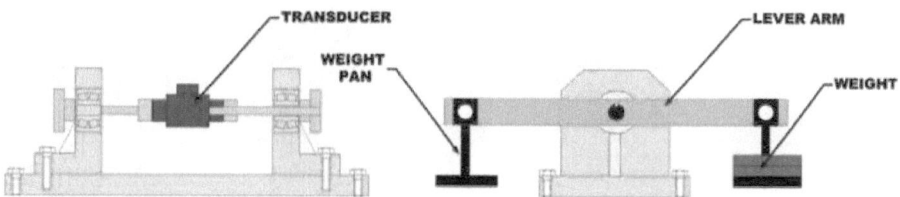

Figure 13.29 Calibration arrangement.

13.7.7 Electrical Calibration

The transducers are provided with electrical calibration facility. A shunt calibration facility is provided in the transducer to permit electrical calibration. The shunt calibration is the known, electrical, and unbalancing of a strain gauge bridge, by means of a fixed resistor that is placed, or "shunted," across one leg of the bridge. The "Wheatstone Bridge" utilized by load cells and torque sensors are typically calibrated using the shunt calibration technique.

Purpose of Electrical Calibration

The electrical calibration is a method of periodically checking the gain or span of a signal conditioner, which is used in conjunction with a strain gauge-based transducer, without exposing the transducer to known traceable, physical input values.

Shunt calibration simulates the mechanical input to a transducer by unbalancing the bridge with a fixed resistor placed across, or in parallel with, one leg of the bridge. For tension shunt calibration, the shunt resistor (*SRt*) is shunted across the [+] excitation (*A*) and [+] signal (*B*) leg of the bridge. For compression shunt calibration, the shunt resistor (*SRc*) is shunted across the [–] excitation (*D*) and [+] signal (*B*) leg of the bridge.

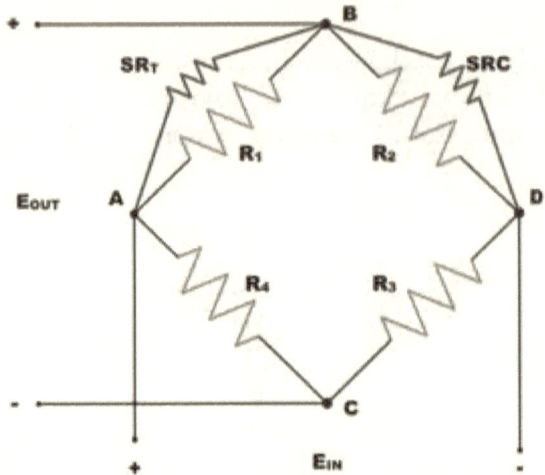

Figure 13.30 Schematics of shunt calibration.

Procedure for Shunt Calibration

The typical procedure followed in industry for shunt calibration is as follows:

1. Connect the transducer to an appropriate strain gauge signal conditioner and allow adequate time for the system to stabilize.
2. Apply a full-scale, N.I.S.T. traceable, mechanical input (or load) to the transducer.
3. Adjust the signal conditioner's gain or span controls, as required, to obtain a full-scale electrical output signal, and/or numeric display that represent the applied, mechanical input quantity.
4 Remove the mechanical input (or load).

5. Place a shunt calibration resistor across an appropriate leg of the Wheatstone bridge as discussed above.
6. Record the value of the signal conditioner's output signal and/or numeric display. This value is the shunt calibration value, or equivalent load.
7. It is important to note that the shunt calibration value is specific for the particular shunt resistor used. This value and the particular resistor are now matched to the transducer and form the basis of the transferable shunt calibration.

13.8 Magnetic Pickup

The accurate speed measurement of dynamometer is important in establishing the engine performance. The speed measurement can be done using following methods.

1. Magnetic pickup with 60-toothed wheel
2. Tachogenerator
3. Encoder

13.8.1 Magnetic Pickup with 60-toothed Wheel

The majority of the dynamometers use a 60-toothed wheel in conjunction with a magnetic pickup to measure the rpm of the engine under test. The output signal of a *magnetic speed sensor* is caused by the excitation of its inductive coil by a passing target such as 60-toothed wheels. As conductor moves past the magnetic sensor, conductor's surface discontinuity (i.e., a gear tooth) excites a voltage in the pickup coil, producing an electrical analog wave. The *frequency and voltage of the analog signal is proportional to actuator velocity*. Each passing discontinuity in the wheel causes the magnetic sensor to generate a pulse. The cyclical pulse train created by a 60-toothed wheel is fed to electronic counters or speed measuring device.

13.8.2 Why 60-toothed Wheel Is Used?

Let us assume that a test engineer needs to measure the speed of the engine connected to dynamometer. As dynamometer is mounted with magnetic pulse pickup which works in conjunction with 60-toothed wheel to measure the speed, it is necessary to connect the magnetic pickup to a counter or microcontroller. The counter/microcontroller needs input of pulse and amplitude provided by magnetic pickup.

As the speed of the dynamometer changes so does the output of magnetic pickp changes in amplitude and in frequency. To find out the rpm, digital counter needs to count the pulses. However the minimum level of amplitude of the signal is the requirement of counter circuit. Since the magnetic pickup works by generating an emf through the rate of change of flux linked into a coil, both the amplitude and the frequency of the output will be proportional to the speed. At low speeds, there may not be sufficient voltage induced to give a measurable signal against noise. Therefore, even if you use the frequency, you need to be aware of the voltage dependency. The output of the magnetic pickup at low speed such as idling speed of the engine is very low in amplitude. This weak signal needs to be treated properly before feeding to the counter/digital speed indicator.

The speed is function of frequency (pulse output).

Mathematically:

$$Speed = f(pulses) \quad \ldots\ldots\ldots\ldots\ldots\ldots\ldots 13.10$$

If dynamometers have 60 teeth gear wheel, then 60 pulses will be generated for each revolution.

$$(Total\ Number\ of\ pulses) = Number\ of\ Teeth \times rpm$$

$$Total\ Number\ of\ pulses = \frac{Number\ of\ Teeth \times Rpm}{60}$$

$$Total\ Number\ of\ pulses = \frac{60 \times Rpm}{60}$$

$$Total\ Number\ of\ pulses = RPM.$$

From above equations, it is clear that if you use gear wheel other than that of 60 toothed, then you need to account it in the counter circuit. For example, if a wheel with 30 toothed is used, then number of pulses generated will be half of the RPM of the dynamometer.

Figure 13.31 Magnetic pickup with 60-toothed wheel. Courtesy of Horiba.

The output signal from magnetic pickup also depends upon the distance between pickup and 60-toothed wheel and magnetic pickup as shown in the following figure.

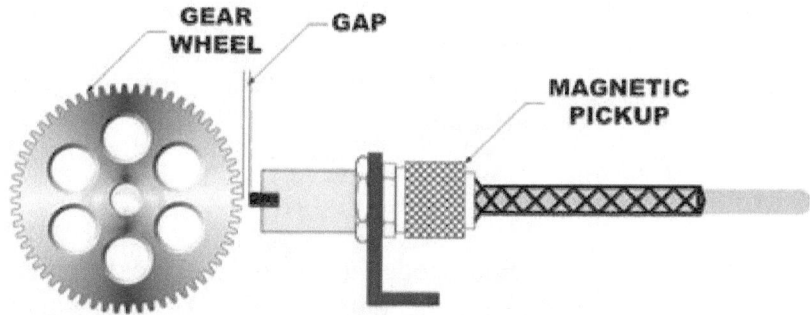

Figure 13.32 Magnetic pickup and toothed wheel.

13.8.3 Tachogenerator

Tachogenerator is a device which generates an electrical output directly proportional to the shaft speed. The voltage thus generated is fed to a speedometer for speed measurement and/or used as feedback signal for controlling the speed of a rotating machine/equipment. In every automatic control system, since the system performance is directly related to the performance, it becomes essential that the *tachogenerator* must satisfy certain minimum requirements such as:

Linear

The rectified output voltage should be strictly linear with respect to the rotational speed.

Output Signal

The ripple content in the rectified output signal must be as low as possible.

Temperature Effect

Variations in ambient temperature should have negligible effect on the TG output.

Long-term Stability

Instability of output cause due to aging of parts, shift in magnetic circuits, etc., should be minimum.

13.8.4 Encoder

The transducer that generates coded reading of a measurement is termed as encoder. The shaft encoders are digital transducers that are used for measuring angular displacements and velocities. Shaft encoders are classified in two categories depending on the nature and the method of interpretation of the output.

1. Incremental encoders.

The output is a pulse signal that is generated when the transducer disk rotates as a result of the motion that is being measured. Incremental encoder works by counting the pulses or by timing pulse width using a clock signal.

2. Absolute encoders.

The absolute encoder has many pulse tracks on its transducer disk. When the disk of an absolute encoder rotates, several pulse trains, equal in number to the tracks on the disk, are generated simultaneously. Absolute encoders are commonly used to measure fractions of a revolution. However, complete revolutions can be measured using an additional track that generates an index pulse, as in the case of an incremental encoder.

Table 13.3 The comparison of analog and digital measurement

	Analog	**Digital**
Accuracy	The pointer of analog indicator points at the reading	A digital meter takes a precise, computer-generated reading and displays it on a screen
Dynamic readings	Analog meters have the advantage that, when fluctuating readings exist, the analogue meter is able to measure these fluctuations. The analogue meter's needle will swing from one position to another constantly in order to represent the fluctuation	When a fluctuation exists, a digital meter is unable to represent the fluctuation; instead it either records an error or calculates one reading.

Ease of Reading	Analog indicator pointer will point out the measured reading; however, reader judgment plays a major role in reading the pointer.	The fact that a digital meter performs the calculation and displays the reading, rather than relying on a reader's ability to calculate the correct reading, makes digital meters user-friendly
Measurement trend	Analog indicator can indicate the trend in measurement, i.e., rising or falling	Digital meters cannot indicate the trend. When parameter under measurement is changing fast, digital meters will show rapidly changing digits making it difficult in establishing the trend of measurement

It is a good practice to install one analog and one digital instrument for important and rapidly changing parameters. This will ensure that the reader understands the trend in measurements as well as get accurate measurements.

13.9 Environmental Measurements

The environmental measurements of the test cell are measurement of test cell temperature, ambient pressure, and relative humidity.

13.9.1 Ambient Air Intake Temperature

The cold air intakes operate on the principle of increasing the amount of oxygen available for combustion with fuel, because cooler air has a higher density (greater mass per unit volume).

Standard air is defined as clean, dry air with a density of 0.075 pounds per cubic foot, with the barometric pressure at sea level of 29.92 inches of mercury and a temperature of 70°F.

Standard Temperature and Pressure

Standard temperature and pressure, abbreviated STP, refers to nominal conditions in the atmosphere at sea level.

- STP—Standard Temperature and Pressure—is defined by IUPAC (International Union of Pure and Applied Chemistry) as air at 0°C (273.15 K, 32°F) and 10^5 Pascal (1 Pa = 10^{-6} N/mm^2 = 10^{-5} bar = 0.1020 kp/m^2 = 1.02×10^{-4} m H_2O = 9.869×10^{-6} atm = 1.45×10^{-4} psi (lbf/in^2)).
- STP—commonly used in the Imperial and USA system of units—as air at 60°F (520°R) and 14.696 psia (15.6°C, 1 atm).

This is essentially the freezing point of pure water at sea level, in air at standard pressure.

Normal Temperature and Pressure

Normal Temperature and Pressure is commonly abbreviated as NTP refers to normal conditions in the atmosphere at room conditions.

- *NTP—Normal Temperature and Pressure—is defined as air at 20°C (293.15 K, 68°F) and 1 atm (101.325 kN/m^2, 101.325 kPa, 14.7 psia, 0 psig, 29.92 in Hg, 760 torr). Density 1.204 kg/m^3 (0.075 pounds per cubic foot)*

Thus the difference between NTP and STP is in temperature: in case of STP, it is 0 deg. C; while in case of NTP it is 20 deg. C.

Advantages of cold air intake include the following:

- Increased horsepower and torque.
- Improved throttle response and fuel economy in most cases.

Table 13.4 STP and NTP by IUPAC

International Union of Pure and Applied Chemistry (IUPAC)				International Union of Pure and Applied Chemistry (IUPAC)			
STP				NTP			
	SI	Imperial	Absolute		SI	Imperial	Absolute
Temperature	0°C	32°F	273.15°k	Temperature	20°C	68°F	293.15°k
Pressure	101.325 kPa	14.696 psi	1.0 atm	Pressure	101.325 kPa	14.696 psi	1.0 atm
Air Density				Air Density	1.204 kg/m^3	0.075 lb/ft^3	

Table 13.5 STP by NIST

National Institute of Standards & Technology			
NIST			
STP			
	SI	Imperial	Absolute
Temperature	20°C	32°F	273.15°k
Pressure	101.325 kPa	14.696 psi	1.0 atm
Air density			

13.9.2 Barometric Pressure

Why we need to measure the barometric pressure? The power calculated after testing the engine needs to be corrected by applying suitable correction factors. The formulae for the correction factor involve the barometric pressure for the location of the engine testing site. Barometric pressure is synonymous to atmospheric pressure.

Atmospheric Pressure

It is approximated by the hydrostatic pressure caused by the weight of air above the measurement point. Low pressure areas have less atmospheric mass above their location, whereas high pressure areas have more atmospheric mass above their location. Similarly, as elevation increases, there is less overlying atmospheric mass, so that pressure decreases with increasing elevation.

A column of air one square inch in cross section, measured from sea level to the top of the atmosphere, would weigh approximately 14.7 lbf (65 N).

Average *sea level pressure* is **101.325 kPa** (1013.25 mbar, or hPa) or **29.921 inches** of mercury (in Hg) or **760 millimeters (mmHg)**.

(1 millibar = 1 hectopascal).

How to Measure the Atmospheric Pressure?

The instrument that measures atmospheric air pressure is called a barometer. The first barometer was invented by **Evangelista Torricelli.**

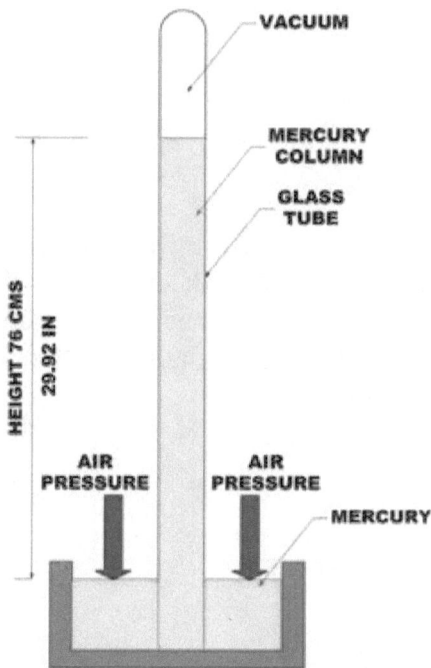

Figure 13.33 Torricelli's barometer.

Aneroid Barometer

Aneroid barometer is one that measures atmospheric pressure via the expansion and contraction of a sealed hollow cell which is partially depleted of air. *Aneroid means using no fluid.*

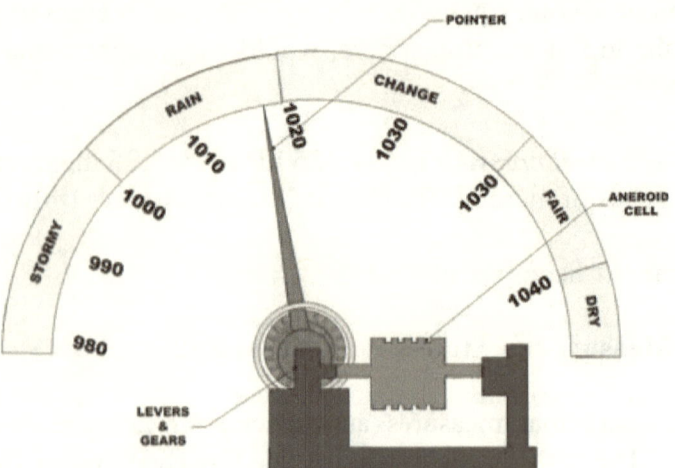

Figure 13.34 Aneroid barometer.

13.9.3 Relative Humidity

The *"Absolute humidity"* is the quantity of water in a particular volume of air. The most common units are grams per cubic meter.

If all the water in one cubic meter of air were condensed into a container, the container could be weighed to determine absolute humidity. The amount of vapor in that cube of air is the absolute humidity of that cubic meter of air. More technically, absolute humidity is the mass of water vapor per cubic meter of air.

$$AH = \frac{M_W}{V_a}, \quad\quad\quad\quad\quad\quad\quad\quad\quad\quad\quad\quad 13.11$$

where
AH = absolute humidity
M_w = mass of water vapor
V_a = cubic meter of air

The absolute humidity changes as air pressure changes.

Relative humidity is the ratio of the air's actual water vapor content compared with the amount of water vapor required for saturation (at that temperature and pressure). In effect, it's a measure of how close to being saturated the air is.

If the amount of water vapor is constant:

Increasing temperature decreases RH.
Decreasing temperature increases RH.

13.9.3.1 Measurement of Relative Humidity

The word psychro means cold and meter means to measure.

Sling Psychrometer

Sling Psychrometer is used to measure both the dry bulb and wet bulb temperatures at time. These temperatures are a measure of humidity content in air.

The sling psychrometer consists of two thermometers for the measurement of wet and dry bulb temperatures. The thermometer whose bulb is bare and is directly in contact with air is used to measure the temperature of surrounding air is called as dry bulb thermometer. The other thermometer whose bulb is covered with a cotton which is moist with distilled water and is in direct contact with surrounding air is called wet bulb thermometer and measures the wet bulb temperature. These two thermometers are mounted on a frame which can be swung.

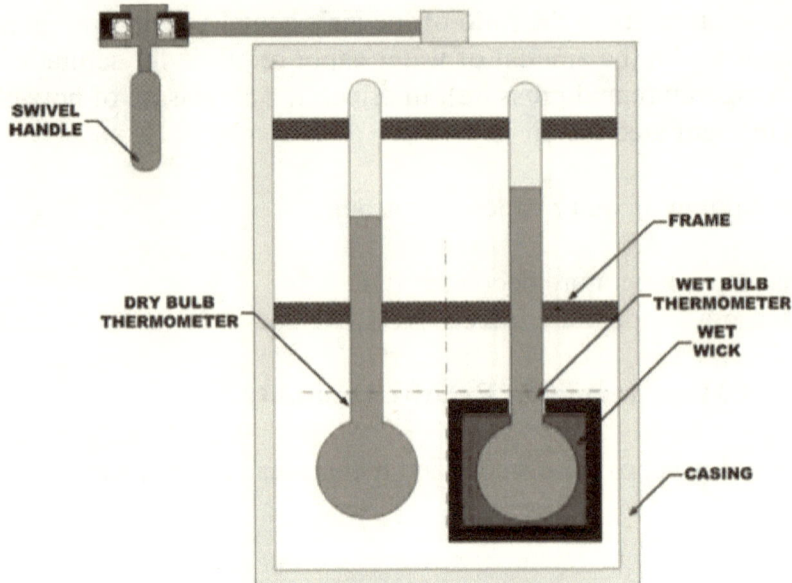

Figure 13.35 Sling psychrometer.

How the Sling Psychrometer Works?

The psychrometer frame which houses the dry bulb and wet bulb thermometers is rotated at 5 m/s to 10 m/s to get the necessary air motion. The accurate measurement of wet bulb temperature is obtained only if air moves with velocity around the wet cotton. The very reason the psychrometer is being rotated is to get the moving air around the wet bulb thermometer. The thermometer whose bulb is covered with the wet cotton comes in contact with the moving air; the moisture present in the wick starts evaporating and a cooling effect is produced at bulb. Thus temperature indicated by the thermometer is the wet bulb thermometer which will naturally be lesser than the dry bulb temperature. The thermometer whose bulb is bare contacts with the air indicates the dry bulb temperature.

Dew Point

The temperature of air would have to be cooled in order for saturation to occur. The dew point temperature assumes there is no change in air pressure or moisture content of the air.

Limitation of Sling Psychrometer

The evaporation process at the wet bulb will add moisture to the air which may reduce the temperature of the surrounding air.

1. It cannot be used in automation requirement situations.
2. It cannot be used for continuous recording purpose.
3. The dust and dirt accumulation on the moist cotton may reduce the rate of evaporation and give wrong readings of the wet bulb temperature.

Digital Relative Humidity Indicator

The sling RH indicator will only provide the wet and dry bulb temperature, and then one needs to look up in psychometric chart to establish the relative humidity. These readings need to be fed manually to data acquisition system to process it further.

However, for the test automation, this data about relative humidity should be available in real time. The digital relative humidity indicator with proportional output which can be fed to data acquisition system will serve the purpose. The following figure shows the schematic of digital relative humidity indicator.

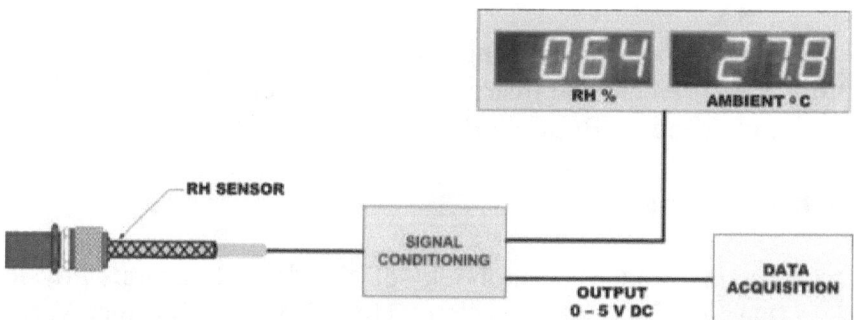

Figure 13.36 Schematics of digital relative humidity indicator.

Closure

The front-end elements in measurement technology are sensors and transducers. The sensors and transducers are critical because they directly measure what is happening on the engine and dynamometer. Knowing precisely how the engine behaves is important to development as well as testing team. Sensors monitor anything that can affect engine performance, that is, pressures, speed, temperatures, braking loads, and torques. The sensors perform a diversity of functions involving sensing, capturing, communicating, and distributing the information sensed. The commonly used sensors and transducers make it possible to automate the engine testing procedures and control system.

Bibliography:

1. PICO Technology, *Thermocouple Application Note.*
2. RTD Platinum Resistance Thermometers—criteria for temperature sensor selection of T/C and RTD sensor types—White paper No. 8500-917-A11 E000, Acromag Inc: USA.
3. Introduction To Strain & Strain Measurement White paper No. 8500-699-A02G000, Acromag Inc: USA.
4. Comparison_of_Thermocouple_and_RTD_Temperature_ Sensors-White paper No. 8500-906-A10L000 Acromag Inc: USA.
5. *Torque Transducer,* Magtrol Inc: USA.
6. *M*easurement and Instrumentation Principles by Alan S Moris published by Butterworth Heinmann 2001

Further Reading

1. Jacob Fraden, Handbook of Modern Sensors: Physics, Designs, and Applications. Third Edition, Springer-Verlag, Inc, 2004.
2. *Modern Sensors hand book Edited by Pavel Ripka and Alois tipek published by ISTE 2007.*
3. Douglas M.Considine, *Process/Industrial Instruments and Controls Handbook,* (4th) (Edn.). McGraw-Hill, 1993.

CHAPTER 14

Automation of Testing Process and Measurements

14.1 Introduction

In the manufacturing context, the term automation generally conjures up images of robots performing repetitive tasks in a large, complex assembly line. More specifically, in the engine testing process, one immediately visualizes automated engine handling, automated hoists, or auto loading of the engines to test beds. There was a time when testing engines were synonymous with manual testing. Over the decades, automated processes have become synonymous with large scale production of engines and thereby automated engine testing.

Manual testing can be described as a situation where a person initiates each test, interacts with it, and interprets, analyzes, and reports the results. The engine test automation is much more than computers' launching test programs.

The automation of the test cell is the apparent evolution of automation of engine testing segment in the manufacturing sector. Safety and control systems were the original charters for all automation. Our discussion is about the test cell and how automation can be successfully implemented in engine testing, low-volume manufacturing environment with a cost-effective test cell. Test automation has many benefits, but it is not a silver bullet. It has its costs and is not always the right answer. It is incumbent upon those in charge of test cells to find a balance between automated and manual testing. Automation has many advantages. Manual testing is expensive. It requires people to click buttons and observe results.

The automation mainly aims at

1. Producing accurate and consistent test results.
2. Optimum utilization of the test cell
3. Increasing the test cell throughput
4. Reducing its contribution to overall engine development cost and time-to-market.
5. Engine specific test cycles.

Despite its advantages, automation is not a panacea. First, it is not free. Automation is expensive to create. If you are able to amortize that cost over a lot of runs, the incidental cost becomes low. On the other hand, if the test is something that will only be run a few times, automation may be more expensive than manual testing. Decision makers must consider the high initial cost before committing their organization to automated testing of the engines. The second drawback of automation is that there are things that simply cannot be automated well. With the increased sophistication of test automation techniques, automated testing can take a larger role than in the past. It cannot, however, replace real experience testers and their diagnostic ability.

14.2 Different Levels of Automation

Engine testing originally started in open area and subsequently a concept of the closed test cell evolved. As engine production became automated, turning larger output, necessity of testing engines faster and quickly became the top priority of automotive industry. Once again, automation of engine testing alludes to complex engine handling and mounting fixtures with a loading and unloading station.

The test cell can be automated in stages incorporating the modular systems which will enhance the functionality of the entire test cell. The test cell composition can include one or more sub-systems that are designed for an optimized integration and the modules that make it work fully automatically. For example, in the modular systems such as throttle controller, signal conditioning can be added step by step as per requirements.

The automation benefits are highlighted:

- Automation can be applied to a high-mix, low-volume test environment;
- Automation is cost-effective for engine testing operations; however, it depends upon the degree of the automation process.
- The test operator can focus on other production and test issues; If required, a lights-out third or weekend shift can be added with *no* manpower costs;
- Test lots can be queued and tested on holidays without supervision or overtime pay;
- Through automation, the engine test cell gains efficiency and productivity;
- One test operator can "manage" multiple automated engine test systems.
- The productivity of the automated test cell requires less commitment to test fixtures for larger production runs.

The important benefit that is readily recognized is the stability and repeatability that automation brings to engine testing process.

Analyzing the return on investment (ROI) for an automated engine test cell has several factors to consider. If one engine testing cell is used daily for two shifts without any idle times (break times), then it will help in reducing the payback period of recovering the investment made in automation.

Any shift added to this simple equation without adding manpower would enhance the time it would take to recoup the initial investment in automating the engine test cell.

The cost and other benefits:

- As the productivity and efficiency of the automation is realized, more test cells will be automated.
- Potentially, the entire test department becomes more effective as time is spent on troubleshooting and manufacturing process enhancement.

- Less time is spent verifying and troubleshooting false test callouts.
- Management can respond to commitment in test excellence and consistency of test.
- No down time due to lunch or shift changes; automation never takes a break or calls in sick.
- The installation often becomes the highlight of customers' facility tours.

14.3 Strategies for Automation

1. Alleviating stress and boredom caused by repetitive jobs and then increasing employee satisfaction.
2. Elimination of human errors occurring due to monotonous work.
3. Environmentally hazardous work.
4. Work requiring high concentration.
5. Work with possibility of accidents.

The most important benefit of the test cell automation is time saving in test set up. Test preparation and engine handling has been one of the key factors during engine testing. The test preparation outside the test cell, especially in assembly area itself, saves lot of time in making necessary connections such as water supply, lubricating oil connection, and other required piping and wiring. The second important factor is how quickly engine can be connected to dynamometer. Quick coupling methodologies help in reducing the preparing time. These factors are responsible for the utilization of the test bed to its optimum capacity.

14.4 Classification of Test Cell Automation

The automation of test cell can be divided into 3 parts:

1. Automation of control system.
2. Automation of engine mounting.
3. Automation of engine auxiliary equipment.

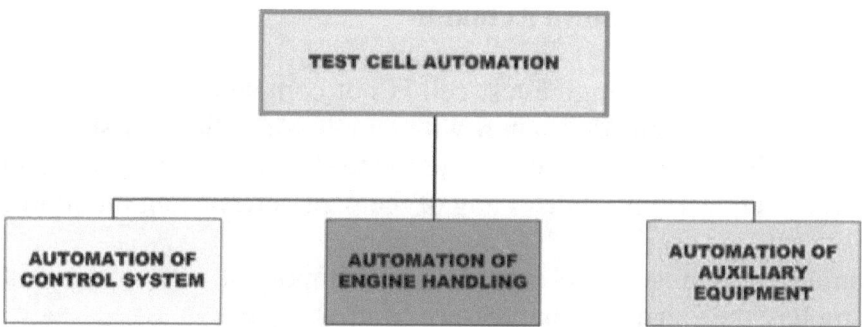

Figure 14.1 Test cell automation.

14.5 Automation of Control System

The automation of control system consists of automation of dynamometer controller, throttle controller, and auxiliary systems.

14.5.1 Engine Throttle Controller

The engine throttle controller plays an important role in test automation. The throttle controller provides easy control of throttle lever either in remote manual mode or remote auto mode. The throttle controllers are built with following safety:

1. Power Fail and Emergency Stop—returns throttle to closed position, using backup power source.
2. Over speed/under speed—returns throttle to minimum position.
3. Throttle jam—attempts to return to minimum throttle position.

The modern throttle controllers have ability to be controlled externally, and they can be interfaced with programmable controller to run the engine in various sequences.

The ability of the electronic throttle controller to listen to commands from the control system and give the feedback about the position of throttle lever makes it first step in the automation process.

14.5.2 Engine Shutdown Actuator

The automation of engine test cell is not complete until and unless you have incorporated actuators with remote controller. The shutdown actuator is one of the actuators required to shut the engine down in the event of an alarming state of any of the parameters being monitored.

Similarly exhaust valve actuator plays important role in actuating remotely the exhaust back pressure control valve.

14.5.3 Computerized Automation of Testing

The computerized automation of testing is achieved through dynamometer data acquisition and control system. It is commonly known as "Engine Data Acquisition and Control System" and abbreviated as *EDACS*. The dynamometer and throttle controller, signal conditioning, and other accessories are interfaced with Engine Data Acquisition and Control System. The dynamometer data acquisition and control system has following modules which enables flawless automated testing of a particular engine.

Admin/User Permissions

This section allows creating and managing user accounts and groups, and setting permissions to perform any system operation on by-user or by-group basis. This section assigns roles to different testers as to fully configure dashboard UI layout, channel display, styling/coloring alarm setting, and sequence generation.

Automated Sequence Generation

Automated sequence editor allows user/tester to configure the test sequence to create full throttle performance testing or part throttle testing or any specific sequence demanded by development team.

The Sequence Editor module of the EDACS is used to create and manage automated test procedures. The EDACS system can be controlled in one of three ways:

1. Fully manual operation: Dynamometer is controlled by user commands from the Dashboard User Interface.
2. Fully automated operation: A predefined test sequence is executed as a sequence of steps initiated by one click from the user.
3. Semi-automated operation: A predefined test procedure is running in the background, and operator can select to measure any specific parameter such as blow-by and oil consumption.

On Screen Display

The system displays real-time data display in the form of analog or digital indication or plotting of any dynamometer data channels. Most of the software programs provide the user the ability to zoom, toggle, between screens, configure display channels.

Data Logging

The data logging module enables the system to log desired data channels as configured by test engineer. The module also helps to read and present the data into engineering units and makes it available for advanced postanalysis applications.

Alarm Management and Safety Controls

The user can decide and configure the alarm channels and configure the status such as warning or shut down of the system. The following figure shows the block schematics of the complete control system.

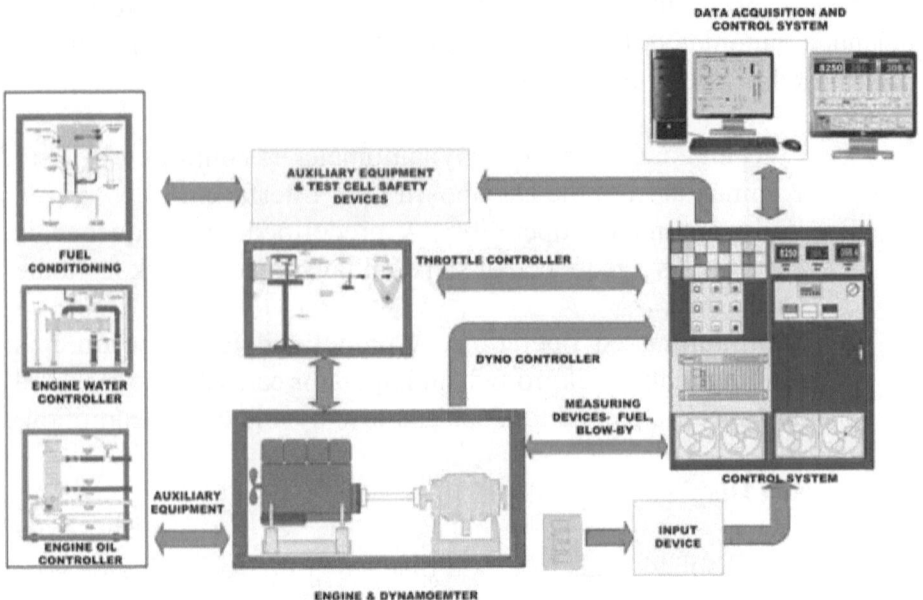

Figure 14.2 Engine data acquisition control system.

The dynamometer controller, throttle controller, engine start-stop controller, signal conditioning unit, and auxiliary equipment controllers are interfaces with the engine data acquisition and control system. Each unit can be controlled and monitored through the central console. Each unit in automation will have the capability to be interfaced with control and data acquisition system. The safety system incorporated monitors the health of the parameters controlled and measured. The parameters crossing the safety limits will generate the alarm, and control action is executed by the control system depending on how that parameter is configured that is generate a warning signal so that operator looks into it or shut down the entire system if condition of the parameter is severe which is likely to cause damage to engine and dynamometer. The following figure shows the signal flow from the control and data acquisition panel to the test cell.

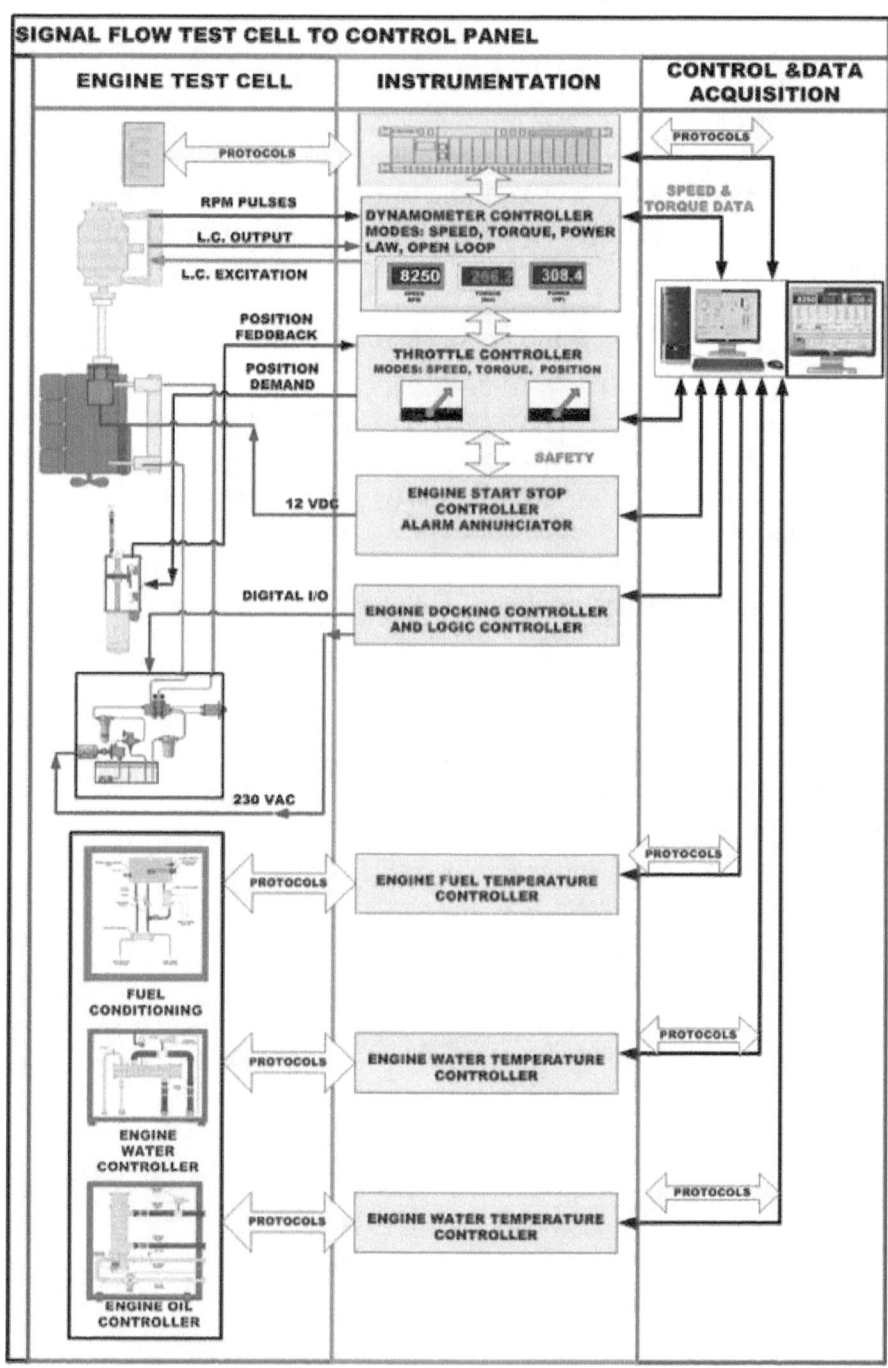

Figure 14.3 Signal flow from test cell to control panel.

14.6 Automation of Engine Handling

The engine handling involves rigging/derigging of the engine, movement of the engine from engine preparation area to test cell, automated docking of palette or trolley, and automated connection of the driveshaft. Similarly after the testing is over, handling process involves undocking, decoupling the driveshaft, movement of the engine to storage or repair section depending on test results.

14.6.2 Automatic Driveshaft Connection

Large amount of time is consumed in connecting and disconnecting the engine from the dynamometer. The process of automatic driveshaft connection reduces the overall testing time. The automatic shaft docking system provides the following advantages:

1. Rapid and easy change of the engine.
2. Time saving in connecting and disconnecting of driveshaft.

The most important requirement of automatic shaft docking system is that should provide the backlash free torque transfer for the smooth running of the engine. The following figure shows the automatic shaft docking sequence.

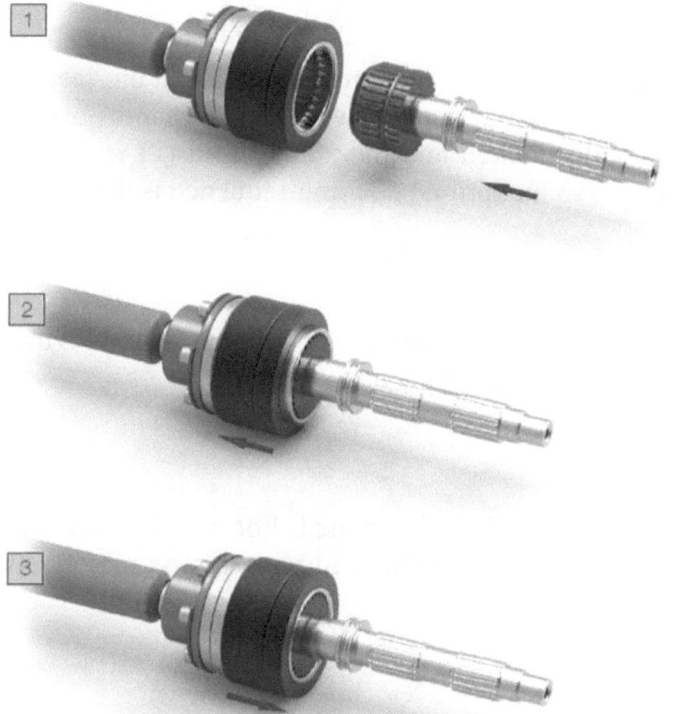

Figure 14.4 Automatic shaft docking.
(Courtesy of tectos gmbh, Glacistrasse, Graz, Austria)

The shaft docking can be employed by using different types of splined shafts. The driveshaft is split in two halves, and one half is mounted on engine which is rigged on palette, trolley, or automated guided vehicle (AGV). The second half of the driveshaft is fixed to dynamometer flange coupling. The following figure explains the automatic connection of driveshaft.

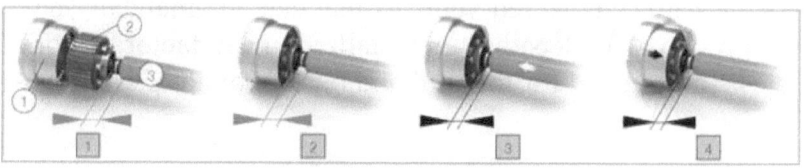

Figure 14.5 Automatic shaft docking.
Courtesy of tectos gmbh, Glacistrasse, Graz, Austria.

Step 1

The engine adaptor (1) is mounted on engine at the service area. The spline head (2) is fixed to the driveshaft (3). In this position, the joint is automatically locked and the spline head (2) is aligned with the driveshaft (3) so an optimal docking sequence is achievable.

Step 2

During the docking sequence, the docking system with the driveshaft (2 + 3) slides into the engine adapter (1).

Step 3

The automatic joint lock is opened. For an ideal connection of the engine, the driveshaft is preloaded with approximately 300 N to the engine.

Step 4

In operation, the constant velocity (CV) joint shaft enables axial, radial, and angular relative movements of the engine without impairment of the test bed function.

14.6.3 Automated Guided Vehicles

AGVs—The material handling institute defines automated guided vehicles as driverless vehicle equipped with on board automatic guidance device capable of following programmed path. Occasionally, AGVs are referred to as mobile robots owing to their reprogram ability. AGVs are battery powered vehicles that can automatically navigate through a network of paths (preinstalled on the factory floor). The vehicles are designed for forward and reverse motion.

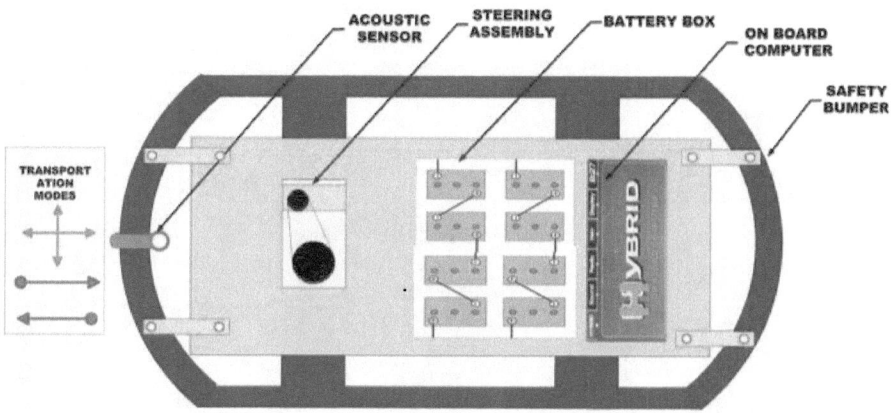

Figure 14.6 Schematics of AGV.

Navigation Guidance for AGVs

The steering is achieved by on board computer executing a prespecified motion trajectory, which is communicated to it from a central area controller, normally via RF. The vehicles are equipped with acoustics sensors to detect obstacles (for collision avoidance as well as with safety bumpers that can detect contact and stop their motion).

The navigation guidance for the AGVs can be in two forms:

1. Active tracking of guidepath
2. Passive tracking of guidepath

Both of the methods rely on noncontact tracking of guidepath installed on or in the floor of the manufacturing plant. Optical passive tracking is the most economical and flexible method, where the guidepath is defined by a painted or taped-on strip. An optical detection device mounted underneath of the vehicle follows the continuous guidepath and guides the vehicles to its destination.

The following figure shows the AGV control block schematic:

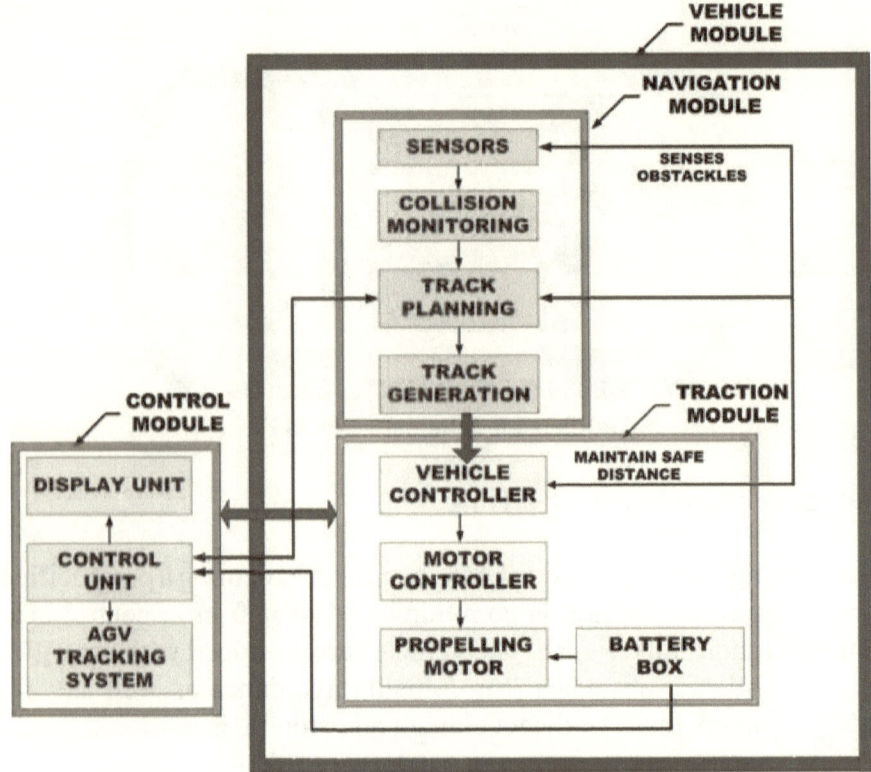

Figure 14.7 AGV control schematics.

The typical engine mounting arrangement on AGV is shown in the following figure. The engine is prerigged and male part of driveshaft is fitted on engine flywheel so that while docking the shaft gets automatically connected to the other half of the driveshaft that is mounted on the dynamometer coupling.

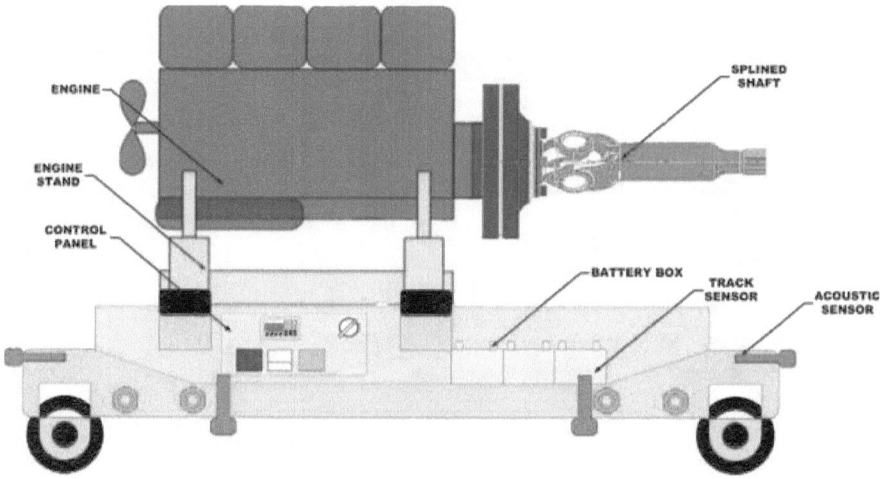

Figure 14.8 Engine mounted on AGV.

14.7 Docking Trolleys and Carts

The docking trolleys and carts facilitate the operator to move the engines quickly in and out from the test cell. The use of trolleys allows you to prepare one engine outside the test area while another engine is being tested in the test cell. The trolleys are equipped with the flexible hoses and pipes which are connected to the engine while they are prepared for testing. Once the trolley arrives in the testing station for docking, the automated docking will dock the trolley eliminating the manual connection of the engine. The services such as oil and water are connected by using the flexible hoses with quick coupling. The following figure illustrates the schematics of the trolley docking system.

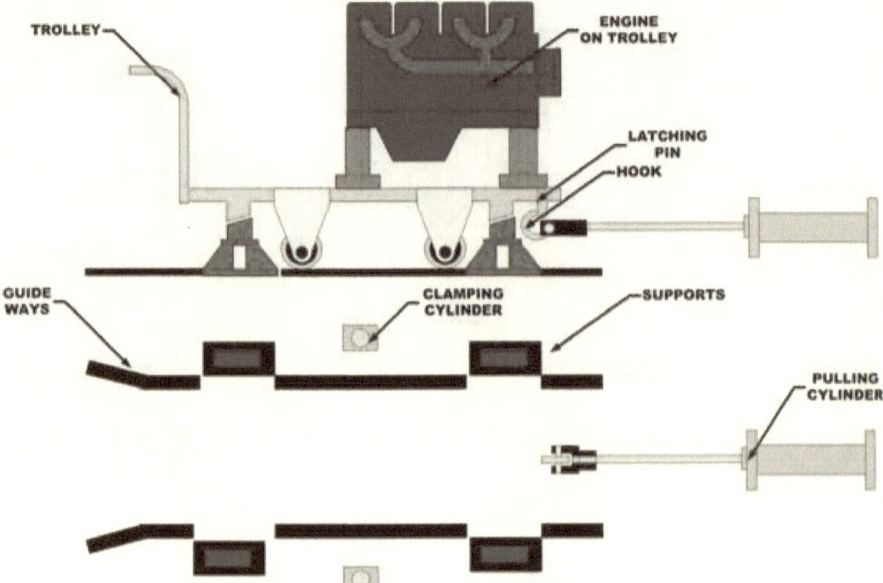

Figure 14.9 Schematics of trolley docking system.

The trolleys are generally designed in such a way that height of engine over trolley is same as the height of the center of dynamometer. However, in some cases it is possible to have different engines with height adjusting pedestals. On these pedestals, vibration mounts are used to connect to engine. The trolley is locked in its position by the pneumatic clamping system when the engine is running. The following figure shows the schematics of the pneumatic circuit.

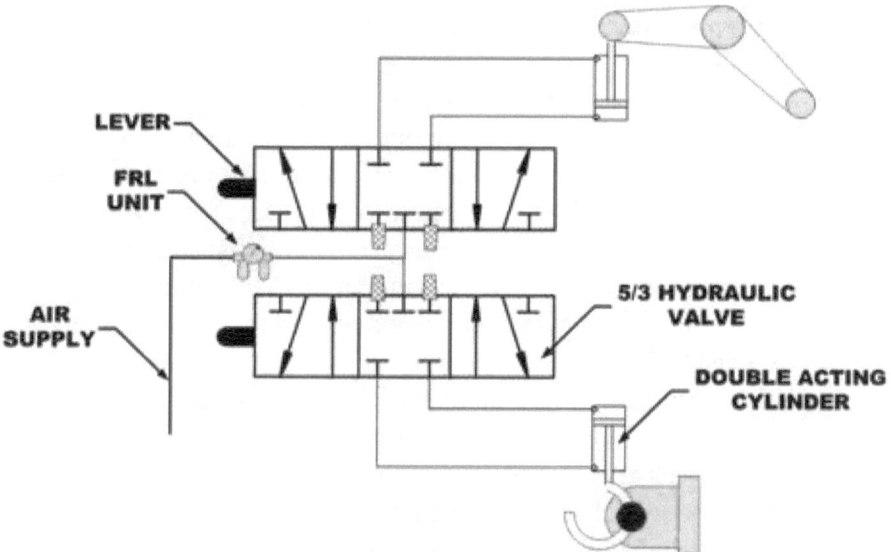

Figure 14.10 Pneumatic circuit.

The pneumatic system incorporates two direction control valves. The direction control valves employed are with 5 ports and 3 fixed positions. The symbolic presentation of direction control valve is shown below.

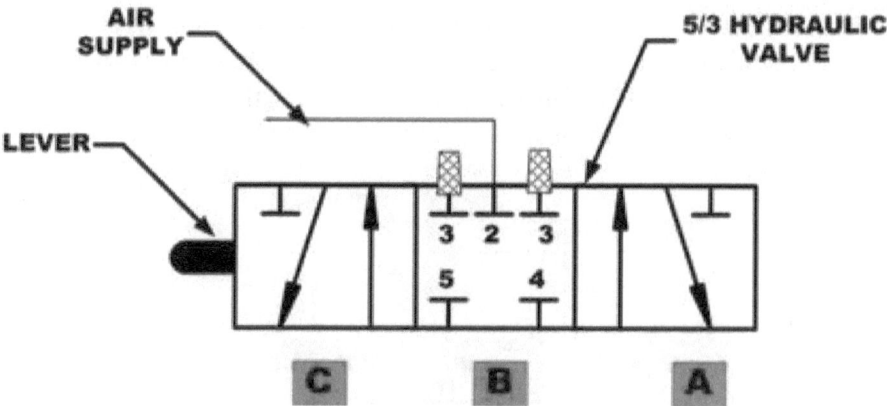

Figure 14.11 Direction control valve.

The docking procedure is described in following steps:

1. When the lever is in center position (B), the ports are closed and air is not supplied to the operating cylinders.
2. When the lever is moved to right, valve is in position (A), the inlet port (2) will be connected to cylinder through the outlet port (4). Direction is shown as arrow in the above figure. The air enters at rod end of cylinder and makes the cylinder to retract. The air at piston end is exhausted to atmosphere by the exhaust valve (3).
3. When the lever is moved to position (3), the inlet port (2) will be connected to cylinder through the outlet port (5). Direction is shown as arrow in the figure. The air enters at the piston end of cylinder and makes the cylinder to extend. The air at rod end is exhausted by the exhaust valve (1).

14.8 Overhead Conveyor and Pallets

The overhead-mounted conveyor eliminates any floor usage and allows lot of space for the workers to move around and do their job. The engine is staged or prerigged on the pallet with complete connections done and terminated to the match plate mounted on the palette. When this prerigged engine arrives on the docking station, a position sensor will sense the engine's position and will activate the docking sequence. The following figure shows the match plates on palette and docking station.

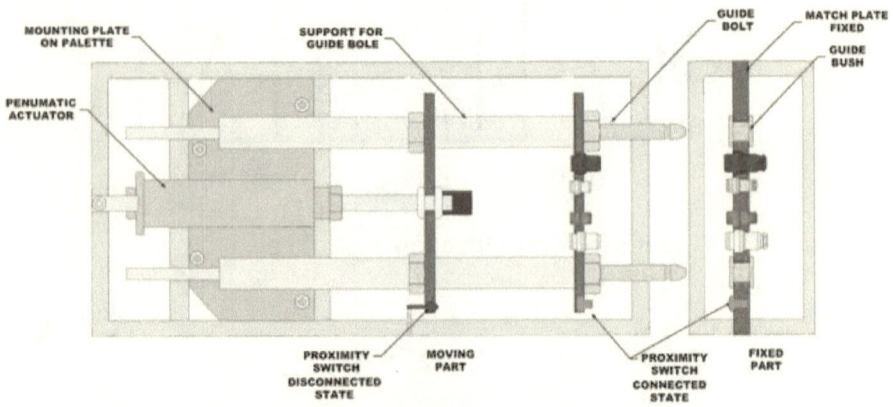

Figure 14.12 Schematics of match plate.

This docking sequence (steps) in the automation of engine handling is described below.

Engine Preparation

This is also called engine rigging on pallet. The fully assembled engine from production line is mounted on a pallet and fixed. The connections between engine and pallet are made and terminated at the match plate on palette.

Preparing engine supplies

Once engine is rigged to the palette, it is sent to test cell via filling station. The filling station bar code reader will identify the type of engine and automatically start filling right amount of fuel, oil, and coolant.

Arrival to Test Cell

Upon arrival to test cell area, the engine will be in waiting room, waiting to be docked in the empty test cell. As soon as the test unit is available, the engine is diverted into this test cell.

Docking and Testing

The pallets will have a companion floor-mounted docking frame on which the pallet gets docked. Once the docking is complete, the match plate on the docking pallet will match the other match plate on the docking station thereby completing all the connections such as engine water supply, engine lubricating oil connection, and all sensor electrical connections. The automatic driveshaft connection is via a spline assembly. The pallet assemblies include compliant engine mounts to isolate engine vibration. The test cell with automation system is usually equipped with fuel, oil, and water conditioning units as well as engine exhaust extraction system and ventilation systems. After finishing the test sequence, the engine is automatically disconnected.

Failed Engine Requiring Repair

If an engine fails the test, then it is rejected and sent for repair/investigation area. The repaired engine goes through the above steps again.

Derigging

If the engine is accepted after it passes the test, the engine is transported out of the test cell to derigging area. The engine is cleaned and sent to vehicle assembly area or to the storage room. The empty pallet is returned to the engine rigging area by transport system installed in the plant such as conveyor system.

The following figure shows a flow chart summarizing above sequence.

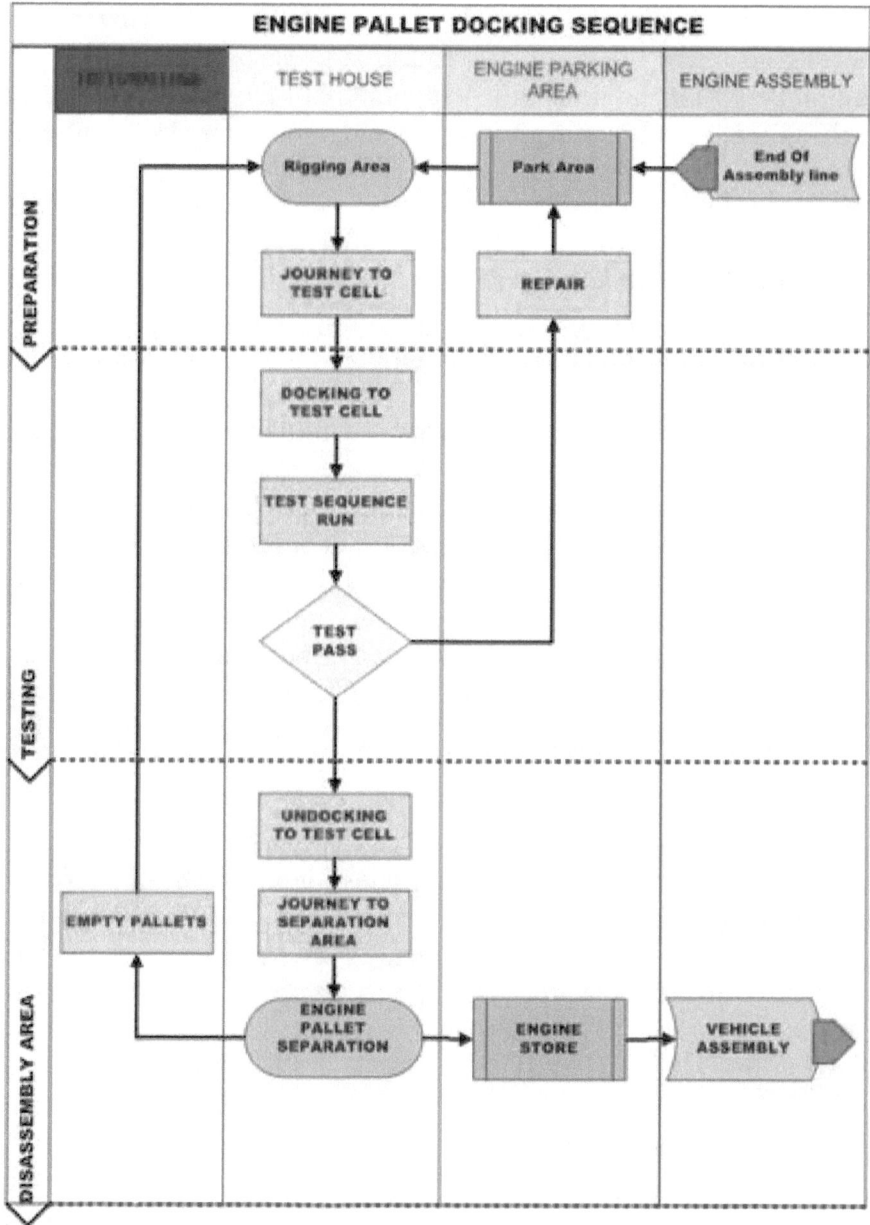

Figure 14.13 Flow chart for docking sequence.

14.9 Automation of Auxiliary Equipment

The test cell is equipped with various auxiliary equipments to facilitate the engine testing in more systematic and scientific manner. The equipments which maintain the temperature of the engine parameters such as water, lubricating oil, fuel, and intake air are mentioned below.

1. Engine water temperature controller
2. Engine oil conditioning unit
3. Engine fuel temperature conditioning unit
4. Engine air intake conditioning unit.

14.9.1 Why Do We Need Temperature Controllers?

The equipments mentioned in preceding paragraph provide more stable conditions to conduct the testing experiments and thereby reduce the test cell utilization time. They are needed in any situation requiring a given temperature be kept stable. This can be in a situation where an object is required to be heated, cooled or both and to remain at the target temperature (set point), regardless of the changing environment around it. There are two fundamental types of temperature control:

1. Open loop

Open loop is the most basic form; it applies continuous heating/cooling with no regard for the actual temperature output. It is analogous to heating or water on gas stove. On a cold day, you may need to turn the heat on to full to warm the water to say 60°C. However, during the summer, the same setting would leave the water warmer than the desired 60°C.

2. Closed loop control.

In a closed loop application, the output temperature is constantly measured and adjusted to maintain a constant output at the desired temperature. Closed loop control is always conscious of the output signal and will feed this back into the control process. Closed loop control is analogous to a radiator with thermostat in the automobile.

14.9.2 Elements of Temperature Controller

A basic temperature controller provides control of heating and cooling processes of engine parameters. In a typical application, sensors measure the actual temperature. This sensed temperature is constantly compared to a user set point. When the actual temperature deviates from the set point, the controller generates an output signal to activate temperature regulating devices such as heating elements for example electrical heaters or cooling water valves to bring the temperature back to the set point.

Input

In engine testing application, sensors measure the actual temperature. This sensed temperature is constantly compared to a user set point (user demand). The sensor measuring the temperature parameter may be either thermocouple or RTD sensor. Depending on the application and its maximum range of temperature measurement limit the type of temperature sensor is selected. The input can be voltage or current signals in the millvolt, volt, or milliamp range from other types of sensors such as pressure transducer and pressure or temperature transmitter.

Controllers are designed typically with a feature to detect when an input sensor is faulty or absent. This is known as a sensor break detect. Undetected, this fault condition could cause significant damage to the engine being controlled.

Outputs

The controllers have an output signal and this signal can be used to control the cooling water supply valve, initiate an alarm, or feed the output to programmable logic controllers (PLCs). The output signal can also be used to record the engine parameters such as oil temperature by interfacing a chart recorder.

The outputs provided with temperature controllers include relay outputs, solid state relay drivers, and linear analog outputs signals. The temperature controller normally provides relay output, and it is generally a single-pole double-throw commonly abbreviated as SPDT relay with a DC voltage coil. The controller energizes the relay coil, providing isolation for the contacts. This lets the contacts control an external voltage source to power the coil of a much larger heating contactor or the motor contactor.

Some controllers provide the analog outputs such as the 0-10 V signal or a 4-20 mA signal. These signals are calibrated so that the signal changes are linear. For example, if a controller is sending a 0 percent signal, the analog output will be 0 V or 4 mA. When the controller is sending a 50 percent signal, the output will be 5 V or 10 mA. When the controller is sending a 100 percent signal, the output will be 10 V or 20 mA.

Set Point

A set point is a target value set by an operator which the controller aims at keeping steady. For instance, a set point lubricating oil temperature of 80°C means that a controller will aim to keep the temperature at this value, within the tolerance limit guaranteed by the manufacturers.

Alarm

This is used to indicate when a parameter to be controlled has reached some alarming condition which may harm the engine under test. The alarms can be configured in a several ways. For instance, a high alarm may indicate that a temperature has reached higher state than some set value. Likewise, a low alarm indicates the temperature has dropped below predetermined set value. For example, in an engine, water temperature crosses the set point; a high alarm prevents a heat source from damaging the engine by pulling the shutdown lever of the engine thereby forcing the engine to stop.

PID

PID, which stands for proportional, integral, derivative, is an advanced control function that uses feedback from the controlled process to determine how best to control that process.

14.9.3 How Controller Works

The controllers control, hold, or maintain some variable or parameter at a set value or demand value from the test engineer. There are two variables required by the controller; actual input signal and desired set point value. The input signal is also known as the process value. The input to the controller is sampled many times per second, depending on the controller.

This input value is compared with the set point value. If the actual value doesn't match the set point, the controller generates an output signal change based on the difference between the set point and the process value and whether or not the process value is approaching the set point or deviating farther from the set point. This output signal then initiates a response to correct the actual value so that it matches the set point. This action is repeated till a set point value is achieved.

The control action taken depends on the type of controller. For instance, if the controller is an ON/OFF control, the controller decides if the output needs to be turned on or turned off, or left in its present state. PID control determines the exact output value required to maintain the desired temperature. The output signal will be in the range from 0 to 100 percent. This output signal is in turn calibrated with the controlling device such as valve opening from 0 to 100 percent.

14.10 Engine Oil Temperature Controller (EOTC)

The engine oil temperature control (EOTC) device is used to control and maintain the engine oil at constant predetermined temperature. The oil temperature control unit circulates oil through oil temperature controller and the engine and helps to precisely, automatically, and reliably maintain it at a set temperature.

Rapid recirculation of the cooling fluid provides a close and uniform temperature relationship between the engine and from engine oil lines. This does, of course, depend on the configuration of the heat exchanger installed in the oil cooling process and any restrictions within piping. This recirculation combined with the heating element and cooling capability gives fast and accurate response to bring the oil to set point temperature or to changes in the settings when needed.

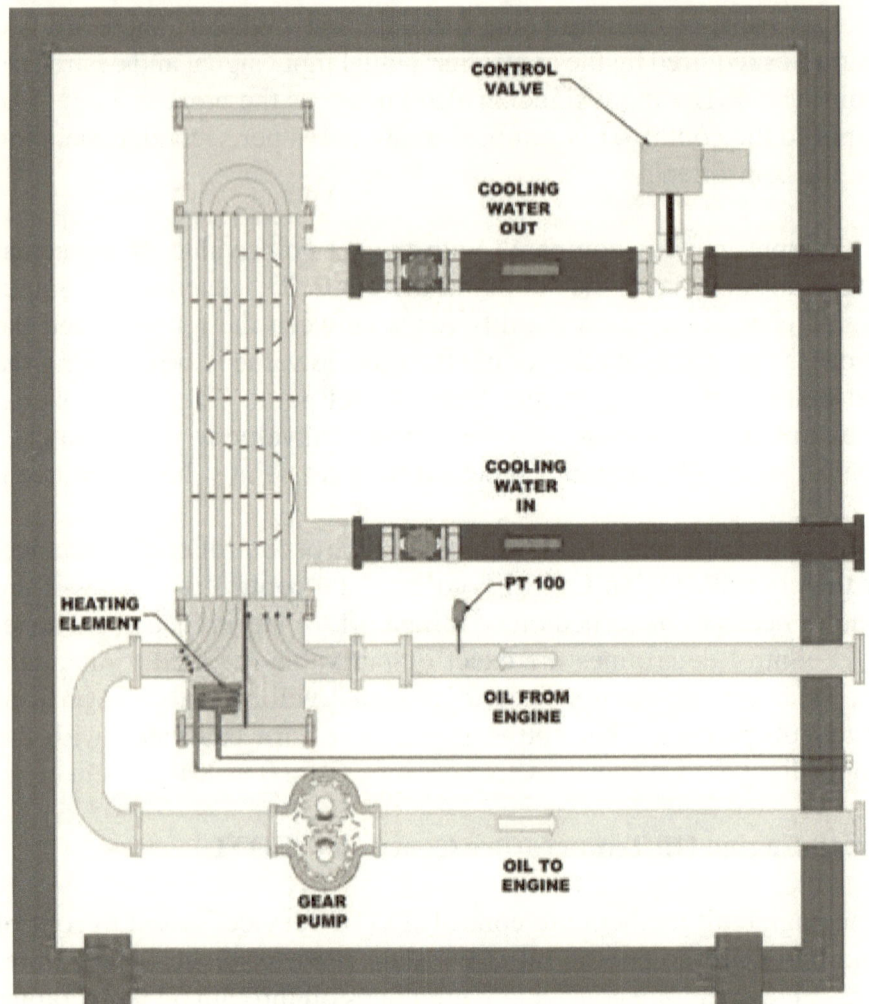

Figure 14.14 Schematics of engine oil temperature controller.

Elements of Oil Temperature Controller

Heater

The fluid is heated by the specially designed three-phase low-watt density electrical immersion heater and regulated by the controller.

The standard heater has a steel sheath for low-watt density and good heat transfer.

Pump

The pump used is generally a positive displacement pump, and it is failure-free design. The pump used in the EOTC is well suited for use with a variety of commercially available heat transfer fluids. The pump has only two internal moving parts and a specially designed seal to give years of trouble-free service, even at high temperatures.

Heat Exchanger

The heat exchanger is generally shell and tube type heat exchanger and works as main media to exchange the heat between the oil and cooling fluid. The tubes are constructed of copper-nickel for durability and optimal heat transfer. The modular construction of the shell and tube heat exchanger allow the tube bundle to be easily removed for periodic cleaning. Additionally, check valves are installed on the water supply and drain lines to prevent water from back flowing into the heat exchanger from a closed drain or into the water supply piping.

The controller automatically regulates cooling by opening and closing the cooling solenoid. This allows the proper amount of oil to pass over the tubes of the heat exchanger.

Valve Motor Drive

A special type of general-purpose controller is the valve motor drive controller. These controllers are specifically designed to control valve motors. Special tuning algorithms give accurate control and fast output reaction without the need for slide wire feedback or excessive knowledge of three-term PID tuning algorithms. The valve motor drive controller controls the position of the valve fully closed (0 percent) to fully open (100 percent) depending on the energy needs of the engine under test at any given time.

1. Advantages of EOTC in the test cell.
   Fuel cost savings. This results from quicker warm up times and operating at optimum oil temperature.
2. Increased engine life.
   Oil is maintained at optimum operating temperature. This prevents prolonged engine operation with low oil temperatures resulting in crankcase oil dilution and contamination leading to internal engine corrosion and accelerated wear.

14.14 Engine Water Temperature Controller (EWTC)

The primary function of the engine water temperature controller (EWTC) in test cell is to simulate the engine cooling system in automobile. The engine water temperature control systems replicate the normal operating conditions of an engine in the test cell.

The application of EWTC in test cell facilitates:

1. Maintaining the engine temperature as demanded by the test.
2. Helping in preheating the engine if the test demands.
3. Draining the system faster for quick engine changes after the completion of the test.

The following figure shows the schematics of EWTC.

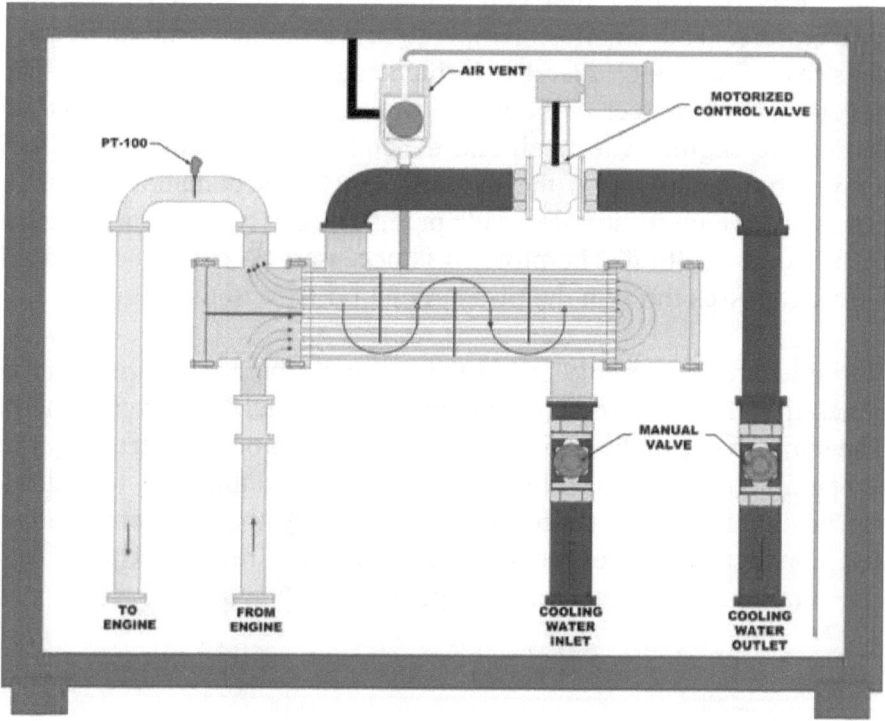

Figure 14.15 Schematics engine water temperature controller.

The engine water temperature controller, engine oil temperature controller, engine fuel conditioning equipment, engine air intake conditioning system and the throttle controller help in automation of the engine testing and getting the consistent test results. This helps in eliminating the variance caused by human involvement in the testing process. The automation process also eliminates the human monitoring process of the test. The test cell composed of the above equipments works in a perfectly integrated way.

14.12 Engine Fuel Conditioning Unit

Hydrocarbon-based fuels, such as gasoline, kerosene, and diesel fuel, are more efficient if they are heated prior to combustion. The efficiency and engine performance are increased while the exhaust emissions are decreased as the temperature of the fuel is raised. Heating the fuel increases the vaporization of these molecules. The measurement of fuel consumption is very important factor during engine development. The measurement of fuel consumption is influenced by the fuel temperature. The smallest temperature variations lead to density changes of the fuel thereby influencing fuel measurement. The development engineer will demand the most stable conditions while testing the engine in test cell and hence the need to control the fuel temperature arises. Also the modern legislation demands less CO_2 emissions which in turn demands precise and accurate measurement of fuel consumption. Therefore, it is necessary to condition the fuel before it passes through the measurement device to the engine.

While testing the engine inside the test cell, the fuel temperature is generally related to ambient temperature or the test cell environment. The temperature of the fuel going to the engine is also governed by the temperature of the return line. The fuel conditioning unit will allow the development and testing engineer to vary the fuel temperature and conduct the experiments to establish the relationship of fuel consumption versus fuel inlet temperature.

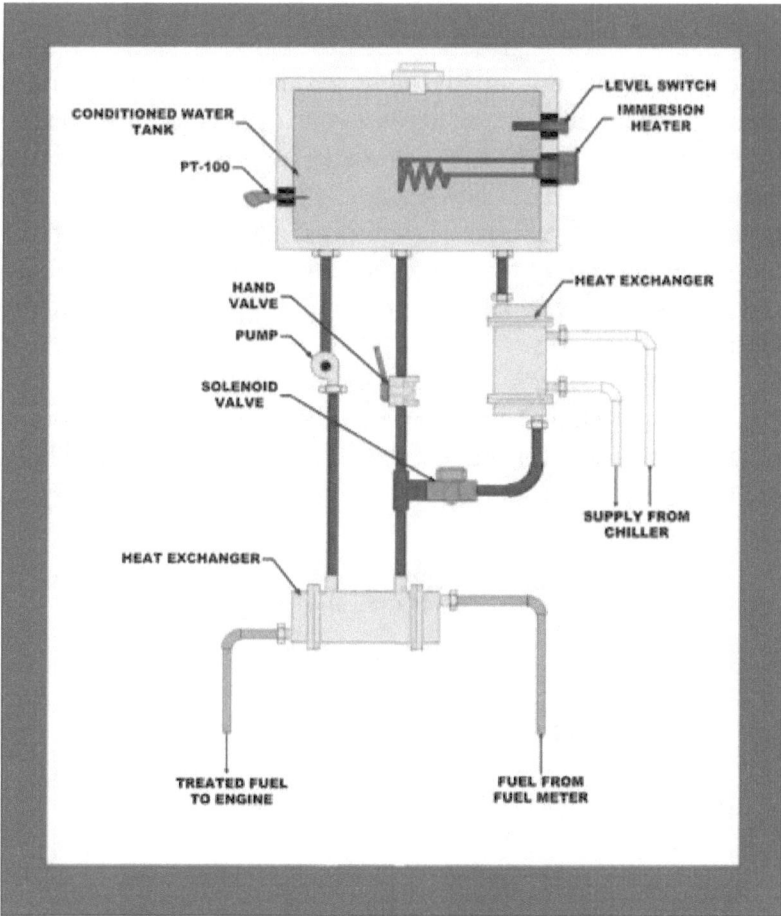

Figure 14.16 Schematics of engine fuel temperature controller.

The conditioned water stored in the header tank is used as primary coolant for controlling the temperature of the fuel going to the engine. The heat is exchanged inside the heat exchanger and fuel is cooled or heated. The fuel to be conditioned is fed to heat exchanger from the fuel meter, and once the desired temperature is achieved, it is supplied to the engine.

Closure

The modern internal combustion engines are getting more and more complex and hence need to increase the quality of test performed on engine test bed. The test bed needs to be most efficient, reducing calibration time and cost. The automated test bed offers the quality of reproducibility of results. The automation also helps in reducing the engine development time and thereby costs of development. The automated test bed will reduce the overall project cost of engine development.

The automation system stores the data of all the engine models and is a centralized repository for the test data and can be exchanged between different engine tests. It helps in easy data storage and distribution.

CHAPTER 15

Basics of Dynamometer Data Acquisition and Control System

15.1 Introduction

Data acquisition system is an essential part of the test cell control system. The modern test cells are getting more and more complicated with advancement in technology. Too many measuring channels and cluster of modern instrumentation make it necessary to have some kind of device which will automatically acquire the data and process it and make the life of testing engineer easy. In the test cell environment, the testing engineer needs to knowand should have the fundamental knowledge of inter disciplinary systems such as mechanical, electrical and electronics, and instrumentation. Data acquisition is widely used in many areas of industry. Data acquisition is used to acquire data from sensors and other sources under computer control and bring the data together and store and manipulate it. The simplest form may be a technician manually logging information such as the temperature of an oven. However, this form of data acquisition has its limitations. Not only is it expensive because of the fact that someone has to be available to take the measurements, but being manual it can be subject to errors. Readings may not be taken at the prescribed times, and also there can be errors resulting from the manual fashion in which the readings are taken. As can be imagined the problems become worse if a large number of readings need to be taken, as timing may become more of an issue, along with the volume of work required. To overcome this, the simple answer is to use computer control to perform the data acquisition.

15.2 What Is Data Acquisition?

Data acquisition is the process of sampling signals that measure real world physical conditions and converting the resulting samples into digital numeric values that can be manipulated by a computer. Data acquisition systems (abbreviated with the acronym DAS or DAQ) typically convert analog waveforms into digital values for processing.

Data acquisition systems, as the name implies, are products and/or processes used to collect information to document or analyze some phenomenon. Data acquisition and control allows us to automate the long hours of routine data collection, freeing us for tasks that can make a difference.

15.3 What Data Acquisition Does?

Data acquisition gathers data that would otherwise be difficult to obtain manually. This data can help us better understand a process of engine testing. Improvements can take the form of higher quality, more efficiency, or lower costs, resulting in savings of time and money. The data acquisition and control system offers the following benefits when it is employed for the task of the engine testing.

Advantages of Data acquisition and control system (DACS):

- Reduces time needed to manually collect data.
- Reduce errors that can occur when inputting data to a computer by hand.
- Operates unattended for desired period of time.
- Makes high-speed collection rates possible.
- Provides tighter control than is possible manually.

15.4 Where Do We Use Data Acquisition and Control System?

The data acquisition system finds its application in engine testing and vehicle testing.

Characterization: Measure the response of a product or process relative to all variables. Typically used in engineering to test new designs, new engines built or engine modified.

Monitoring: Measure the response of an engine and dynamometer parameters or measuring process relative to key variables. Alarms are invoked when the key variables go outside predefined/pre-set limits.

Control: Measure the response of key variables and take corrective action when the variables go outside prescribed limits.

15.5 The Elements of Data Acquisition Systems

1. Sensors that convert physical parameters to electrical signals.
2. Signal conditioning circuitry to convert sensor signals into a form that can be converted to digital values.
3. Analog-to-digital converters, which convert conditioned sensor signals to digital values.
4. Software.
5. Computer.

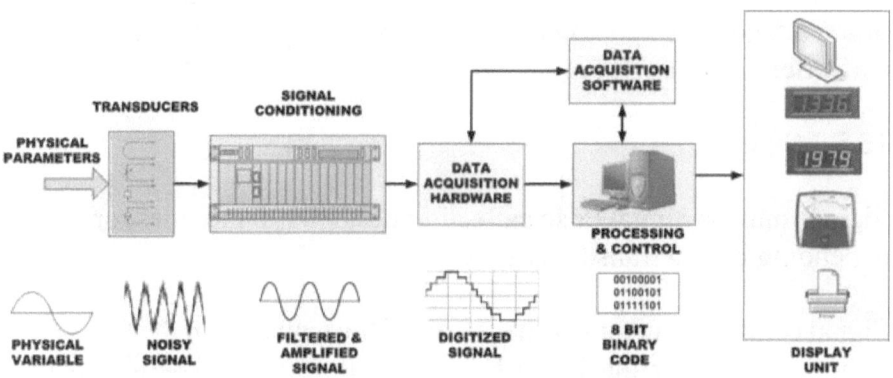

Figure 15.1 Schematics of data acquisition system.

Data acquisition applications are controlled by software programs developed using various general purpose programming languages used by applications to access and control the data acquisition hardware.

Data acquisition begins with the physical phenomenon or physical property to be measured. Examples of this include temperature, light intensity, gas pressure, fluid flow, and force. Regardless of the type of physical property to be measured, the physical state that is to be measured must first be transformed into a unified form that can be sampled by a data acquisition system. The task of performing such transformations falls on devices called *sensors*.

15.5.1 Sensor

It is a type of *transducer*, is a device that converts a physical property into a corresponding electrical signal (e.g., a voltage or current) or, in many cases, into a corresponding electrical characteristic (e.g., resistance or capacitance) that can easily be converted to electrical signal.

The ability of a data acquisition system to measure differing properties depends on having sensors that are suited to detect the various properties to be measured. There are specific sensors for many different applications. DAQ systems also employ various signal conditioning techniques to adequately modify various different electrical signals into voltage that can then be digitized using an Analog-to-digital converter (ADC).

15.5.2 Signals

Signals may be digital (also called logic signals, sometimes) or analog depending on the transducer used.

Signal conditioning may be necessary if the signal from the transducer is not suitable for the DAQ hardware being used. The signal may need to be amplified, filtered, or demodulated. In addition, many transducers require excitation currents or voltages, bridge completion, linearization, or high amplification for proper and accurate operation. Various other examples of signal conditioning might be bridge completion, providing current or voltage excitation to the sensor, isolation, and linearization. For transmission purposes, single-ended analog signals, which are more susceptible to noise, can be converted to differential signals. Once digitized, the signal can be encoded to reduce and correct transmission errors.

The input signals may be analog or digital:

1. Analog
   - Signal is continuous.
   - Example: strain gauge. Most transducers produce analog signals.
2. Digital
   - Signal is either ON or OFF
   - Example: light switch.

15.5.3 Data Acquisition Hardware

DAC hardware is what usually interfaces between the signal and a PC. It could be in the form of modules that can be connected to the computer's ports (parallel, serial, USB, etc.) or cards connected to slots (e.g. Apple Bus, ISA, MCA, PCI, PCI-E, etc.) in the mother board. Usually the space on the back of a PCI card is too small for all the connections needed, so an external connection box is required. The cable between this box and the PC can be expensive due to the many wires and the required shielding.

DAC cards often contain multiple components (multiplexer, ADC, DAC, TTL-IO, high-speed timers, RAM). These are accessible via a bus by a microcontroller, which can run small programs. A controller is more flexible than a hard-wired logic, yet cheaper than a CPU so that it is permissible to block it with simple polling loops. For example, waiting for a trigger, starting the ADC, looking up the time, waiting for the ADC to finish, moving value to RAM, switching multiplexer, getting TTL input, and letting DAC proceed with voltage ramp. Many times reconfigurable logic is used to achieve high speed for specific tasks, and digital signal processors are used after the data have been acquired to obtain some results. The fixed connection with the PC allows for comfortable compilation and debugging. Using an external housing, a modular design with slots in a bus can grow with the needs of the user.

15.5.4 Data Acquisition Software

DAQ software is needed in order for the DAQ hardware to work with a PC and the instrumentation in the test cell. The acquisition software is used to acquire, store, and process the data in logical format. The device driver performs low-level register writes and reads on the hardware, while exposing a standard API for developing user applications. Software allows the test engineer to configure, modify the testing sequence, and create his own test methodology.

15.6 Data Acquisition Terminology

- **Analog-to-Digital Converter (ADC)**

    An electronic device that converts analog signals to an equivalent digital form. The analog-to-digital converter is the heart of most data acquisition systems. The converter determines how often conversion takes place. The higher the sampling rate, the better the system is.

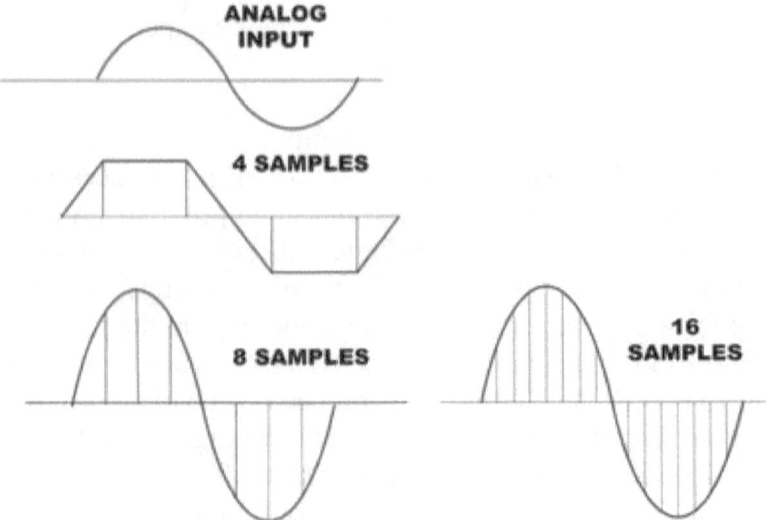

Figure 15.2 Analog to digital conversion.

- **Analog-to-Digital Converter Throughput**

    Effective rate of each individual channel is inversely proportional to the number of channels sampled. For example,

    — 100 KHz maximum.
    — 16 channels.

    100 KHz/16 = 6.25 KHz per channel.

- **Digital-to-Analog Converter (D/A)**

    An electronic component found in many data acquisition devices that produce an analog output signal.

- **Digital Input/Output (DIO)**

- Digital I/O is discrete signals which are either one of two states. These states may be on/off, high/low, 1/0, etc. Digital I/O is also referred to as binary I/O.

- **Differential Input**

    Refers to the way a signal is wired to a data acquisition device. Differential inputs have a unique high and unique low connection for each channel. Data acquisition devices have either single-ended or differential inputs, many devices support both configurations.

- **General Purpose Interface Bus (GPIB)**

    This standard bus is used for controlling electronic instruments with a computer. This is also called IEEE 488 in reference to defining ANSI/IEEE standards.

- **Resolution**

    The smallest signal increment that can be detected by a data acquisition system. Resolution can be expressed in bits, in proportions, or in percent of full scale. For example, a system has 12-bit resolution, one part in 4,096 resolutions, and 0.0244 percent of full scale.

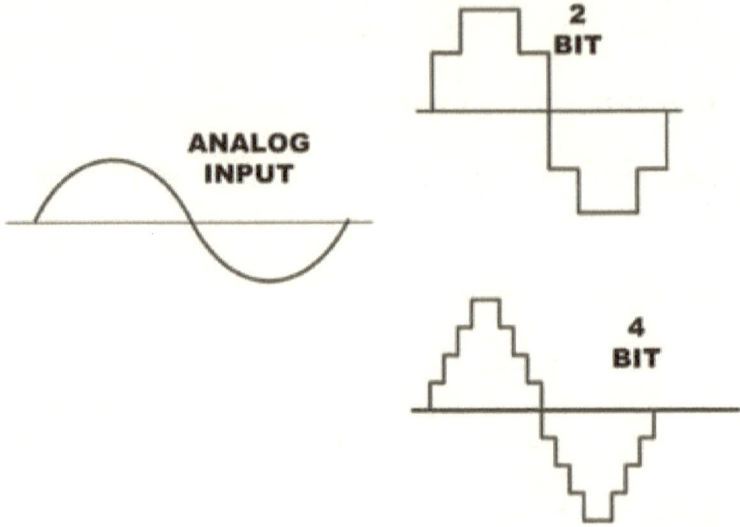

Figure 15.3 Resolution.

- **RS-232**

    A standard for serial communications found in many data acquisition systems. RS232 is the most common serial communication; however, it is somewhat limited in that it only supports communication to one device connected to the bus at a time and it only supports transmission distances up to 50 feet.

- **RS-485**

    A standard for serial communications found in many data acquisition systems. RS-485 is not as popular as RS232; however, it is more flexible in that it supports communication to more than one device on the bus at a time and supports transmission distances of approximately 5,000 feet.

- **Sample Rate**

    The sample rate is the speed at which a data acquisition system collects data. The speed is normally expressed in samples per second. For multichannel data acquisition devices, the sample rate is typically given as the speed of the analog-to-digital converter (A/D). To obtain individual channel sample rate, you need to divide the speed of the A/D by the number of channels being sampled.

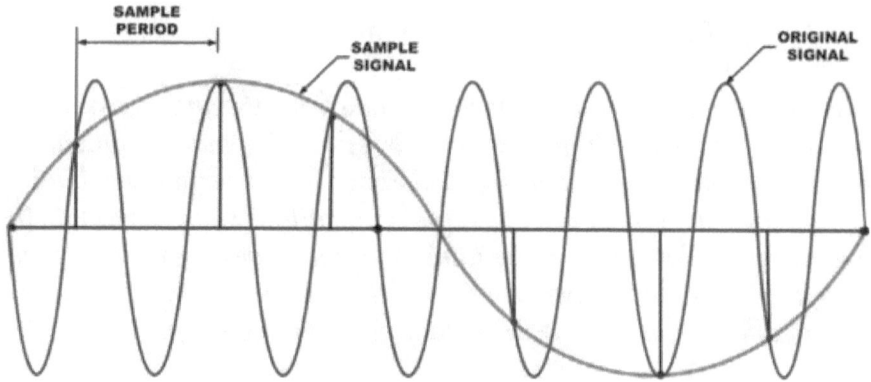

Figure 15.4 Sample rate.

- **Single-ended Input (SE)**

    Refers to the way a signal is wired to a data acquisition device. In single-ended wiring, each analog input has a unique high connection but all channels share a common ground connection. Data acquisition devices have either single-ended or differential inputs. Many support both configurations.

- **Data Parameters**

    Data parameters are the data points in the instrument that may be measured and stored. The list of data parameters is arbitrary, but it essentially consists of readings that appear in test measurements on the display, as well as some internal parameters, such as the concentration just prior to calibrating the instrument. Most data parameters have measurement units associated with them, such as mV, PPB, cc/m, etc., although some have no units. The DAS is not designed to permit these units to be changed.

This restriction is mainly due to the fact that the instruments themselves don't support changing the measurement units of most of its readings. If a computer is being used to interface to the DAS, that computer can perform unit conversions. Data parameters have a few user-configurable properties, which are summarized in the table below.

15.7 Types of Data Acquisition Systems

Data acquisition is a very important area of the test and measurement industry. Data acquisition systems are required in many applications from electronics manufacturing to chemical engineering, mechanical manufacture as well as more diverse applications such as monitoring geographical data from mountains and volcanoes as well as many other interesting and diverse uses. In the case of engine and vehicle testing, data acquisition is important part of the testing.

There are various types of data acquisition systems such as

1. Wireless Data Acquisition system
2. Serial Communication Data Acquisition Systems
3. USB Data Acquisition Systems
4. Data Acquisition Plug-in Boards

15.7.1 Wireless Data Acquisition Systems

The wireless data acquisition system finds the application where one needs to acquire the data from remote places, for example, the monitoring of the bridge structure in remote place. This type of data acquisition also finds application in getting information from a vehicle on the road through wireless transmission. Sometimes it is necessary to collect specific data from the car in order to increase knowledge of car and drivers' performance. In such cases, a wireless data acquisition system finds its application.

Wireless data acquisition systems can eliminate costly and time-consuming field wiring of process sensors. These systems consist of one or more wireless transmitters sending data back to a wireless receiver connected to a remote computer.

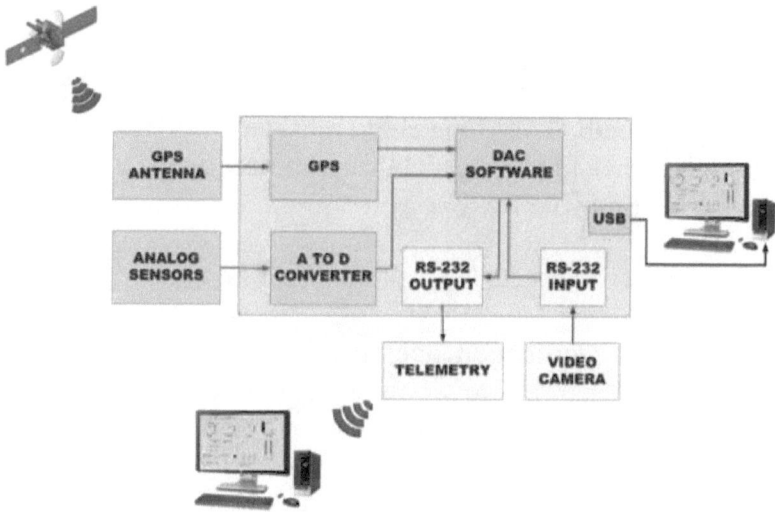

Figure 15.5 Schematics of wireless DAC.

15.7.2 Serial Communication Data Acquisition Systems

Serial communication data acquisition systems are a good choice when the measurement needs to be made at a location which is distant from the computer. There are several different communication standards. RS-232 is the most common but only supports transmission distances up to 50 feet. The RS-232 serial interface communications standard has been in use for very many years and is one of the most widely used standards for serial data communications as a result of it being simple and reliable.

RS-232 Interface Basics

The interface is intended to operate over distances of up to 15 meters. This is because any modem is likely to be near the terminal. Data rates are also limited. The maximum for RS-232C is 19.2 k baud or bits per second although slower rates are often used. In theory, it is possible to use any baud rate, but there are number of standard transmission speeds used.

Common Data Transmission Rates

50, 75, 110, 150, 300, 600, 1200, 2400, 4800, 9600, 19200, 38400, 76800

Normally, speed used is 192200 bits/sec. Above this noise that is picked up, especially over long cable runs, can introduce data errors. Where high speeds and long data runs are required, then standards such as RS-422 may be used.

RS-232 Connections

The RS-232 C specification does not include a description of the connector to be used. However, the most common type found is the 25 pin D-type connector.

RS-232 Applications

The RS-232 standard provided an ideal method of connection and therefore it started to be used in a rather different way. However, its use really started to take off when personal computers were first introduced. The RS-232 standard provided an ideal method of linking the PC to the printer.

15.7.3 USB Data Acquisition Systems

The Universal Serial Bus (USB) Interface is well established as an interface for computer communications. The USB is a new standard for connecting PCs to peripheral devices such as printers, monitors, modems, and data acquisition devices. As a result, USB data acquisition modules and devices are now widespread on the market. The use of USB has grown in all sectors of the computer market as a result of its convenience. In line with this, the data acquisition sector has also used the interface for small and cost-effective devices.

Advantages of USB Data Acquisition:

1. Since USB connections supply power, only one cable is required to link the data acquisition device to the PC, which most likely has at least one USB port.
2. USB data acquisition modules can be connected and disconnected without the need to power down the computer.
3. USB ports are standard on most PCs these days making it an almost universally available method of connection.
4. It is possible to expand the connectivity using a USB hub so that several USB data acquisition devices can be connected.

Many USB devices may be plugged directly to the computer. In some cases, it is required to use the extender cable. However, the length is limited to 5 meters.

USB data acquisition cards (PCB Boards) are another alternative available. These can be connected to virtually any kind of PC effortlessly. This type of data acquisition system can be extended relatively easily by adopting USB hub to enable further units to be connected.

15.7.4 Data Acquisition Plug-in Boards

Computer data acquisition boards plug directly into the computer bus. Advantages of using boards are speed (because they are connected directly to the bus) and cost (because the overhead of packaging and power is provided by the computer). Boards offered are primarily for IBM PC and compatible computers. Features provided by the cards can vary due to number and type of inputs (voltage, thermocouple, and on/off), outputs, speed, and other functions provided. Each board installed in the computer is addressed at a unique Input/ Output map location. The I/O map in the computer provides the address locations the processor uses to gain access to the specific device as required by its program.

15.8 Acquisition Channel Definitions

Data Channel

Data channel is an information route and associated circuitry that is used for the passing of data between systems or parts of systems. In an interface that has a number of parallel channels, the channels are usually separately dedicated to the passing of a single type of information, for example, data or control information.

Measured Data Channel

Measured data channel displays the data acquired and processed from the sensor or transducer.

Calculated Data Channel

Calculated data channel displays the data calculated by program using one or more acquired data from measured data channel and information from parameterization. In the context of dynamometer power is calculated channel. There is general understanding that dynamometer measures the power. In reality, the dynamometer measures the load or the absolute force. With help of lever arm or dynamometer arm on which load cell is mounted, a torque is derived. The physical unit of torque is derived quantity.

Torque = Force (W) × distance from the center of the rotation(R)

$$= W \times R \text{ kgm,}$$

where

W = weight in kg
R = distance in meter

Therefore, power is calculated using relevant formula using appropriate system of measurement such as SI, MKS, and FPS.

15.9 Alarm Annunciation

The alarm annunciator is a device incorporated into the control system which gives audiovisual information about a parameter which has crossed the set/prescribed limit. Alarm takes place when something happens in test cell, such as a particular parameter being measured shows abnormal reading. Annunciator is used to call attention to abnormal process conditions. A limit alarm trip can be a life saver, triggering the response needed to maintain normal, safe operations.

High Alarm

A status change (alarm condition) of a single high alarm occurs when the input rises above the trip point.

High/High Alarm

This alarm accepts one input but has two high relays, each with its own trip point. When the input rises above Trip Point 1 (the lower trip point), the first set of contacts will change status to serve as a warning. Should the input rise above Trip Point 2 (the higher trip point), the second set of contacts change status, which may initiate an emergency shutdown. With four relay outputs, you can provide three levels of warning and then an emergency shutdown.

Low Alarm

A status change (alarm condition) of a single low alarm occurs when the input falls below the trip point. A typical application of a low alarm is warning of a low tank level to avoid problems with a pump running dry.

Low/Low Alarm

A dual low alarm accepts one input but has two relays, each with its own independent trip point. When the input falls below Trip Point 1, the first set of contacts will change status merely to serve as a warning.

Should the input fall below Trip Point 2, the second set of contacts change status, possibly initiating a shutdown of the process. A typical application includes monitoring the low extreme temperature of a cryogenic tank to avoid overcooling.

High/Low Alarm: A dual high/low alarm accepts one input and has two relays, each with a separate trip point.

Alarm Types

Most alarm trips can perform high/low functions. Other types of alarm trips are mentioned below:

Rate-of-Change of Alarm

Used to detect changes in the measured value in units per minute or second, a rate-of-change alarm monitors an input for a change in value with respect to time. The alarm is set to trip when the input rate of change exceeds a user-selected rate (Delta) over a user-selected time period (Delta Time).

Input Fault Alarm

On some alarm trips, you can set one or more of the relays to trip when an input is interrupted such as in the instance of a sensor break.

Self-Diagnostic Alarm

Some limit alarm trips continuously monitor their own status during operation and trip if they are not operating properly.

Average and Differential Alarms

This trip when the average of two or three input signals exceeds a preselected high or low trip point. A differential alarm trips when the difference between two input signals, such as two RTD temperature sensors, exceeds a specific value.

Window Alarm

It's activated when the process variable is outside the low/high trip point ranges.

On/Off Control

A limit alarm trip can also be used as a simple on/off controller such as those required in level applications (pump/valve control) when filling or emptying a container or tank.

Dead Band

The alarm trip fires its relay at the trip point and the relay resets when the process variable reaches the dead band point. Without dead band, if the process variable was hovering and cycling above or below the trip point, the relay would be chattering on and off, leading to premature failure. By setting the dead band just 1 or 2 percent away from the trip point, you can avoid excessive relay wear.

Latching versus Nonlatching

A latching alarm is one where the relay cannot automatically reset. Once the relay trips, it remains in the alarm condition until an operator manually resets the relay (usually through a pushbutton). Latching alarms are most commonly employed when you want to force an operator to acknowledge the alarm condition.

Time Delay

In many applications, a momentary overrange signal may not warrant an alarm trip. Some alarm trips can be set with an alarm response time delay that stops the alarm from going into an alarm condition unless the trip point has been exceeded for a specific time. This can be used to stop false or premature alarms.

Transmitter Excitation

Some limit alarm trips offer the advantage of being able to provide 24 V DC power to a two-wire (loop-powered) transmitter. This saves the cost of specifying and installing an additional instrument power supply.

15.10 Configuration of Alarms as Failsafe and Nonfailsafe

Configuring an alarm trip as either failsafe or nonfailsafe is a primary safety consideration. In a safety application, the foremost concern should be the alarm trip's action in the case of power failure. An alarm trip with a relay that de-energizes if the input signal exceeds the trip point is called failsafe. It's "failsafe" because, even if power to the alarm trip fails, the unit's relay de-energizes as if it were in the alarm condition. Failsafe relay action is chosen for the vast majority of alarming applications.

In a nonfailsafe alarm trip, the unit's relay is de-energized when the input signal is in the normal condition and energized when an alarm occurs. In this configuration, the alarm trip will not provide a warning if there is a power failure. Should a loss of power and an alarm condition coincide, the alarm would go undetected.

The characteristics of failsafe/nonfailsafe and normally open/normally closed relay action can be integrated to provide specific alarming characteristics. To illustrate, consider an application where a light needs to be turned on when a high alarm trip point is reached. If the relay is nonfailsafe, it is de-energized when in normal state, and it is energized when in alarm state. Therefore, when the trip point is exceeded, the relay energizes, and the normally open (NO) side of the contact closes, turning on the light. Note that the light has to be wired to the NO side of the contact so that when the high trip occurs, the relay energizes and the circuit closes.

If the relay is failsafe, by definition it is energized when in normal state and de-energized when in alarm state. When the trip point is exceeded, the relay de-energizes, and the normally closed (NC) side of the contact closes, turning on the light. In this configuration, the light needs to be wired to the NC side of the contact.

15.11 Alarm Trip Choice—Hard or Soft?

When engine is being tested in modern test cell and if lubricating oil signal "falls" or engine water temperature signal "shoots up" then the test cell engineer needs to know these events. A limit alarm trip can trigger the response needed to maintain normal and safe operation. A limit alarm trip monitors a signal such as engine lubricating oil pressure and temperature, engine water temperature, and dynamometer water inlet pressure and compares it against the preset limits. If signal from these parameters crosses the undesirable high or low condition, the alarm activates a relay output to warn of trouble and provide on/off control or institute an emergency shutdown. The limit alarm trips are best known as a sure way to activate a warning flashing light or audible signal such as siren or shut down the system.

Hard Alarms

These types of alarms are hard wired into the system and provide relay outputs. These independent limit alarms are called as hard alarms.

Soft Alarms

These are software independent alarms which are programmed into data acquisition software or programmable logic controller.

Advantages of Hard Alarm

The test cell engineers are using the software-controlled systems; as such they might argue that hard alarms are not necessary. The hard alarm trips complement software set trips alarms. In most of control systems, once an alarm trips, it remains in an alarm condition until the parameter signal re-crosses the trip point and passes out of dead band. The dead band is adjustable that it makes it possible to increase or decrease the range, thus governing what point the relay contact returns to its normal position. Hard wire alarm provides following advantages:

1. Provides the warning signal by providing hard alarm output when parameter under monitoring exceeds its set limit.
2. Provides an independent emergency shutdown system in the event of software failure.
3. Provides redundant warning or shutdown capabilities to back up and compensate for failure of data acquisition system.
4. The simple control systems where data acquisition system and software are absent hard wired alarms are easier to set up and use.
5. Hard wired alarms are used to sense dangerous conditions and shut down the engine and avoid further damage to dynamometer.
6. Hard alarms are good back up or replacement for the soft alarms as soft alarms are intermittent scanning of individual parameters as it is accomplished by the software used in data acquisition and control system. It entirely depends upon the scanning speed of the computer. Each hard wired alarm provides the continuous monitoring/supervision of an individual process signal.
7. Hard wired alarms are easy to setup and eliminate the potential programming errors.

15.12 Types of Relays

The relays used in dynamometer control system are either normally open or normally closed. The word normal in this context means the relay is in de-energized state. When the relay is in the de-energized state, a normally open (denoted by: NO) relay contact does not allow current to flow to the common pole (denoted by C) resulting in open circuit.

Figure 15.6 Single pole single throw (SPST) relay in de-energized state.

When relay is energized, there is closed circuit between NO and the C terminal.

A normally closed (denoted by NC) relay contact allows current to flow to the common ("C") when the relay is in the normal (de-energized) state. When the relay is energized, there is open circuit between the NC and C terminal.

There are three common types of alarm relay configurations:

1. Single pole single throw (SPST)

   In this type of relay, there is only one pole. It allows current to flow when the contact closes. If the relay is normally open (NO), then current only flows when the contact trips, that is, it is energized. If the contact is configured normally closed (NC), current will flow until the alarm trips (energizes). The choice of normally open or normally closed is typically selectable.

2. Single pole double throw (SPDT)

   A SPDT contact has one pole and sends the electrical path in one or two directions. The provision of both NO and NC contacts make SPDT relay to be quickly wired for any application.

Figure 15.7 Single pole double throw relay in de-energized state.

3. Double pole double throw (DPDT)

   This type of relay gives a single alarm trip two separate outputs from one relay. The both contacts on DPDT relay change status simultaneously. A DPDT relay makes it possible for an alarm trip to perform tow simultaneous functions. A DPDT relay is used to annunciate and cause an action to occur such as shutting down the engine and giving the alarm of low lubricating oil pressure.

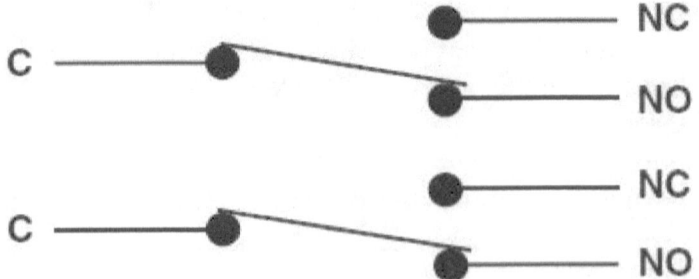

Figure 15.8 Double pole double throw (DPDT) relay in energized state.

Response to Tripping Limits

The response time of the relay is primarily dependant on contact change over time.

15.13 Audiovisual Indications for Tripping

The visual annunciator systems are utilized to display engine and dynamometer parameter status by lighting individual windows of the annunciator, identifying specific parameter or function. The inputs are derived from dry or live contacts to indicate the parameter condition or status. Single point or multipoint alarm logic modules operate the window lights based on a preselected custom sequence.

Alarm Management and Control

ISA—International Society of automation ISA 18.11979 Annunciator sequence and specification defines the following sequences.

First-out Sequence

When a group of alarms is initiated, it is often important to know which of them the first to occur was. This is achieved by having the first-out alarm flashing in a different manner compared to the subsequent alarms.

Auto Reset Sequence

In this mode, the contacts returning to normal on an acknowledged alarm cause that alarm to reset. If the alarm contact returns to normal prior to being acknowledged, the alarm will reset immediately on Acknowledge.

No Lock-in Sequence (Nonlatched)

Alarms will reset immediately when the contacts clear (i.e., the alarm is not latched), although while in the alarm state, the Acknowledge, First-out Reset, and Mute pushbuttons operate normally.

Reflash Sequence

In the standard mode, when an alarm has occurred and has been acknowledged, any further changes of state of the alarm contact will not affect the alarm window display. With the reflash sequence selected whenever the alarm reoccurs on an acknowledged alarm point, the alarm is re-initialized, the window will flash, and the horn will sound just as a new alarm.

Ringback Sequence

This mode is used to indicate to operators that the alarm contact has returned back to its normal state hence avoiding having to continually press the reset pushbutton, to see if the plant contacts have returned to their nonalarm state. When an alarm contact returns to normal, the ringback facility will restart the audible, but at a very much slower rate, as shown in the following sequence tables. The visual indication for the window returning to normal will pulse in sync with the horn so it is noticeably different from a normal alarm.

Manual Reset (Sequence Code M)

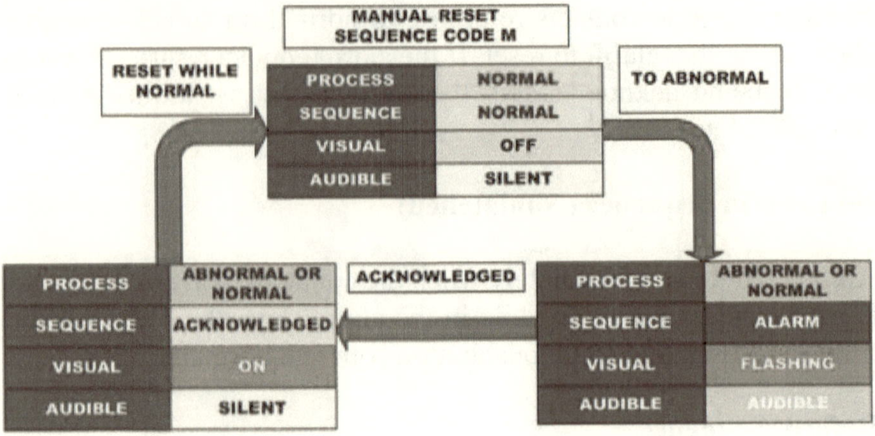

Figure 15.9 Manual rest (sequence code M).

Sequence Description

In this manual mode, momentary alarms are locked in until acknowledged. The audible device is silenced and flashing stops when acknowledged. When process condition comes to normal, then only manual reset of acknowledged alarms is possible.

Automatic Reset (Sequence Code A)

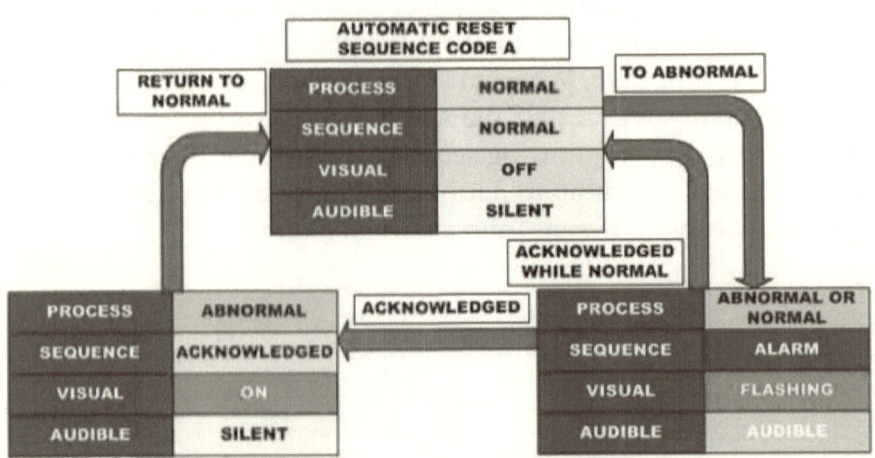

Figure 15.10 Automatic reset (sequence code A).

Sequence Description

In this type of sequence, there is lock-in of momentary alarms until acknowledged. The audible device is silenced and flashing stops when acknowledged. Automatic reset of acknowledged alarms is possible when process conditions return to normal.

No Lock-in

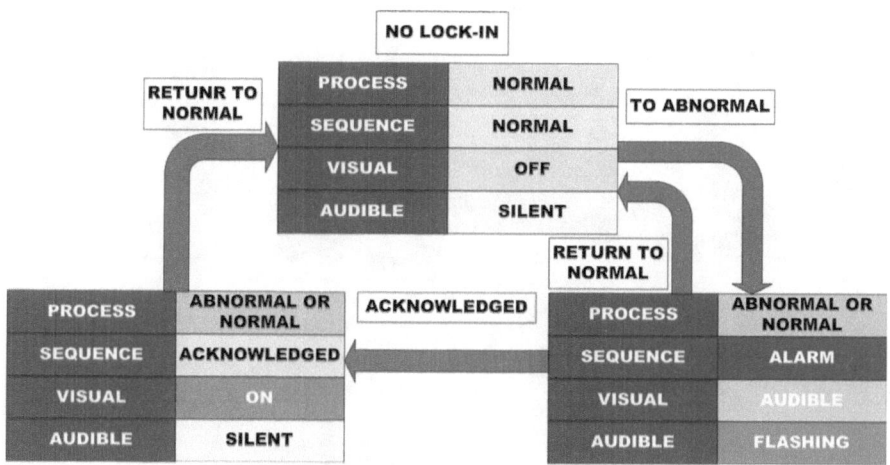

Figure 15.11 No lock-in sequence.

Sequence Description

In this type of sequence, there is no lock-in of momentary alarms. The audible device is silenced and flashing stops when acknowledged. All alarms acknowledged or not return to normal when process conditions return to normal.

Ringback (Sequence Code R)

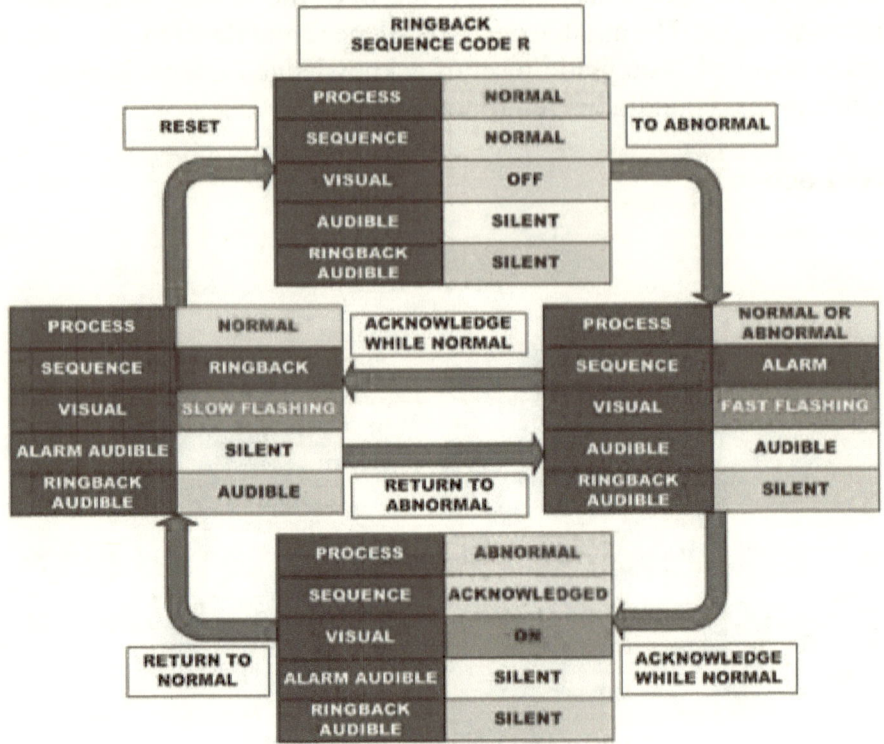

Figure 15.12 Ringback (sequence code R).

In this type of sequence, there is lock-in of momentary first alarm until acknowledged. Ringback visual and audible indicates when process conditions return to normal. It is required to do manual reset of ringback indications.

Automatic Reset First-out (Sequence Code F1A) With No Subsequent Alarm State

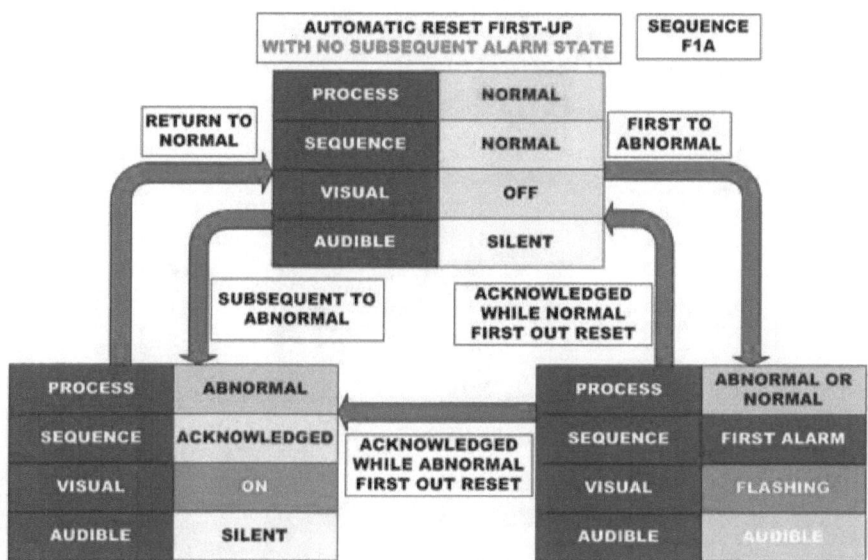

Figure 15.13 Automatic reset first-out (sequence code F1A).

Sequence Description

In this type of sequence, there is lock-in of momentary first alarm until acknowledged. There is no lock-in of momentary subsequent alarms. There is flashing and audible indications for first alarm only. New subsequent alarms go to the acknowledged state. First-out indication is reset and the audible devices silenced when acknowledged. There is "automatic reset" of acknowledged alarms when process conditions return to normal.

Manual Reset First-out (Sequence Code F2M-1) with No Subsequent Alarm Flashing and Silence Pushbutton

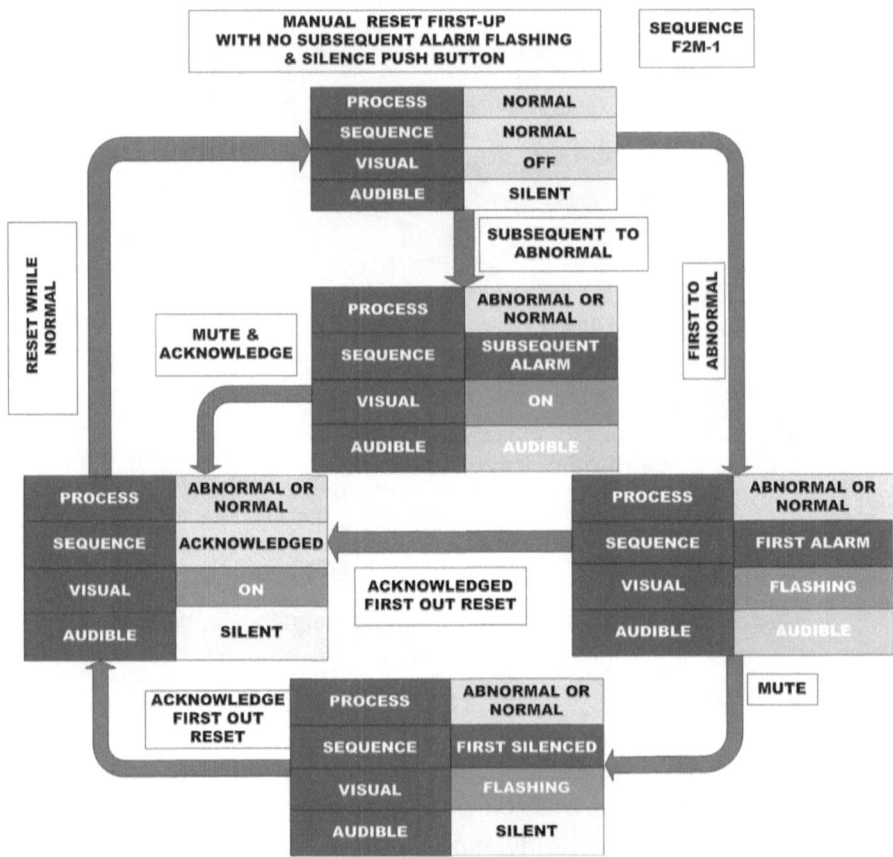

Figure 15.14 Manual reset first-up (sequence code F2M-1).

Sequence Description

When this type of sequence is implemented, then Mute, Acknowledge Reset, and Lamp test pushbuttons are required. Also alarm audible device is required. There is a lock-in of momentary first alarm until acknowledged. Silence pushbutton is used to silence the audible device while retaining first-out flashing indication. There is flashing indication for first alarm only. New subsequent alarms have the same visual as acknowledged alarms. First-out indication is reset and the audible devices silenced when acknowledged. There is manual reset of acknowledged alarms after process conditions return to normal.

Automatic Reset First-out (Sequence Code F3A) with First-out Flashing and "First-out Reset" Pushbutton

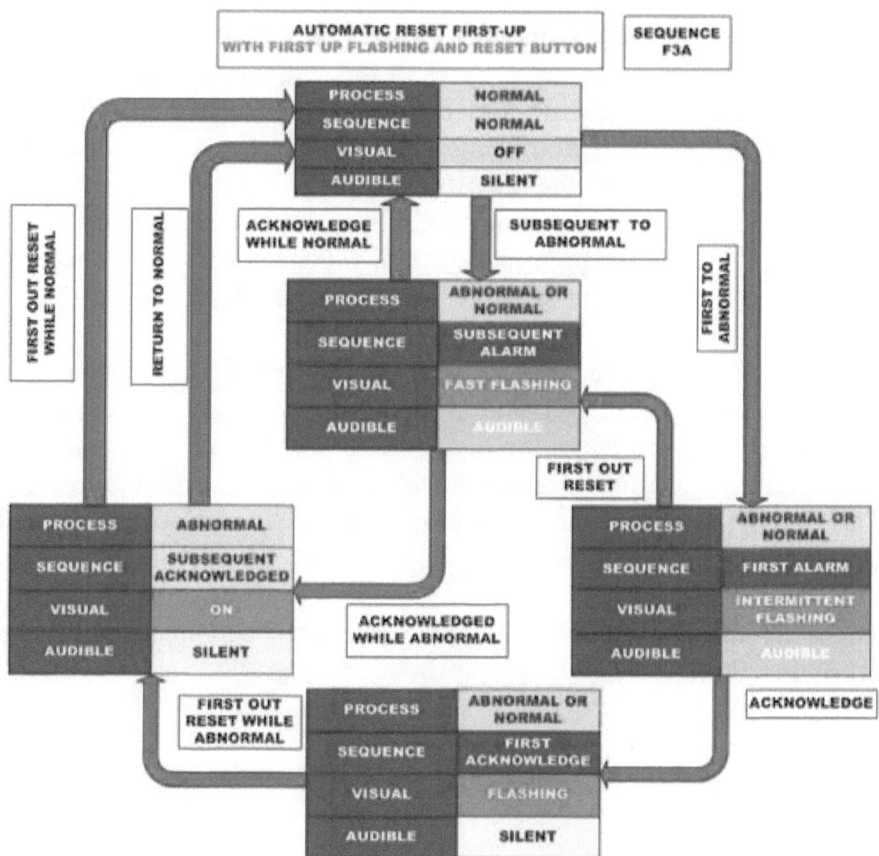

Figure 15.15 Automatic reset first-up (sequence code F3A).

Sequence Description

When this scheme is employed, there is Acknowledge, First-out reset, and Lamp test pushbuttons are provided. Also there is provision of alarm audible device. In this type of sequence, there is lock-in of momentary first alarm until acknowledged and first-out flashing different from subsequent flashing. First-out Reset pushbutton is used to change the first-out visual indication. The automatic reset of acknowledged alarms is possible only when test cell conditions return to normal.

15.14 Engine Data Acquisition and Control System (EDACS)

The engine data acquisition systems for engine and dynamometers consist of basic instrumentation and controls and custom designed software and desktop computer system and interfaces to control various instruments and accessories in the test cell. It integrates data acquisition, engine control, and facility control in a single system. Automatic tests are created using the test sequence editor which has been designed to allow the user to configure simple or sophisticated tests with ease to meet the requirements of the governing regulation. These sequences or programs essentially consist of logical step for automatic running of the engine through predetermined test cycle.

EDACS will facilitate to view live engine and dynamometer data while it is automatically collected for post-processing after the test is completed. Report generation algorithm of the EDACS program can be used to produce multiaxes graphical plots, maps, and tabular results. The four level alarm annunciations on all measured and calculated channels can be configured with the capability of software.

The schematics of the EDACS configured in test cell are shown in the following figures.

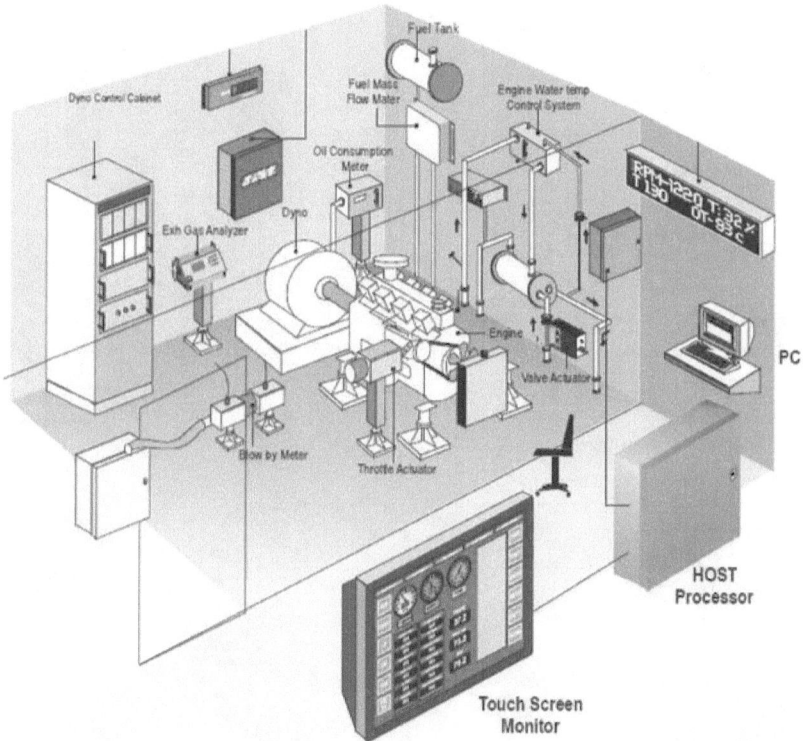

Figure 15.16 Data acquisition and control system for single test cell. Courtesy of Do-India.

Engine data acquisition and control system can control and integrate multiple cells. The configuration of the engine data acquisition system will change depending upon the type of dynamometer and the application. The following figure shows the block schematics of the data acquisition for different dynamometers and application.

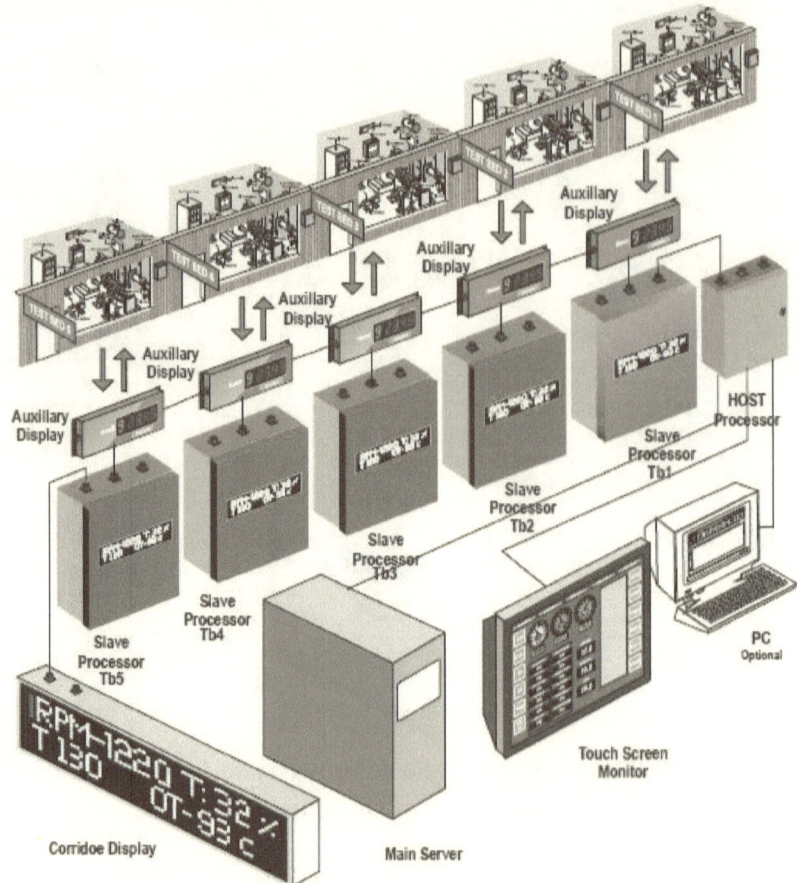

Figure 15.17 Data acquisition and control system for multiple test cell. Courtesy of Do-India.

Data Acquisition system with EC Dynamometer

The data acquisition system interface with the eddy-current dynamometer works in conjunction with eddy current, a controller, and throttle actuator. The following schematics show the various components interfaced in a test cell where EC dynamometer and data acquisition system are used.

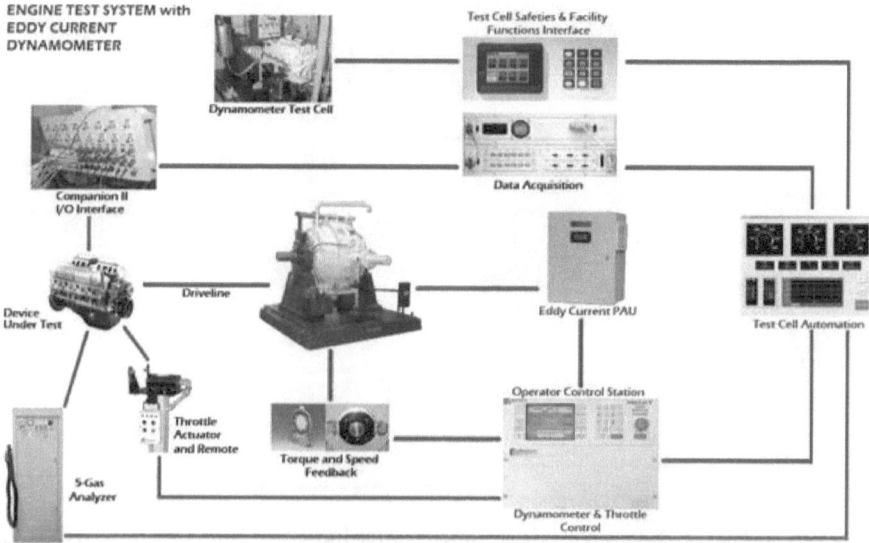

Figure 15.18 Data acquisition and control system eddy-current dynamometer. Image courtesy of Dyne Systems, Inc. and Midwest & Dynamatic Dynamometers, USA Copyright 2010 Dyne Systems, Inc.

Data Acquisition System with AC Dynamometer

The data acquisition system interface with the alternating current (AC) dynamometer works in conjunction with AC driven (AC controller) and throttle actuator. The following schematics show the various components interfaced in a test cell where alternating current (AC) dynamometer and data acquisition system are used.

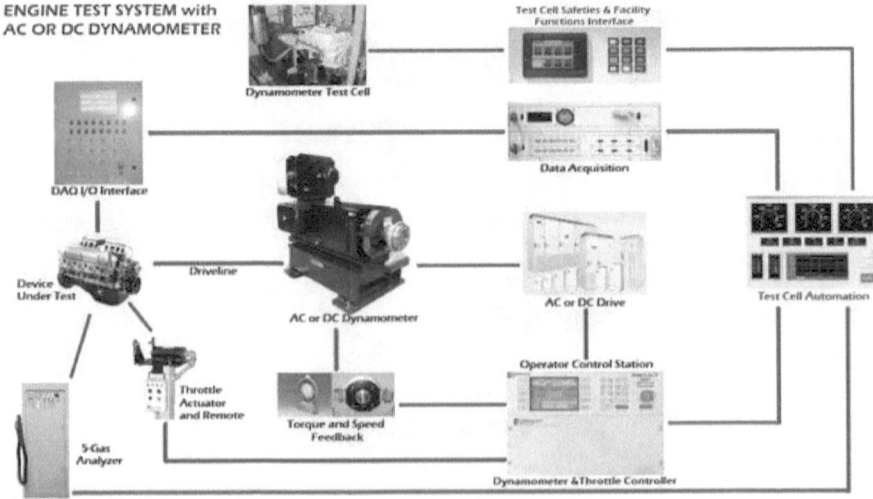

Figure 15.19 Data acquisition and control system alternating current dynamometer. Image courtesy of Dyne Systems, Inc. and Midwest & Dynamatic Dynamometers, USA Copyright 2010 Dyne Systems, Inc.

Data Acquisition System with Transmission Test System

The transmission is a more generalized term. A more specific term will be a gear box of the vehicle. The gear box may be tested independently or coupled with an engine which is being tested. The dynamometer is generally AC or DC dynamometer. The data acquisition system is interfaced with AC or DC drive under consideration. The general schematic is shown in the following figure.

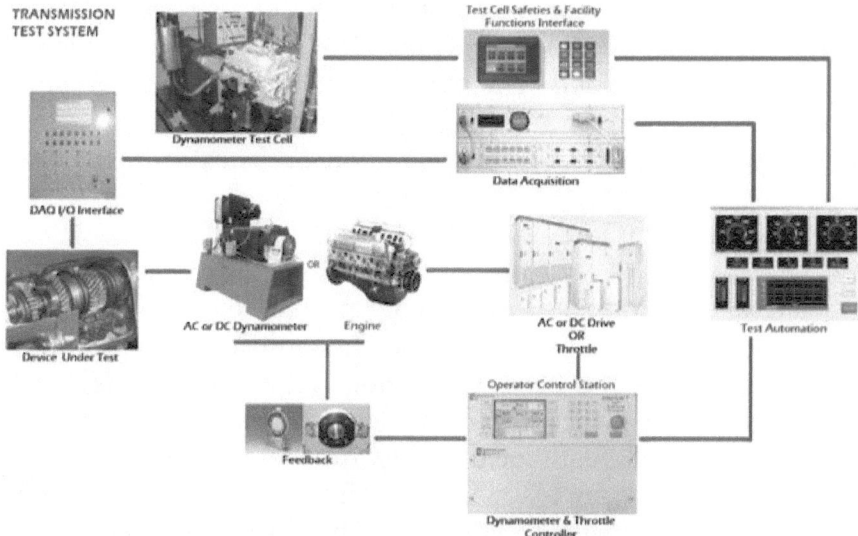

Figure 15.20 Data acquisition and control system transmission test dynamometer. Image courtesy of Dyne Systems, Inc. and Midwest & Dynamatic Dynamometers, USA Copyright 2010 Dyne Systems, Inc.

15.15 Executing the Automated Test as Programmed

The data acquisition system has a module called sequence generator. This sequence generator can be configured by the test engineer to run the desired tests in predefined sequence with predefined measurement of particular parameters such as fuel consumption at particular RPM and blow-by at another RPM. Once the test sequence is configured, the software program will run the test automatically at the will of test engineer.

15.16 Display of Data—Analog and Digital Indicators

During the test, the data acquired may be logged or displayed using analog or digital indicators for the visibility during the test so that data can be monitored.

15.17 Recording and Printing of Data

Generally to gather data for a particular analysis, the engine is run at a multitude of engine operating conditions. For example, the tests are run at incremental RPM points as well as throttle opening points. Sometimes tests are repeated desired number of times for getting repeatable results. The recorded/stored data is available for the printing in desired test format. The desired printing formats are sometimes hardcoded in the software of the data acquisition system.

15.18 Management and Analysis of Data

The data acquisition and data logging provide the data of various types of tests carried out. Once these data are stored, the same can be retrieved for post-processing. Various types of analysis can be performed on these retrieved data. For example, lookup tables in engine controllers are calibrated directly from test data, forcing engine tests to be run at most or all operating points in order to achieve a full calibration. The statistical tools can be used to further analyze the data. The data is often acquired in a structured way. This data can be imported into analysis tool for further processing as desired by the test engineer.

Closure

Data acquisition is a very important area of the test and measurement industry. Data acquisition systems are required in many applications from electronics manufacturing to chemical engineering, mechanical manufacturing as well as more diverse applications such as monitoring geographical data from mountains and volcanoes as well as many other interesting and diverse uses. In the case of engine and vehicle testing, data acquisition is important part of the testing.

The engine data acquisition system plays important role in conducting automated test and logging the data for post-processing. The data acquisition system is essential part of the test cell automation. There are many different measurements that can be made by data acquisition systems. Those mentioned above are just a few of the possibilities, but many more exist. The data acquisition system can be configured to make a custome built system as required by the test engineer.

Bibliography:

1. Test cell pictures from Do-India web site
2. Data acquisition—application notes by Radio electronics.com.
3. Gary Prentice, *Knowing Your Limits: Traversing the Ups and Downs of Alarm Trips*, Application engineer—Moore industries.

Further Reading

1. Jan Axelson, *USB Complete-The Developer's Guide*, (4th) (Edn.).
2. John Park ASD, Steve Mackay, *Practical Data Acquisition for Instrumentation and Control Systems.*

PART III

MODERN TEST CELL CONCEPTS

CHAPTER 16

Engine Test Cell and Its Evolution

16.1 Introduction

As testing of engine started becoming complex, more attention was given to its needs. Engine testing operation from open house became close house. Concept of test house was evolved. In the end of twentieth century, many regulations came into existence to regulate the fuel consumption and the emissions emitted into environment. These regulations demand more sophisticated equipments to be used for the testing of the engine. As engine testing became more complex involving more advance dynamometers, controls, and measuring instruments, a need was aroused to crate dedicated space to accommodate these modern instruments and controls thus giving birth to the concept of the "Test Cell." Testing of engine eventually moved from test house to test cell. There are few oblivious reasons why one should have the test cell.

1. To control noise in your facility. This will protect your employees from hearing damage and allow you to conduct other business (such as phone calls) while running tests.
2. To prevent testing from disturbing neighbors.
3. To provide the appropriate amount of ventilation during the test.
4. To have a controlled test environment that allows you to repeat your test with the same conditions.
5. To provide the safe environment in your test facility.

The test cell development engineer needs to consider designing of individual systems such as water supply to dynamometer, air ventilation, exhaust handling, fuel supply system, fire fighting system, material handling system—hoist/crane, engine docking. *The "Test Cell Design" is subject by itself; however, some important aspects need to be considered by test facility development engineer.*

16.2 Types of Test Cells

There are basically two types of test cell:

1. Modular test cell
2. Custom built

The type of test cell depends upon the individual needs and largely depends on type of tests conducted such as performance, emission, after sales market such as overhauling workshops/repair workshops.

16.3 Noise in Test Cell

One of the important criteria while designing the test cell is to reduce noise level. Sound will leak through the path of least resistance. If you have constructed a test cell of concrete block and have not used a sound rated door or a sound rated window, the sound is probably leaking through these openings.

Hard (nonabsorptive) walls and ceilings cause reverberation. This can cause noise levels inside and outside your test cell to increase approximately to the tune of 10 dBA. This sound is quite loud. You will need to add absorptive panels to the walls or ceiling of your test cell.

If the test cell is constructed with masonry hard surfaced ceilings and walls then it is necessary to add absorption to the walls of the test cell. The test also needs to employ silencers for cell intake and exhaust air to reduce the noise escaping out through test cell ventilation system. The test cell noise is measured in dBA. To harness the sound properly to a designed level, the designer should arrest sound by employing sound proof doors and windows.

Let us see some of the terms used in noise control.

16.3.1 What Is Noise?

Noise is unwanted sound. It does not have to be loud, just unwanted. It can be disturbing. It may interfere with communications or it can be hazardous to your health. Noise is measured in dBA.

16.3.2 What Is dBA?

Sound pressure is measured in decibels (dB). The dBA is a weighted measurement that more accurately reflects the way the human ear hears sound. Humans are less sensitive to low frequencies than high frequencies, so the dBA measurement places a higher importance on high frequencies by "weighing" out low frequencies.

16.3.3 What Is NRC?

Noise reduction coefficient (NRC) is the average sound absorption of the four speech interference frequencies (250 Hertz, 500 Hertz, 1000 Hertz, and 2,000 Hertz). A material with an NRC of 0.95 absorbs approximately 95 percent of the noise that strikes it. That is, it prevents sound from reflecting off it.

16.3.4 What Is Sound Transmission Class?

Sound transmission class (STC) is a single number rating of a material's ability to stop sound from going through it. It is used to rate doors, windows, walls, ceilings, or any other partition between spaces. The higher is the STC rating, the greater is the sound reduction.

Understanding how loud is your engine is most important before tackling the noise control methods. List below shows some of the typical sounds the human ear is exposed in day-to-day life:

1. Jet engine 160 dBA
2. Engine test cell 130-140 dBA
3. Motorcycle test cell 120-130 dBA
4. Human threshold of pain 120 dBA

5. Pneumatic chipper 106 dBA
6. Lawn mower 98 dBA
7. Heavy traffic 90 dBA
8. OSHA eight-hour criteria 90 dBA
9. Typical manufacturing plant 80 dBA
10. Normal speech 70 dBA
11. Typical office area 60 dBA
12. Rustling leaves 20 dBA

The above sound sources are illustrated pictorially below:

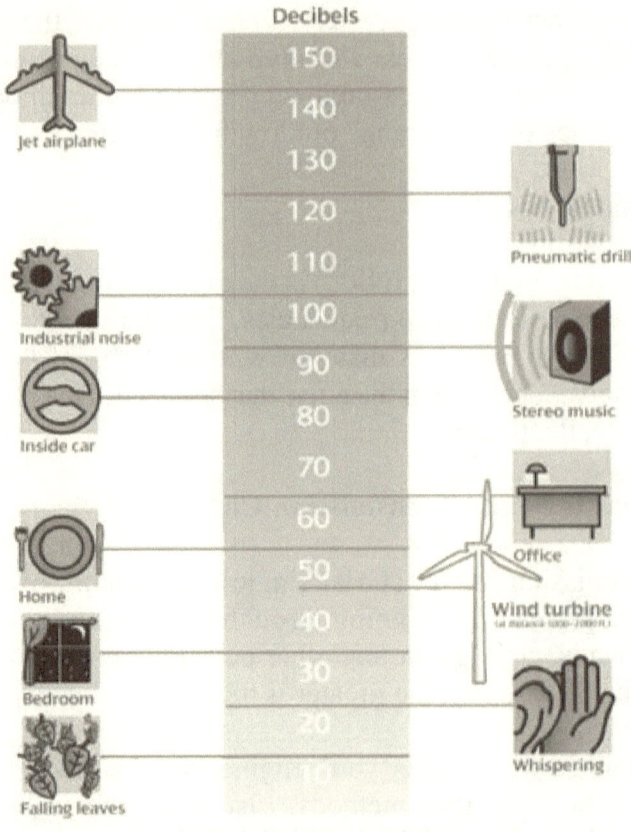

Figure 16.1 Sound level in decibel.
Courtesy of American wind Energy Association.

16.3.5 How to Interpret the dBA?

Every decrease of 10 dBA sounds like the noise is ½ as loud as before. If you reduce the noise by proper sound proofing by say 40 dBA, then the decreased sound sounds like the noise has been cut in half four times.

16.4 Test Cell Ventilation

Test cell ventilation system designed for the test cell should be capable of maintaining desired temperature and pressure in the test cell and provide the combustion air. The cooling air is the amount of air flow required to cool the test cell and remove the radiant heat from the engine.

The combustion air is the air flow required for the combustion of the engine. If separate air intake conditioning equipment is installed, then the supply of combustion air will be from intake air conditioning supply and it will not be a part of the ventilation equipment.

The test cell ventilation system typically comprises of:

1. The main blower or the fan
2. Hoods and dampers and air ducts
3. Control system to control

    - Standard operation of circulating the air in test cell.
    - Purge mode—Used to purge or vent the cell smoke and fumes in the event of engine failure.
    - Ventilation ducts

The ventilation system size is normally governed by the maximum size/capacity of the engine to be tested in the test cell and test cell volume. The design of ventilation system considers the following:

1. Determining the capacity or sizing of the total air flow which is sum total of:

- Cooling air

The cooling air is the air required to take away the heat radiated by the engine in the test cell. The test cell is subject to radiation heat from other equipments such as fans, lighting, and blowers which in turn increase the overall temperature of the test cell. The ventilation system should be capable of maintaining the test cell temperatures demanded by the test procedure under consideration. Test cells dedicated to testing petrol/gasoline engines and carrying out extended full power tests will experience a red-hot exhaust manifold which needs to have a cold air blast on the engine exhaust manifold to avoid the dangers arising due to manifold becoming red hot. The application of insulation on exhaust pipes, silencers, and jacket water pipes will reduce the amount of heat radiated by auxiliary sources into the test cell.

- Combustion air

The combustion air requirement is an essential parameter for the design of test cell ventilation system. The test cell designer should collect the air consumption of those engines which will be tested in the test cell. Many engine manufacturers specify a thumb rule for diesel engine which states as "0.1 m³ of air/min/brake kW (2.5 ft³ of air/min/bhp." Generally the air consumption or the combustion air requirement is specified at reference conditions by the engine manufactures. To convert both mass airflow from reference conditions to site conditions, use the following correction formulae:

$$M_{Test\ cell} = M_R \left(\frac{T_{Test\ cell}}{T_R} \right), \quad \ldots\ldots\ldots\ldots\ldots 16.1$$

where

$M_{Test\ cell}$ = mass flow rate at test cell conditions

M_R = mass flow rate at reference or standard conditions

$T_{Test\ cell}$ = test cell ambient temperature

T_R = air temperature at reference or standard conditions

The requirement of combustion air may be computed in volumetric units as well.

$$Mass = \frac{Volume}{Density}, \quad\quad\quad\quad\quad\quad\quad\quad\quad\quad 16.2$$

- Ventilation air

The ventilation air flow requirement is based on desired test cell temperature to be maintained. The total ventilation air required is sum of cooling air and combustion air. The cooling air requirement can be calculated from the basic formula from thermodynamics:

$$H = V.\rho.C_p.\Delta T, \quad\quad\quad\quad\quad\quad\quad\quad\quad\quad 16.3$$

where

H = heat radiation from engine and auxiliary equipment in test cell (kW)
V = ventilation air required to remove the heat. (M³/min) or BTU.
P = density of air at test cell air temperature.
C = specific heat of air (0.017 kW × min/kg × °C), (0.24 Btu/LBS/°F).
ΔT = permissible temperature rise in engine room (°C), (°F).

2. Determining the capacity of fan/blower:

The selection of the ventilation fan/blower is based on the following factors:

- Volume of the ventilation air.
- Ventilation air pressure required.
- Space limitation or physical size of the fan that can be accommodated in given space.

Ventilation blower types are:

- Axial fan/blower.
- Centrifugal fan/blower.
- Propeller type fan/blower.

The ventilation system design normally employs tow fans for supply and exhaust which gives optimum control on air distribution in the test cell. The fan sizing is based on the understanding of fan performance characteristics and ventilation system design parameters. The main governing parameters are volume and pressure requirements of the ventilation system for the test cell. The exact determination of volume and pressure will help in choosing optimum fan size for the test cell.

3. Ventilation duct routing:

It is important to see that ventilation system will not create negative pressure in the test cell. Normally maximum pressure differential from barometric air pressure approximately 30 Pa is maintained. The ventilation system ducting should follow guidelines mentioned below:

1. Incoming air inlets should be located as far from the sources of heat as practical and as low as possible inside of the test cell.
2. Ventilation air should be exhausted from the engine room at the highest point possible, preferably directly over the engine. A fume hood should be installed above the engine which will assist in removing fumes, vapors, and heat. Hood should be connected to the ducting which leads to the outside of test cell open to atmosphere.
3. Ventilation air inlets and outlets should be positioned to prevent exhaust air from being drawn into the ventilation inlets. The recirculation of the air should be avoided.
4. Ventilation air inlets and outlets should be positioned to prevent pockets of stagnant or recalculating air, especially in the vicinity of the engine air inlet.

5. Where possible, individual exhaust suction points should be located directly above the primary heat sources. This will remove heat before it has a chance to mix with engine room air and raise the average temperature. It must be noted that this practice will also require that ventilation supply air be properly distributed around the primary heat sources.
6. Avoid ventilation air supply ducts that blow cool air directly toward hot engine components. This mixes.

Schematics of Ventilation System 1

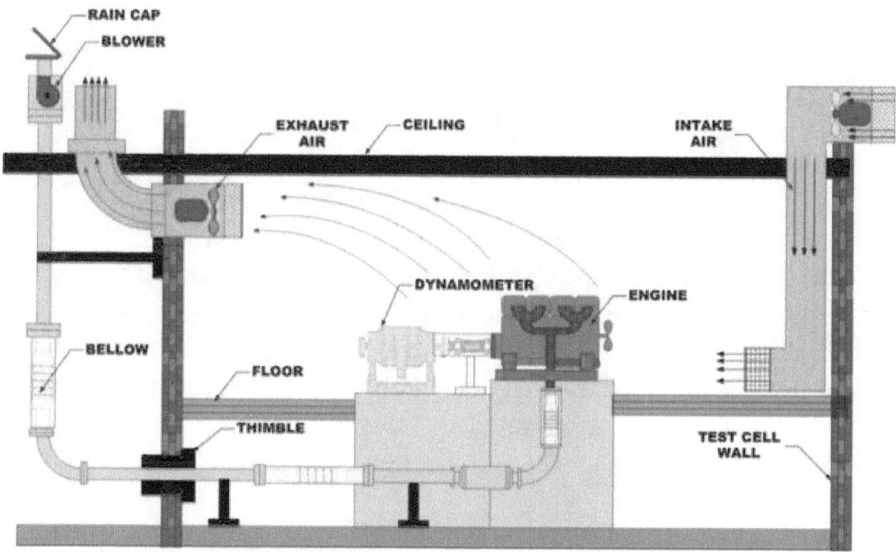

Figure 16.2 Schematics of ventilation system 1.

In ventilation system 1, the outside air is brought through the ducting inside the test cell. The ducting opens at almost floor level. The discharge of air is at bottom level of the dynamometer and the engine bed. The exhaust fan is mounted at the highest point in the test cell and should be above the heat source.

Schematics of Ventilation System 2

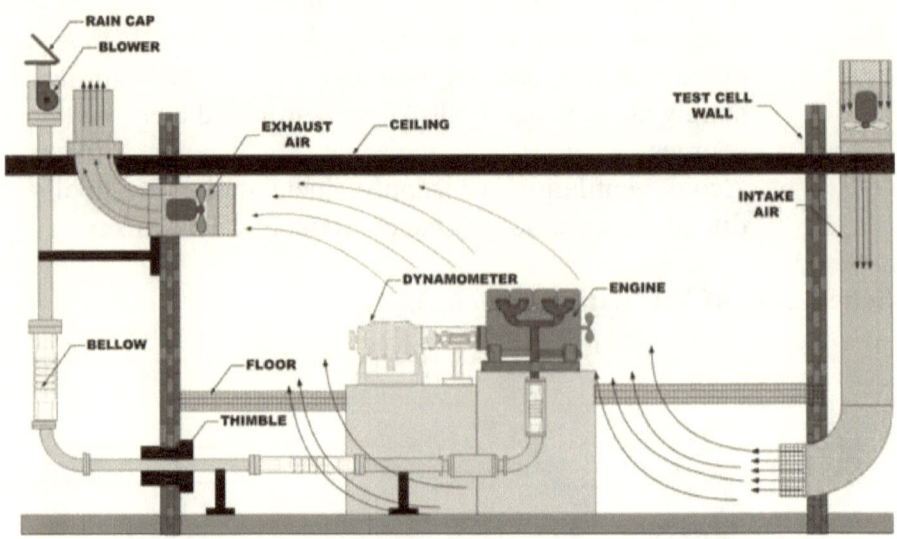

Figure 16.3 Schematics of ventilation system 2.

The air is brought into the test cell through the ventilation ducting and discharged below the floor and directed upward toward the engine and dynamometer through the perforated grating on floor. Regular inspection is required to inspect whether the perforated grating is clean and air flow is unrestricted.

Schematics of Ventilation System 3

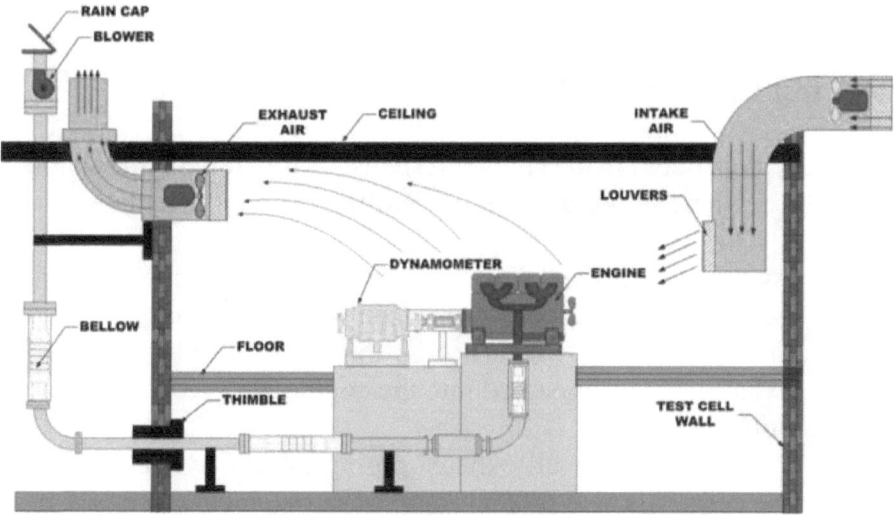

Figure 16.4 Schematics of ventilation system 3.

If ventilation scheme described in preceding paragraphs is not feasible, then the schematics of ventilation system 3 should be used. The ventilation air is discharged into the test cell as low as possible.

Schematics of Ventilation System 4

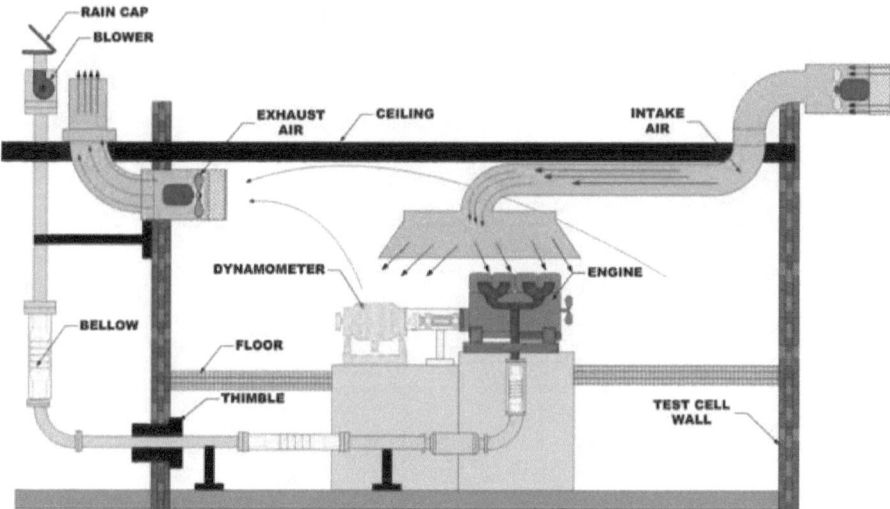

Figure 16.5 Schematics of ventilation system 4.

The ventilation system shown in above figure is less efficient than other types described earlier. The ventilation scheme has routing factor of 2.5.

16.5 Test Cell—Engine Exhaust Handling

This is dealt in detail in the chapter Exhaust Extraction. In this chapter, we will see some important tips in handling the engine exhaust.

1. Avoid exhaust leaks in test cell. Besides reducing engine performance, carbon monoxide (Co) exhaust is deadly to humans. It is worth noting that even a small volume of exhaust fumes, breathed into the engine, will reduce its power output.
2. Provide double-wall exhaust piping to reduce heat load into the cell. This also protects against burns, melted wires, etc.
3. Failure to provide adequate exhaust system cooling to the underside of the vehicle/engine during extended trails on dynamometer testing can lead to a vehicle fire!
4. As far as possible use flexible stainless steel bellows to connect exhaust manifold to the piping extracting exhaust outside the test cell.
5. Use mufflers to damp the sound.
6. Provide suitable tapings to exhaust pipe to facilitate measurement of exhaust gas emissions.

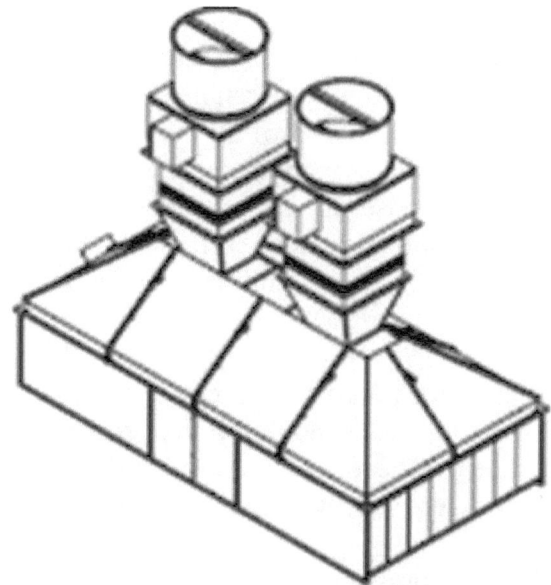

Figure 16.6 Double fan canopy. Courtesy of Taylor Dynamometer.

16.6 Water Supply Systems

A successful water system will supply the correct quantity of water of adequate quality and temperature for each application.

Hydraulic or the eddy-current dynamometer needs cooling water supply. In case of hydraulic dynamometer, water plays dual role: 1. Power absorption and 2. Cooling of dynamometer.

The dynamometer's ability to dissipate heat is a function of how long a load will be applied. Therefore, the maximum power ratings given are based on continuous operation under load, as well as a maximum of 5 minutes over load.

We have seen in chapter 8 Economics of test cell about water supply requirements for hydraulic dynamometer. While designing test cell, one need to take into account the following:

Influence of water system on working of the hydraulic dynamometer

The water supply system should supply the water under steady head, that is, constant pressure. The common header supplying the water to the test cells in test house should be designed in such way that it will maintain the required pressure even if all the test cells are operating those housing the hydraulic dynamometer. Let us say two test cells are operating in a test house drawing water from common header which may maintain the pressure head. When all other remaining test cells start operating, the header should be capable of supplying the required quantity of water without pressure drop.

The fluctuations in water pressure will create the load stabilization problem. Load variation is dependent on the pressure fluctuation in the water supply. The overhead tank is considered as one good source of constant water pressure supply to the dynamometer.

The fluctuation in water pressure is sometimes annoyance and setting of P.I.D. becomes cumbersome task as pressure fluctuation affect dynamometer load. A water pressure regulator and/or an accumulator will correct most problems with inconsistent water pressure.

A water accumulator located near the test cell dampens the pressure fluctuations of the water and eliminates the pressure spikes associated with water system. However, this may add to the cost of the installation.

A dynamometer which is starving for the water quantity may tend to increase the water temperature and water vaporization inside the dynamometer, and this will cause dynamometer control problems. It is normal practice not to exceed the dynamometer water outlet temperature to 60°C.

Water pressure fluctuation does not pose a problem if the power absorption unit is eddy-current dynamometer and need to have pressure regulator and the accumulator is eliminated. However, the water supply system remains the same if the eddy-current dynamometer is of the same size to that of hydraulic dynamometer.

16.6.1 pH Value of Water

The pH value of neutral water is 7.07. The pH value of acidic water is less than 7.07 and alkaline water is greater than 7.07. The soft water is acidic and tends to corrode the parts of dynamometer, whereas alkaline water is hard and tends to form the scale inside the dynamometer.

Water with pH value of between 8 and 8.5 is acceptable. Water of high acidity tends to promote electrolytic erosion. Theoretically, water of high alkalinity is preferable, as this would promote the formation of a protective hard scale deposit on power absorption elements such as stator and rotor, but in practice it has been observed that this advantage is nullified by buildup of scale on the surface of control system parts such as butterfly valve.

Traditionally, magnesium plugs are fitted where dissimilar materials are connected in common water path.

16.6.2 Purity of Water

A suitable water filter should be installed in the supply line to filter the impurities in supply. The water supplied should be free from the impurities such as sand and dirt which has the tendency to erode the power absorption elements such as stator and rotor of dynamometer.

16.6.3 Water Quantity

The quantity of water to be circulated depends upon amount of heat to be removed from the dynamometer. As we have seen in chapter 01 that

$1 \text{ KW} = 1000$ watts/sec.
$1 \text{ watt} = 1 \text{ Nm/sec} = 1$ joule.

Therefore, 1 Kw/hr = 1000 watts × 3600 secs = 3600000 joules (watts/hr) of heat is generated per kilowatt per hour.

In FPS system, 2545 Btu/ bhp/hr.
In MKS system, 633 kcal/cv/hr.

All of this heat generated due to power absorption is carried away by circulating water. Quantity of water supplied to dynamometer must normally be sufficient to prevent the outlet temperature rising above 60°C (140°F). The lower temperature than 60°C reduces the rate of scale deposition on the heated surfaces.

Therefore using **SI** system,

$$Q = \frac{3600000}{w.C_p.(t_2 - t_1)} \quad \text{..................16.4}$$

where

Q = quantity of water required

w = mass of 1 kg water in kg (1 kg)

C_p = coefficient of heat (4200 joules/kg)

t_1 = inlet temperature of water to dynamometer, say 38.5°C

t_2 = outlet temperature of water from dynamometer = 60°C

Now solving above equation, we get,

$$Q = \frac{3600000}{w.C_p.(t_2 - t_1)}$$

$$Q = \frac{3600000}{1 \times 4200(60 - 38)}.$$

$$Q = 39.86 \; ltr/Kw.hr.$$

Now using **MKS** system,

1 HP absorbed per hour produces 633 kcal, therefore,

$$Q = \frac{633}{w \cdot C_p \cdot (t_2 - t_1)},$$

where

Q = quantity of water required

w = mass of 1 kg water in kg (1 kg)

C_p = coefficient of heat (1 kcal/kg)

t_1 = inlet temperature of water to dynamometer say 38.5°C

t_2 = outlet temperature of water from dynamometer = 60°C

Now solving above equation, we get,

$$Q = \frac{633}{1 \times 1 (60 - 38)},$$

$$Q = 29.44 \; ltr/Kw \cdot hr.$$

Now using **FPS** system,

1 HP absorbed per hour produces 2545 Btu, therefore,

$$Q = \frac{2545}{w \cdot C_p \cdot (t_2 - t_1)},$$

where

Q = quantity of water required

w = mass of 1 gallon water in pounds (10 lb)

C_p = coefficient of heat (1 Btu/lb)

t_1 = inlet temperature of water to dynamometer say 100°F

t_2 = outlet temperature of water from dynamometer = 140°F

Now solving above equation, we get,

$$Q = \frac{2545}{1 x 1 (140 - 100)}.$$

$$Q = 4.24 \; gallons/bhp.hr.$$

The following figure shows typical schematics of water supply system to test cell.

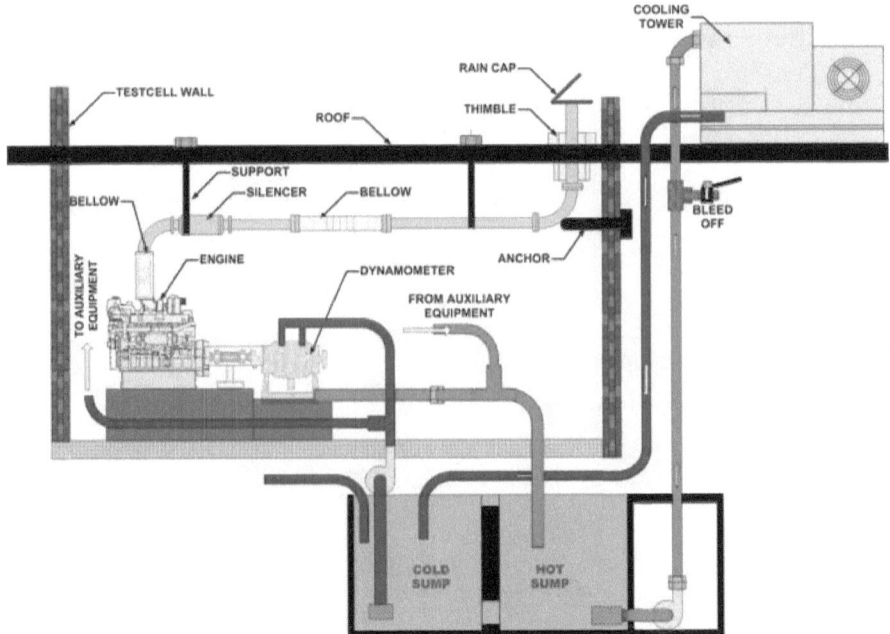

Figure 16.7 Test cell cooling water schematics.

16.6.4 Cooling Towers

All cooling towers operate on the principle of removing heat from water by evaporating a small portion of the water that is recalculated through the unit. The heat that is removed is called the latent heat of vaporization. Each one pound of water that is evaporated removes approximately 1,000 BTUs in the form of latent heat.

Following factors should be considered while designing the water piping system:

1. The water distribution system is composed of the pipes, valves, holding tanks, and pumps, which supplies water to its end use points at the required flow rate and pressure. Water connections should be located to limit number bends and joints. Pipes must be sized, so pressure drops due to frictional are not excessive.
2. When the water source and pressure tank cannot deliver the required flow rate, an intermediate storage and two-pump system can be used. Intermediate storage also facilitates water reuse, which can significantly reduce total water quantity requirements. This pump fills an intermediate storage with water for peak use periods. A second pump draws the water from the intermediate storage and forces it into a pressure tank. Size the second pump to provide the peak flow rate. Intermediate storage can be plastic, concrete, or steel tanks. Protect the storage from contamination. The intermediate storage may be elevated to avoid the use of pressure tank and second pump.
3. The types of pipe generally used in water systems are galvanized steel, copper, and plastic. Galvanized steel pipe is suitable for all piping inside buildings. Plastic and copper are preferred for underground installations. Highly mineralized water greatly reduces the life of steel pipe.
4. Before beginning layout and sizing of the piping system, prepare a schematics of cooling water system using Figure 16.7. This diagram can be used to explore options for reuse and to develop a layout of the piping system that minimizes the length of pipe runs and number of fittings. The peak flow rates for each section are used for pipe sizing.
5. General design rules of thumb for pipe sizing are that friction loss should not exceed 5 psi from the pressure tank to the service entry and should not exceed 10 psi from the pressure

tank to any isolated fixture. This criterion is generally met if peak flow water velocities are limited to 4 feet per second.
6. The pipes in the distribution system should meet or exceed the minimum requirement of the national plumbing codes and local codes.

16.7 Compressed Air System

Compressed air needs are defined by the air quality, quantity, and level of pressure required by the end uses in your engine test cell. Analyzing needs carefully will ensure that a compressed air system is configured properly. The test cell will need compressed air supply for the various purposes as discussed below:

1. Dynamometer

   - If a dynamometer is hydraulic unidirectional and using a pneumatically operated turn table to make the use of dynamometer from other end of driveshaft.
   - If dynamometer is high speed dynamometer and has oil mist (micro fog) lubricated bearings.
   - If pneumatically operated shutdown actuator is used in the system or throttle actuator units fitted with override units.

2. Engine stand

   - If trolley docking is pneumatic.
   - Cardan shaft height adjustment jack is pneumatically operated.
   - Pneumatic engine starter.

16.7.1 Air Quality

The compressed air quality ranges from plant air to breathing air. Test cell needs following air quality only.

Table 16.1 Air Quality

Air quality	Application
Plant air	Actuators for load throw off, pneumatic cylinders for docking, shutdown actuators. Test cell doors
Instrument air	Micro fog/oil mist lubrication

Normally factory supply of air is enough for most of the applications listed above in test cell. Normally a F. R. L. (Filter-Regulator Lubricator) unit is provided at the inlet of each equipment, which is pneumatically operated. The following figure indicates a typical layout of compressor and air piping.

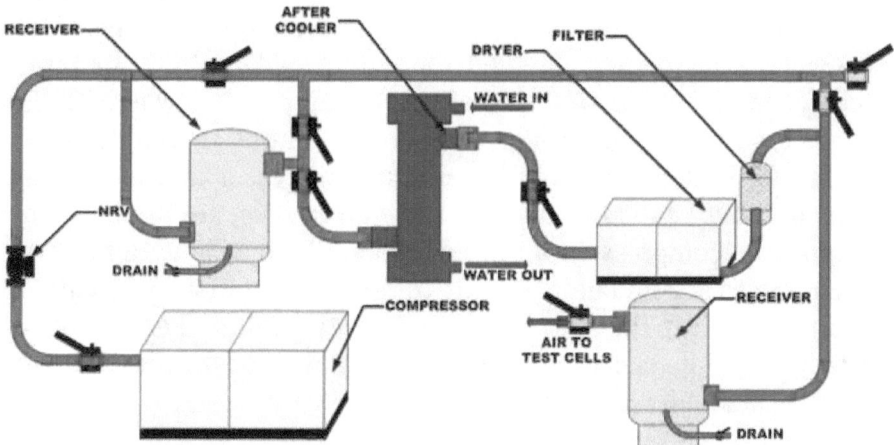

Figure 16.8 Schematics of air supply to test cells.

Quality is determined by the dryness and contaminant level required by the end uses and is accomplished with filtering and drying equipment. The higher the quality, the more the air costs to produce. Higher quality air usually requires additional equipment, which not only increases initial capital investment, but also makes the overall system more expensive to operate in terms of energy consumption and maintenance costs.

16.7.2 Air Quantity

The quantity of air required will depend upon the individual equipment needs. If there are many pneumatically operated equipments in each test cell and test house has battery of test cells, then a separate compressor installation dedicated to the test house may be worth considering.

The test cell design engineer needs to consider the above facts for planning of source of compressed air.

16.7.3 Air Pressure

The industrial compressors are normally rated for 7 kg/cm^2 or 100 psi pressure range. Majority of pneumatic equipments used in test cell are covered by this pressure range. The pressure criteria plays major role when a test cell is devoted for testing of large diesel engines used for heavy generating sets or marine application where engine is started with the help of high pressure air. In this case, a dedicated air starting system with high pressure compressor and air storage tanks, normally called as high pressure air bottles, are used. The air compressors compress the air in stages and fill up the huge air bottles, which acts as accumulator. The air compressors compress usually up to 35 bar and keep the air bottles filled up all the time. The number of air bottles and its capacity depend on the power of the diesel engine under consideration.

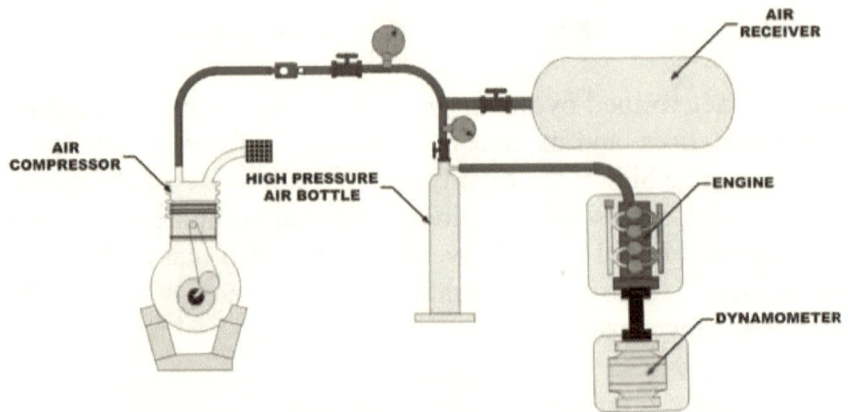

Figure 16.9 Schematics of air starting arrangement.

The smaller engines are equipped with air starting motor which engages its pinion to the flywheel mounted starter gear ring. The motor is powered by high pressure compressed air flow.

16.8 Fuel System

The fuel oil system is the heart of engine test cell. System design depends on type of fuel required by the engine under test and balance of the following points:

- Reduced space and weight
- Increased reliability
- Lower wear part
- Good cold starting capability
- Increased safety in fuel handling and safety

While designing the fuel system, test engineer must also consider the following points:

1. Cetane number—it is the measure of ignition quality of fuel. It affects the cold starting warm-up, combustion roughness, and exhaust smoke density.
2. Viscosity—this affects the fuel flow through the pipes to the engine.
3. Flash point—safety and precaution while handling and storage.
4. Pour point—lowest temperature requirement for easy flow and depends on certain number.
5. Maintenance consideration.

Fuel journey starts from the main storage tank for the premises to the overhead tank of the test cell to the daily tank to the fuel conditioning to fuel measuring system of the test cell to the engine. Typical layout is shown in the following diagram.

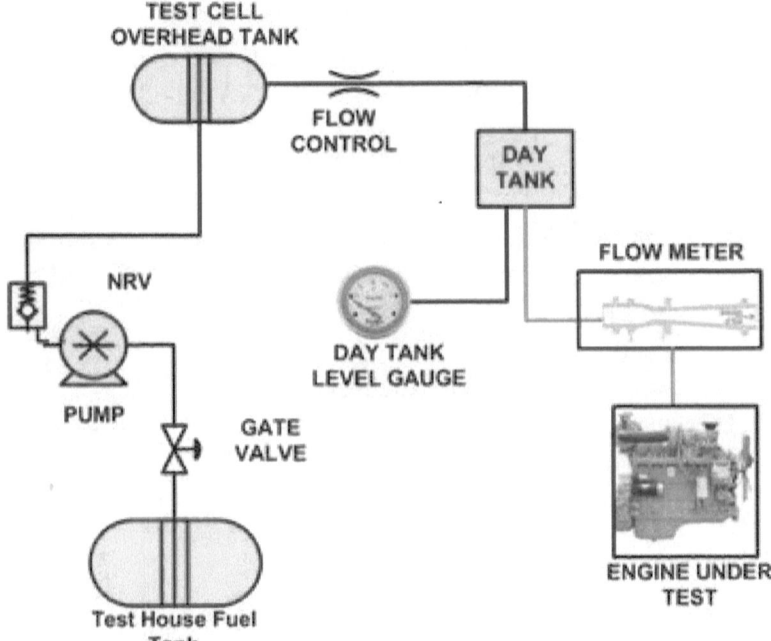

Figure 16.10 Fuel supply schematic.

16.9 Engine Oil Supply System

The engines coming to test cell are normally prepared for testing and as such they will be prefilled with oil. However, if the engines are required to be filled with oil at test stand, then this requirement must be taken care of while designing the test cell. A similar arrangement to that of fuel oil can be installed to supply oil.

Following should be considered while designing the oil supply system:

1. Provision for oil metering should be incorporated so that accurate filling can be done.
2. Proper oil drain should be arranged and care should be taken that oil is not mixed or contaminated with water drain.
3. Provision should be made to recycle the drained/used oil.

4. Oil cooling arrangements should be made for prolonged testing of engine on test bed. If required, oil temperature controller should be installed. However, care should be taken that oil is not cooled excessively. Excessive cooling of oil other than designed temperature will make the test bed conditions favorable to the engine than it will experience on the road in reality.

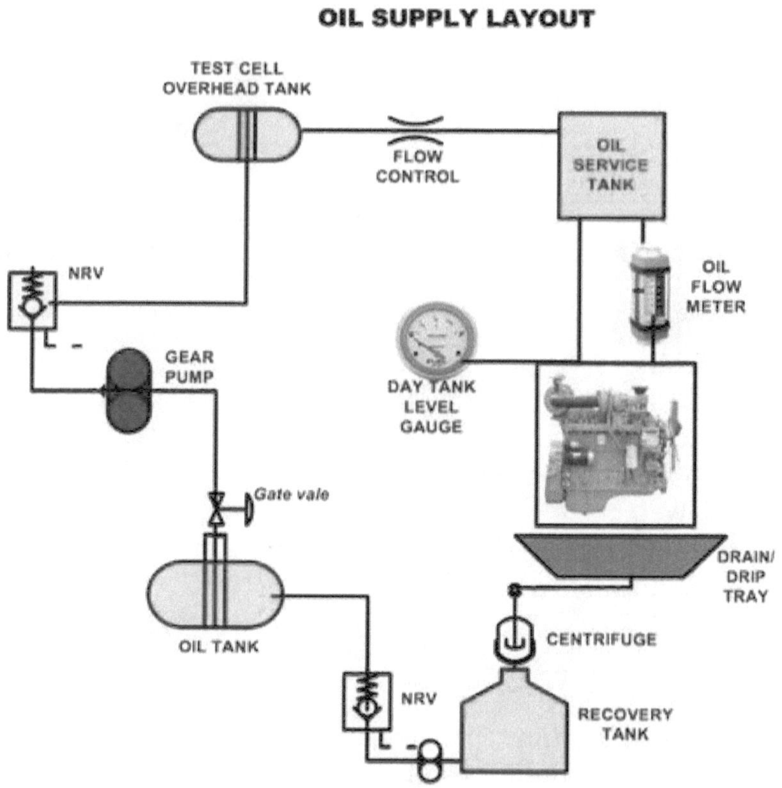

OIL SUPPLY LAYOUT

Figure 16.11 Oil supply schematic.

16.10 Engine and Dynamometer Foundation

16.10.1 Foundation

The function of a foundation is not only to support the weight of the machine/equipment, but also to keep the vibration levels and dynamic displacement of the isolation system within acceptable limits.

Designing foundations supporting diesel engines or dynamometer that can produce static and dynamic loads require sound engineering procedures for a reliable result. An incorrectly designed foundation is extremely difficult to correct once installed.

Following engineering disciplines are required to be taken into consideration while designing a foundation for dynamometer and engine.

1. Theory of vibrations.
2. Soil mechanics.
3. Structural analysis.
4. Dynamic analysis.

The requirements for the designing work of foundation for the dynamometer and the I.C engine are classified into three main groups:

1. Machine (dynamometer and engine) properties

    - Unbalanced forces
    - Operating speed
    - Weight
    - Center of gravity of dynamometer and engine
    - Allowable deflection

2. Soil parameters

    - Load bearing capacity
    - Shear modulus of soil
    - Density of soil
    - Composition of soil at various depths

3. Environmental requirements

    - Degree of isolation required
    - Isolation required at what frequencies.

The traditional rules observed in the past of making the foundation 3 to 5 or even 10 to 12 times the weight of the equipment/machine it supports are applicable only when the foundation will be isolated by the soil and where the soil dynamic properties are known.

The calculations for the stiffness of a foundation yield the static and dynamic behavior and stress concentration points that occur. Stresses are related to the geometry of the foundation and the distribution of loads and forces acting upon it.

16.10.2 Resonance

During startup or shutdown of a machine, a temporary resonance condition may be tolerated, where the support structure or even the vibration isolators are in resonance with the machine's operating frequency, especially if significant damping is available. For the purpose of the dynamic analysis data on the operating speed and forces generated by engines or the measured vibration amplitudes and frequencies at which they occur for engines are required.

16.10.3 Concrete

An important part of a foundation's structure and stiffness is the specified concrete strength used in the design. A specified concrete strength is easy to obtain and is often used as the only criteria. However, shrinkage control can be one of the most important factors in providing a successful project. The following are major factors controlling shrinkage:

1) Water/cement ratio (slump) of delivered concrete
2) Aggregate proportioning and size
3) Water reducing additives
4) Site conditions, such as hot, dry climate
5) Curing
6) Control joints and reinforcing

Each of these six factors needs consideration. Slump is controlled by controlling the total water per cubic yard of concrete, while strength is governed by the thickness or consistency. This thickness is determined by the ratio of the weight of water to the weight of cement.

Shrinkage is simply the reduction in volume that takes place when the concrete dries from its original wet condition down to a point where its moisture condition reaches equilibrium with the humidity in the air. Unrestrained shrinkage does not develop cracks.

When designed and cured properly, large foundations result in very low concrete shrinkage while in a controlled environment.

16.10.4 Foundation Isolation

The purpose of isolation is to control unwanted vibration so that its adverse effects are kept within acceptable limits. In other words, if the equipment requiring isolation is the recipient of unwanted vibration, the purpose of isolation is to reduce the vibration transmitted from the support structure to the recipient to maintain performance. Operating frequencies of rotating/reciprocating machines often are very close to the natural frequency of their support structure.

While installing a dynamometer and the engine stand on supporting concrete block foundation that rests directly on soil as means of providing isolation, the soil conditions must be taken into account. It is important to make note that any static and dynamic forces exerted on the foundation also are exerted on the soil, and the load bearing capacity of the soil is a key factor in determining the size of the foundation.

16.10.5 Foundation Bolts

Foundation bolts or holding down bolts are available in various sizes. The correct size of the holding down bolts should be used specified for the dynamometer.

16.10.6 Holding Down Bolts

The following figure shows typical holding down bolt. A variety of foundation blots are available and should be selected as per recommendation of the foundation expert. The following figure shows the foundation bolt with leveling wedges used to level the baseplate.

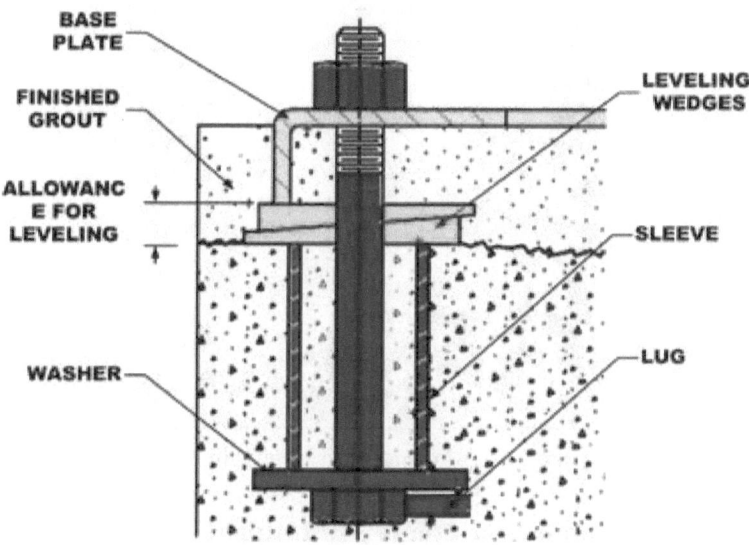

Figure 16.12 Holding down bolts.

16.10.7 Typical Foundation Hole

The foundation hole schematic is shown in the following figure. The top of hole opening is flared to enable the grouting to be poured in the hole.

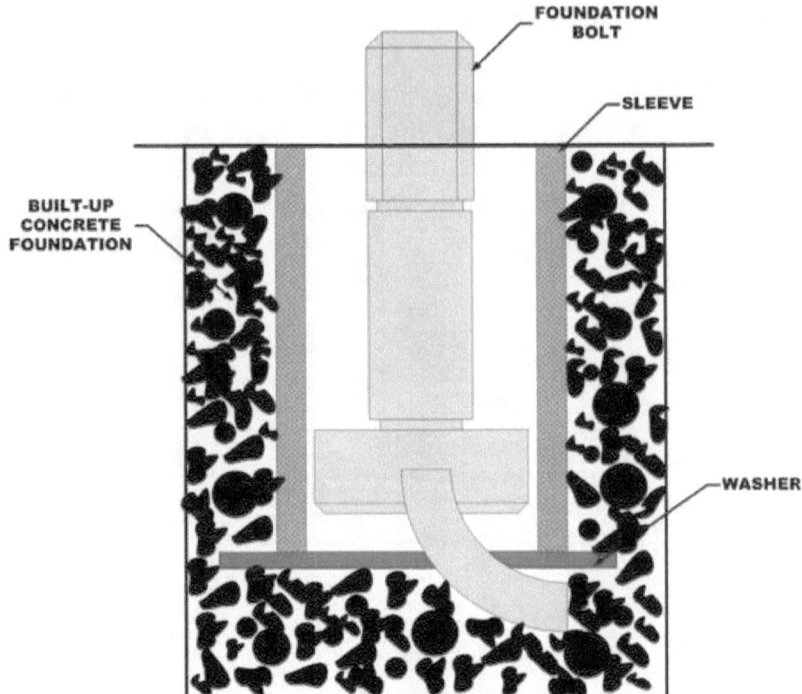

Figure 16.13 Foundation hole schematic.

16.10.8 Dynamometer Baseplate Mounting

This procedure assumes that a concrete foundation has been prepared with anchor or hold down bolts extending up ready to receive dynamometer baseplate.

1. Use blocks and shims under base for support at anchor bolts and midway between bolts to position base approximately 1" above the concrete foundation, with studs extending through holes in the baseplate.
2. By adding or removing shims under the base, level and plumb the dynamometer shaft and flanges.
3. Draw anchor nuts tight against base and observe dynamometer and engine shafts or coupling hubs for alignment. (Temporarily remove shaft guard for checking alignment.)

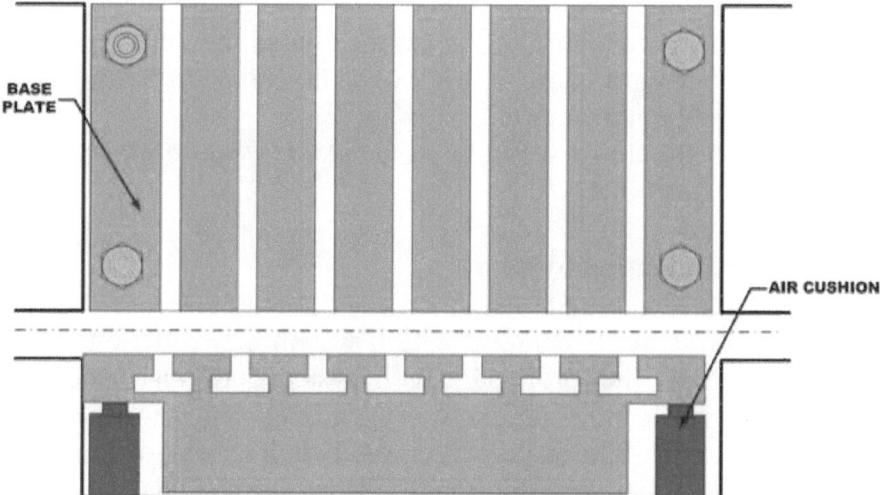

Figure 16.14 Dynamometer baseplate on air cushion.

5. If alignment needs improvement, add shims or wedges at appropriate positions under base, so that retightening of anchor nuts will shift shafts into closer alignment. Repeat this procedure until a reasonable alignment is reached.
6. Check to make sure the piping can be aligned to the dynamometer flanges without placing pipe strain on either flange.
7. Grout in baseplate completely and allow grout to dry thoroughly.

16.11 Lighting System

The test cell and control room should be well lit. The lighting should be enough and sufficient to read all displays and signs clearly.

- Emergency lighting should be provided in the test cell as well as control room.
- Exit signs should be installed where ever necessary and should be clearly visible.
- All switches should be explosion proof to avoid the sparking and danger of fire.

- The switches to turn off the test cell lights should be provided near the control desk, and it will be useful to watch for sparks and red-hot surfaces.
- Lights should be securely mounted so as not to move in the ventilation wind.

16.12 Communication System

There should be proper communication system between the test cell and control room. The cordless intercom system between test cell and control room is one of the alternatives for proper communication. For visual monitoring of the test cell, a closed circuit camera and monitor will provide continuous monitoring of the engine and test cell.

16.13 Fire Fighting System

It is important to know that the causes of fire in engine room. Fire results from the combining of three factors:

1. A combustible substance.
2. A source of ignition.
3. A supply of air.

These three factors are said to be the three sides of the fire triangle. Removing any one or more of these sides will break the triangle and result in the fire being put out. The complete absence of any one of the three will ensure that a fire never starts.

The first side of the triangle, the substance that can burn, is in abundance in test cell—tanks of oil, pipes full of oil, etc. This side can be removed by good housekeeping in the test cell such as efficient garbage management, good maintenance of pipes, flanges, gaskets, and machines, etc.

The second side of the triangle, air, is also in abundance as huge test cell ventilation blowers pump air into the test cell for ventilation, cooling, and for combustion.

The third side of the fire triangle is the ignition source, which can also be present sometimes due to failure of electrical insulation which will result in sparking which will lead to fire. This may happen during the maintenance of test cell which calls for welding and gas cutting of pipes inside the test cell.

The test cell designs engineers must make sure that high temperature areas like exhaust manifolds are insulated and that the insulation is in good condition. Fuel spilling and leaks should be avoided as fire is caused by fuel spilling from a leaking pipeline on to a hot surface and catching fire. Naked flames like butane lighters and cigarettes should be avoided in the test cell and control room because of the presence of oil vapors. Smoking should only be done in designated areas like the engine control room, which is the certified smoking area.

Figure 16.15 Causes of fire. Courtesy of Wikipedia

In the event of a cell fire, you need immediate access to the ventilation fan's shut-off switch. A big red button that is also interlocked with the engine's ignition system and fuel pump is recommended. Test cell should be equipped with fire extinguishers and automatic sprinklers. The following figure shows the mounting of sprinklers in test cell.

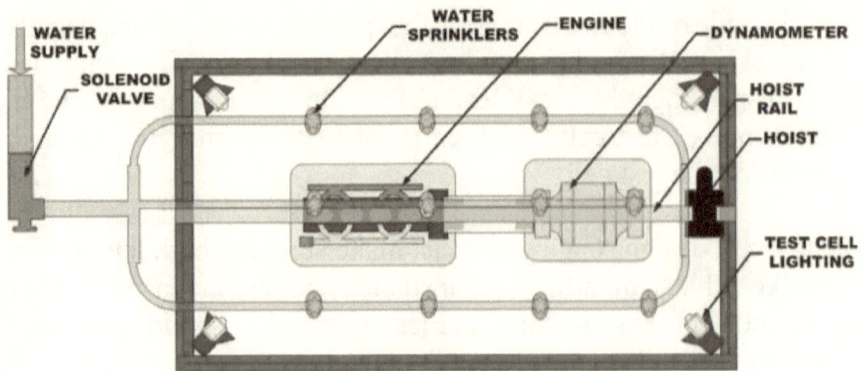

Figure 16.16 Water sprinklers in test cell.

The test cell should have smoke and flash detecting sensors which will activate the water sprinklers in the event of fire breaking out. The sprinkler system sprinkles water in case of fire. In addition to water sprinklers, fire extinguisher is installed in the test cell. In the event of the fire taking place inside the test cell, the fire extinguisher will operate automatically and extinguish the fire.

Closure

The modern test cells are very complex in nature and houses various systems required to test the engine. These systems are interfaced with a Dynamometer control system at a central location. The automation of various systems such as engine docking, water, intake air conditioning, oil temperature controllers and fuel temperature controllers enables the engine to be tested in controlled environment. The performance of the engine can be evaluated under controlled parameters. The nature of the modern test cell demands the inter disciplinary skills to be acquired by test engineer.

References

1. Fire image from http://en.wikipedia.org/wiki/File:Fire_triangle.svg.

CHAPTER 17
Measurements of Modern Engine

17.1 Introduction

The engine under test undergoes different types of testing apart from toque, speed, and power measurement. Modern day regulation and legal requirements demand advanced and accurate testing of following parameters:

1. Fuel consumption measurement.
2. Air consumption measurement.
3. Blow-by measurement.
4. Oil consumption measurement.

17.2 Fuel Consumption Measurement

The measurement procedure of fuel consumption looks very simple prima facie; however, it needs to consider the back flow or return flow of the fuel. Sometimes the back flow is caused by bubbles in the fuel line which may be developed due to engine heat. The fuel flow measurement also depends on the type of flow meter used. Turbulent flow may cause erratic readings in fuel measurement. The density of the fuel is dependent on the temperature which may vary in the test cell environment from as low as −5°C to 70°C. This may mislead the measurement. Fuel consumption can be measured by using flow meters, pipette-type instruments, or gravimetric instruments as discussed below.

17.2.1 Flow Meter

In the field of engine testing, you need to measure the flow of fuel and intake air. Fuel is being the liquid and air being gas both can be measured in volumetric or mass flow rates, such as liters per second or grams per second. A flow meter is an instrument used to measure mass or volumetric flow rate of a liquid or a gas in general a fluid.

17.2.2 Types of Flow Meter Used in Engine Testing

1. Volumetric measurement.
2. Gravimetric measurement.

17.2.3 Selection Criteria of a Flow Meter

The requirements of the particular application play major role factor in selection of the fuel flow meter. Following are some of the factors those govern the selection of the fuel meter.

1. Type of fluid:
   The material of construction of flow meter is largely decided by the type of fuel /liquid.

2. Type of measurement:
   The two types of measurements are more popular in engine testing.

   - Rate measurement
   - Total fuel flow to engine

3. Type of display:
   Following type of displays are in use with engine testing:

   - Local display.
   - Remote display.
   - Rate display (analog)
   - Total flow (normally digital)
   - Output signal for processing in data acquisition system

4. Flow rate:
   The maximum and minimum flow rate through the flow meter is another important factor. You need the flow meter which gives consistent results at extremely minimum and maximum flow rates.

   What is the minimum and maximum fuel flow rate required by range of engines under test?

   What is the minimum and maximum process pressure of the fuel supplied to the engine?

   What is the minimum and maximum fuel temperature on test bed?

   Is fuel and its chemical composition compatible with the flow meter wetted parts?

17.2.4 Flow Measurement Orientation

Selection of the fuel flow meter also depends upon certain intangible factors such as:

1. Reliability of the instrument and its MTBF
2. Service support availability
3. Complexity of the calibration procedure and how often it is required to be done.

The fuel flow meters employed in engine testing bed are highly accurate and expensive, and it is highly recommended that the cost of the installation be considered and added to the cost of fuel meter.

The test engineer should be clearly aware of the requirements of the engine testing. Therefore, time should be invested in fully evaluating the fuel to be measured and the material of construction of the measuring apparatus.

Expected minimum and maximum pressure and temperature values should be given in addition to the normal operating values when selecting flow meters.

Care to be taken in installation of flow meter:

1. Avoid too many bends and joints in piping installation.
2. Proper size of pipeline should be considered.
3. Include proper valves and regulators.

The range of meter to be identified by considering minimum and maximum flows (mass or volumetric) that will be measured in a particular engine test cell considering the range of engines to be tested in that test cell. Once you have decided the range of the required flow, then measurement accuracy is determined.

17.2.5 Types of Fuel Consumption Meters

There are various types of fuel consumption meters used in engine testing. They are described below:

17.2.5.1 Orifice-type Flow Meter

An *orifice meter* is a device used for measuring the rate of fluid flow. It uses the same principle as a Venturi nozzle, namely Bernoulli's principle which says that there is a relationship between the pressure of the fluid and the velocity of the fluid. When the velocity increases, the pressure decreases and vice versa.

An orifice plate is basically a thin plate with a hole in the middle. It is usually placed in a pipe in which fluid flows. As fluid flows through the pipe, it has a certain velocity and a certain pressure. When the fluid reaches the orifice plate, with the hole in the middle, the fluid is forced to converge to go through the small hole; the point of maximum convergence actually occurs shortly downstream of the physical orifice, at the so-called *vena contracta* point. Beyond the vena contracta, the fluid expands and the velocity and pressure change once again. By measuring the difference in fluid pressure between the normal pipe section and at the vena contracta, the volumetric and mass flow rates can be obtained from *Bernoulli's equation.*

Incompressible Flow Through an Orifice

By assuming steady state, incompressible (constant fluid density), inviscid, laminar flow in a horizontal pipe (no change in elevation) with negligible frictional losses, Bernoulli's equation reduces to an equation relating the conservation of energy between two points on the same streamline:

$$P_1 + \frac{1}{2} \cdot \rho \cdot V_1^2 = P_2 + \frac{1}{2} \cdot \rho \cdot V_2^2, \quad \ldots\ldots 17.1$$

or:

$$P_1 - P_2 = \frac{1}{2} \cdot \rho \cdot V_2^2 - \frac{1}{2} \cdot \rho \cdot V_1^2 \ldots\ldots 17.2$$

By continuity equation:

$$Q = A_1 \cdot V_1 = A_2 \cdot V_2, \quad \ldots\ldots 17.3$$

$$\text{or } V_1 = \frac{Q}{A_1} \text{ and } V_2 = \frac{Q}{A_2},$$

$$P_1 - P_2 = \frac{1}{2} \cdot \rho \cdot \left(\frac{Q}{A_2}\right)^2 - \frac{1}{2} \cdot \rho \cdot \left(\frac{Q}{A_1}\right)^2.$$

Solving for Q:

$$Q = A_2 \sqrt{\frac{2(P_1 - P_2)/\rho}{1 - (A_2/A_1)^2}} \quad \ldots\ldots 17.4$$

and:

$$Q = A_2 \sqrt{\frac{1}{1 - (D_2/D_1)^4}} \sqrt{2(P_1 - P_2)/\rho}. \quad \ldots\ldots 17.5$$

The above expression for Q gives the theoretical volume flow rate. Introducing the beta factor $\beta = D_2/D_1$ as well as the coefficient of discharge C_d

$$Q = C_d \cdot A_2 \sqrt{\frac{1}{1-(\beta)^4}} \sqrt{2(P_1 - P_2)/\rho} \quad \text{...............17.6}$$

And finally introducing the meter coefficient C which is defined as

$$C = \frac{C_d}{\sqrt{1-\beta^4}}$$

to obtain the final equation for the volumetric flow of the fluid through the orifice:

$$Q = C A_2 \sqrt{2(P_1 - P_2)/\rho} \quad \text{...............17.7}$$

Multiplying by the density of the fluid to obtain the equation for the mass flow rate at any section in the pipe:

$$m = \rho Q = C A_2 \sqrt{2\rho(P_1 - P_2)}, \quad \text{...............17.8}$$

where
Q = volumetric flow rate (at any cross section), m³/s
m = mass flow rate (at any cross section), kg/s
C_d = coefficient of discharge, dimensionless
C = orifice flow coefficient, dimensionless
A_1 = cross-sectional area of the pipe, m²
A_2 = cross-sectional area of the orifice hole, m²
D_1 = diameter of the pipe, m
D_2 = diameter of the orifice hole, m
β = ratio of orifice hole diameter to pipe diameter, dimensionless
V_1 = upstream fluid velocity, m/s
V_2 = fluid velocity through the orifice hole, m/s
P_1 = fluid upstream pressure, Pa with dimensions of kg/ (m·s²)
P_2 = fluid downstream pressure, Pa with dimensions of kg/ (m·s²)
ρ = fluid density, kg/m³

Following assumptions were made while deriving the above equations:

1. The cross section of the orifice opening is not as realistic as using the minimum cross section at the vena contracta.
2. The frictional losses were considered as zero.
3. The viscosity and turbulence effects may be present; hence, the coefficient of discharge C_d is introduced. The coefficient of discharge can be established as a function of the Reynolds number.
4. An orifice only works well when supplied with a fully developed flow profile. This is achieved by a long upstream length (20 to 40 pipe diameters, depending on Reynolds number) or the use of a flow conditioner. Orifice plates are small and inexpensive but do not recover the pressure drop as well as a venturi nozzle does. If space permits, a venturi meter is more efficient than a flow meter.

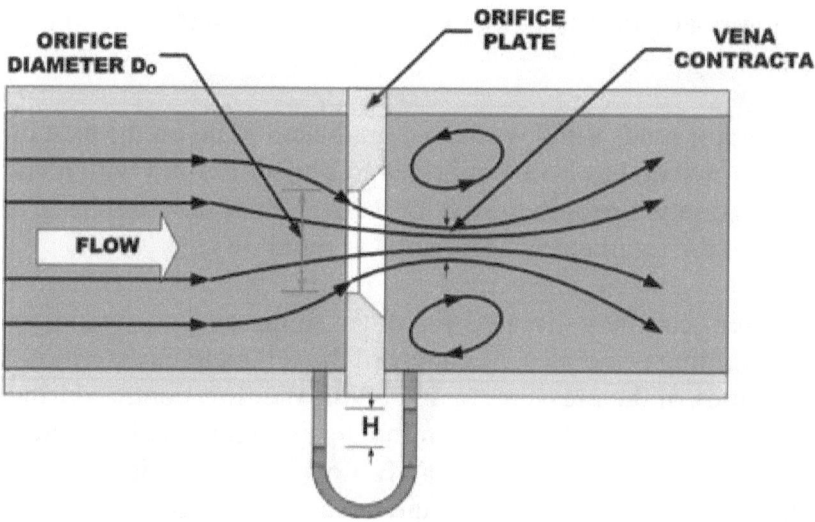

Figure 17.1 Orifice meter.

17.2.5.2 Rotameter

A rotameter is a device that measures the flow rate of liquid or gas in a closed tube. The variable area flow meters feature an upright tapered measuring tube, tapering upward. In this tube, specially designed float moves freely up and down.

The fluid flows through the tube from bottom to top. Due to upward flow, it raises the float until there is an annular gap between the wall of the tube and the float and equilibrium of the forces applied to the float has been achieved.

The flow rate inside the rotameter is measured using a float that is lifted by the fluid flow based on the buoyancy and velocity of the fluid opposing gravity pulling the float down. For gasses, the float responds to the velocity alone, buoyancy is negligible.

The float moves up and down inside the rotameter's tapered tube proportionally to the flow rate of the fluid. It reaches a constant position once the fluid and gravitational forces have equalized. Changes in the flow rate cause rotameter's float to change position inside the tube. Since the float position is based on gravity, it is important that all rotameters be mounted vertically and oriented with the widest end of the taper at the top. It is also important to remember that if there is no flow, the float will sink to the bottom of the rotameter due to its own weight.

The operator reads the flow from a graduated scale on the side of the rotameter, which has been calibrated to a specific fluid with a known specific gravity. Specific gravity or the weight of the fluid has a great impact on the rotameter's accuracy and reliability.

Rotameters can be calibrated for other fluids by understanding the basic operating principles. Accuracy of the rotameter is determined by the accuracy of the pressure, temperature, and flow control during the initial calibration. Any change in the density and weight of the float will have impacts on the rotameter's flow reading. Additionally any changes that would affect the fluid such as pressure or temperature will also have an effect on the rotameter's accuracy. Given this, rotameters should be calibrated yearly to correct for any changes in the system that may have occurred.

Three main forces act on the float.

1. The **buoyancy A**, which is dependent on the density of the medium and the volume of the float. It is constant (at constant density).
2. The **weight G**, which is dependent on the mass of the float.
3. The flow **force S**: The flow force changes transitionally with a change in the flow until a new state of equilibrium has been achieved.

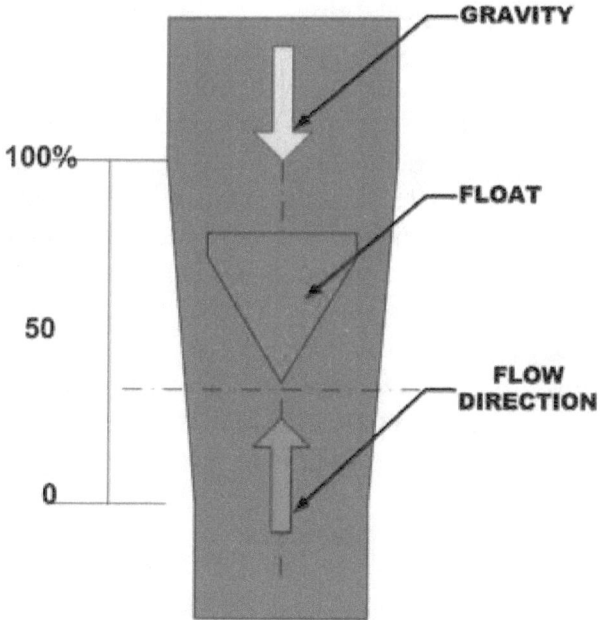

Figure 17.2 Rotameter.

There are several advantages to a rotameter over a more complicated flow meter including:

1. Rotameters can be installed in areas with no power since they only require the properties of the fluid and gravity to measure flow, so you do not have to be concerned with ensuring that the instrument is explosion proof when installed in areas with flammable fluids or gases.
2. Rotameters can be installed with standard pipe fittings to existing piping or through a panel. You do not have to worry

about straight runs of pipe as with a magnetic or turbine flow meter.
3. Rotameters are simple devices that are mass manufactured out of inexpensive materials keeping investment costs low.
4. A glance at a rotameter acts as a sight glass telling the operator that a filter needs cleaning, that there is some other problem causing discoloration of the water, or that the fluid is actually flowing. With a transparent rotameter, they can instantly see if there is any buildup on the float or tube walls.
5. With a properly maintained rotameter, the operator can expect sustained high repeatability.
6. Rotameters offer wide flow measurement ranges or range ability. A typical ratio of 10:1 from maximum to minimum flow rate can be expected. Operators will be able to measure minimum flow rates as low as 1/10 of the rotameter's maximum flow rate without impairing the repeatability.
7. The rotameter's scale is linear because the measure of flow rate is based on area variation. This means that the flow rate can be read with the same degree of accuracy throughout the full range.
8. Pressure loss due to the rotameter is minimal and relatively constant because the area through the tapered tube increases with flow rate. This results in reduced pumping costs.

Disadvantages to the use of rotameters:

1. Because gravity plays a key role in the flow measurement, the rotameter must always be installed vertically with the fluid flowing up through it.
2. The graduated scale on the side of the rotameter will only be valid for the specific fluid and conditions where it was calibrated. The specific gravity of the fluid is primary property to consider; however, the fluid's viscosity and any temperature changes may also be significant. Rotameter floats are generally designed to be insensitive to viscosity, but the operator should verify that any rotameters installed in their system are calibrated to their specific setup prior to relying on the flow measurements provided.

3. It is difficult for rotameters to be adapted for machine reading, although a magnetic float may be used in some instances.

17.2.5.3 Volumetric Fuel Consumption Meter (Pipette Type)

The simplest method is to supply a fixed quantity of fuel and measure the time by stop watch to consume the know quantity of fuel. The volumetric flow will be total volume consumed divided by the time to consume this volume. The photo sensors are used to judge the fuel level exactly and to avoid the error in sighting the level in the glass tube.

The volumetric fuel consumption is popular and inexpensive method to evaluate the gross fuel consumption of the engine. It consists of a pipette with photo sensor and light-emitting diodes to detect the volume accurately and activate the timer. It essentially measures the time in seconds to consume the precalibrated volumes. The following figure shows the volumetric fuel consumption meter.

The fuel meters using photo sensors to sense the level of the fuel in the meter using light beam (which in turn tell volume/weight) may get affected by color of the fuel. The color of the fuel may get changed due to impurities in the fuel line. The position of the photo sensor also plays important role. The level of the fuel in the glass tube depends on its meniscus forming and placement of photo sensor to judge the level exactly.

The volumetric fuel meter consists of a burette with two glass bulbs which are normally of two different volumes say A and B. This will give three volumes to be used for measurements such as

1. Volume A
2. Volume B
3. Volume A + B

The burette is connected by three-way cock which is operated by solenoid. When you start the measurement, the lamps in the photoelectric system light up and make solenoid stop the flow of fuel going directly to the engine and open the connection from burette to the engine. When the fuel level is reached to the top most level, the solenoid stops the fuel supply to the engine as well as to the burette. Now fuel starts flowing through the burette.

Once the level crosses the first photo sensor, the timer is activated and initiates the measuring of the time. When fuel level crosses the second photo sensor mounted at the end of the bulb, the new signal generated will stop the Timer, thus giving the time to consume the volume contained by the bulb and open the solenoid valve to fill the pipette for next measurement.

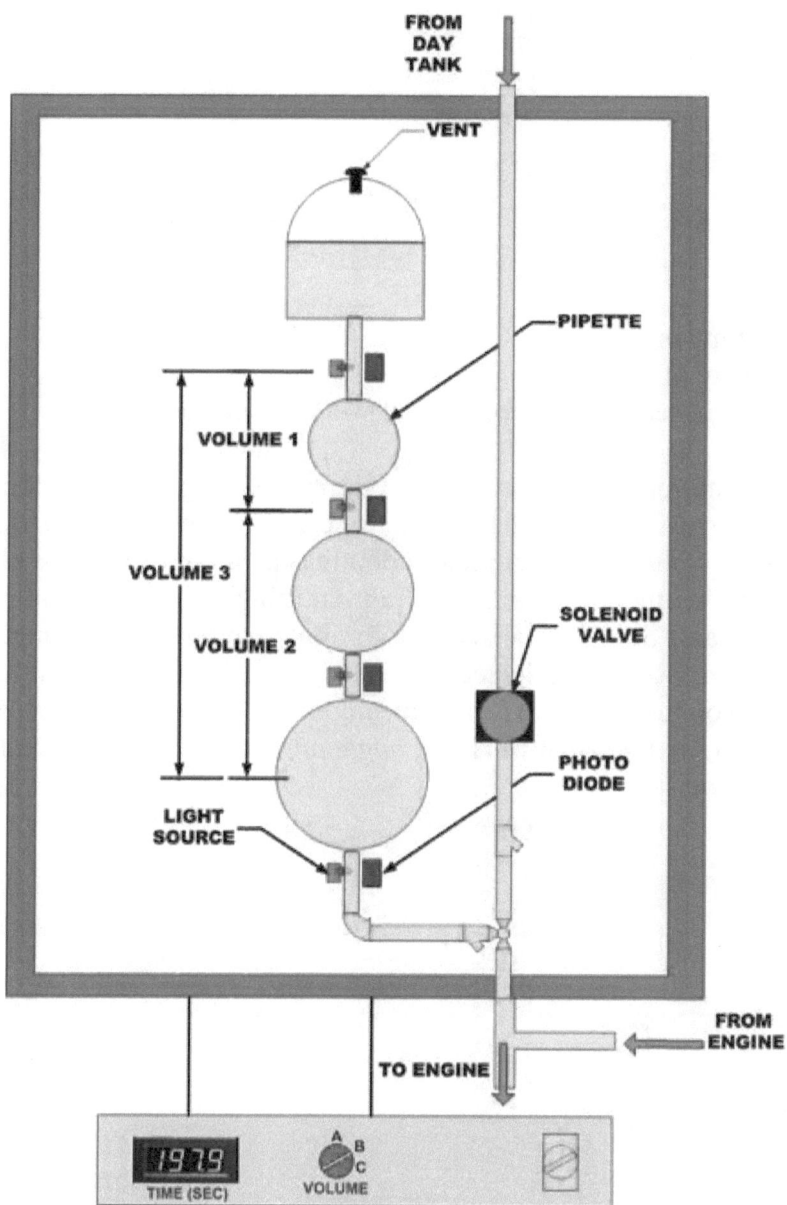

Figure 17.3 Volumetric fuel consumption meter.
Courtesy of Associated Electrodyne Industries Pvt. Ltd.

Specific fuel consumption is computed by using following formula:

$$Fuel\ CConsuption\ (FC) = \left(\frac{Volume.Density}{\frac{BHP}{hr}}\right).$$

$$Fuel\ CConsuption\ (FC) = \left(\frac{3600(V.\rho.T)}{BHP}\right) gms/bhp/hr, \ldots 17.9$$

where
V = volume in CC
P = density of fuel
T = Time in seconds for consuming volume selected

Some of the volumetric fuel consumption measuring units are adopting different way of mounting of photo sensors. The following figure shows the volumetric fuel meter adopting the mounting of the photo sensor in inclined position. The light-emitting sensor is mounted at an angle and light is projected at an angle called incident angle to the fuel surface and reflected at reflection angle. The photodiode is mounted on the pipette at reflection angle. The *law of reflection* states that $\theta_i = \theta_r$, or in other words, the angle of incidence equals the angle of reflection.

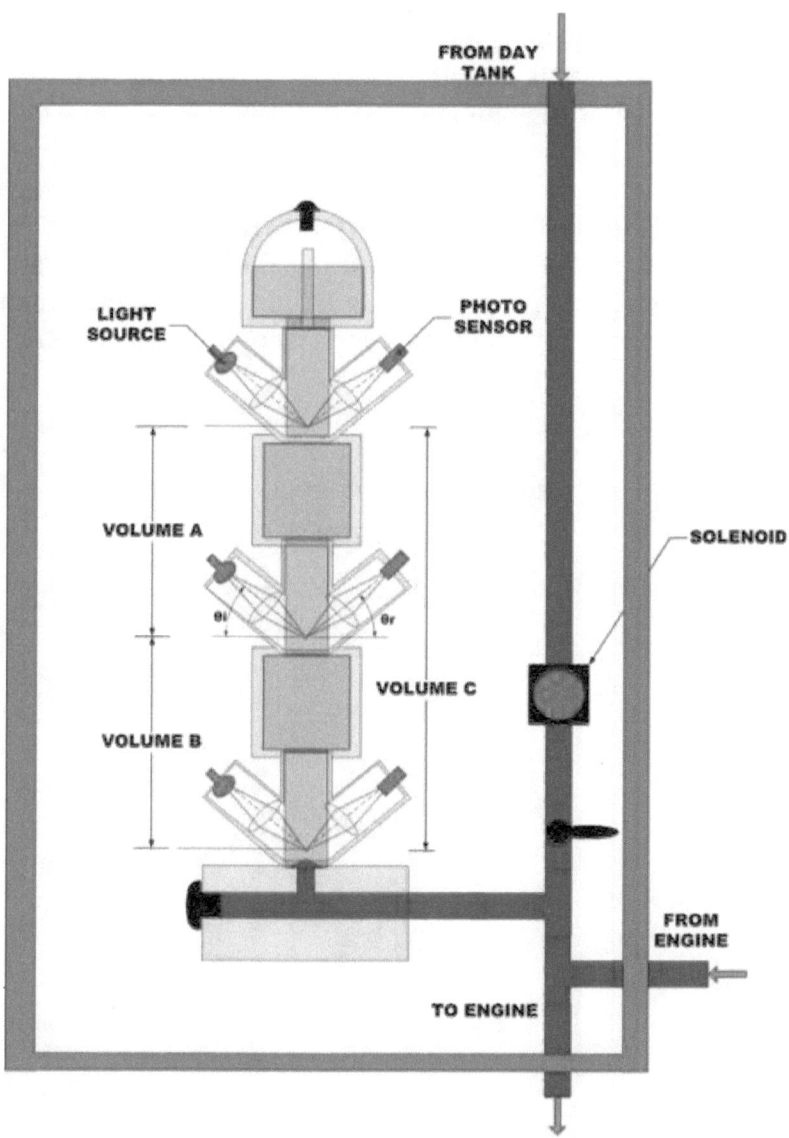

Figure 17.4 Schematics of volumetric fuel consumption meter.

17.2.5.4 Gravimetric Fuel Measurement

The gravimetric fuel consumption meter directly measures the fuel consumption in weight unit. The gravimetric fuel consumption meter employs a load cell or the pressure transducer to sense the weight of the fuel. As the weight is actually measured, the effect of temperature and density variation is eliminated in gravimetric measurement.

The gravimetric fuel meter employing pressure transducer to measure the weight of the fuel actually uses the fuel head to measure the weight of the fuel. Area of the measuring flask being constant, the force on transducer is proportional to height of the fuel column in the measuring flask. The following figure shows the gravimetric fuel meter using pressure transducer.

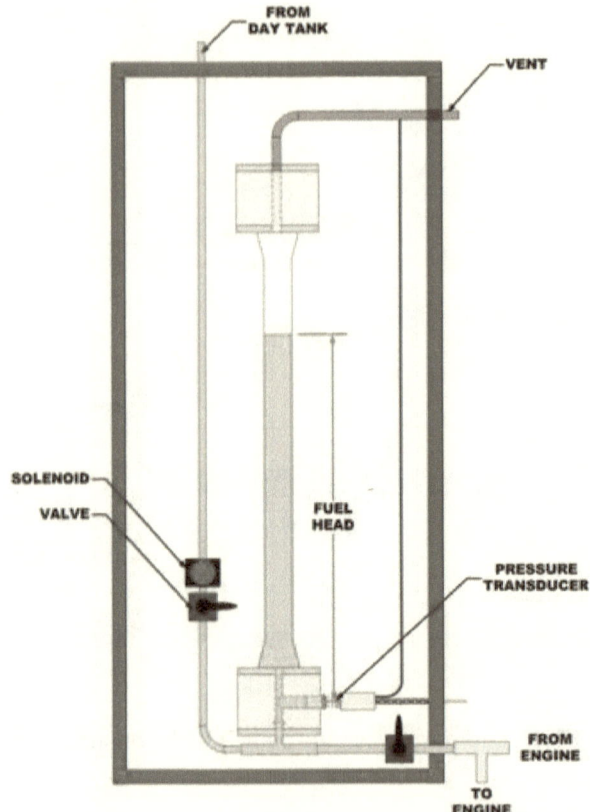

Figure 17.5 Schematics gravimetric fuel meter using pressure transducer. Courtesy of Courtesy of Saj Test Plant Pvt. Ltd

Some of the gravimetric fuel meters are based on sensing the weight directly by using the load cell. The fuel flow assembly includes a fuel supply pipe and engine feed and returns pipes. Recirculated fuel is returned to the weighing container through the fuel flow assembly and enters the container above a baffle. The baffle prevents vapor bubbles entering the engine fuel supply pipe. When weighing is in progress, the solenoid valve is closed, and thus the weight of fuel used can be measured. When metering is not in progress, the valve regularly operates and the fuel level is kept above a preset level in measuring flask.

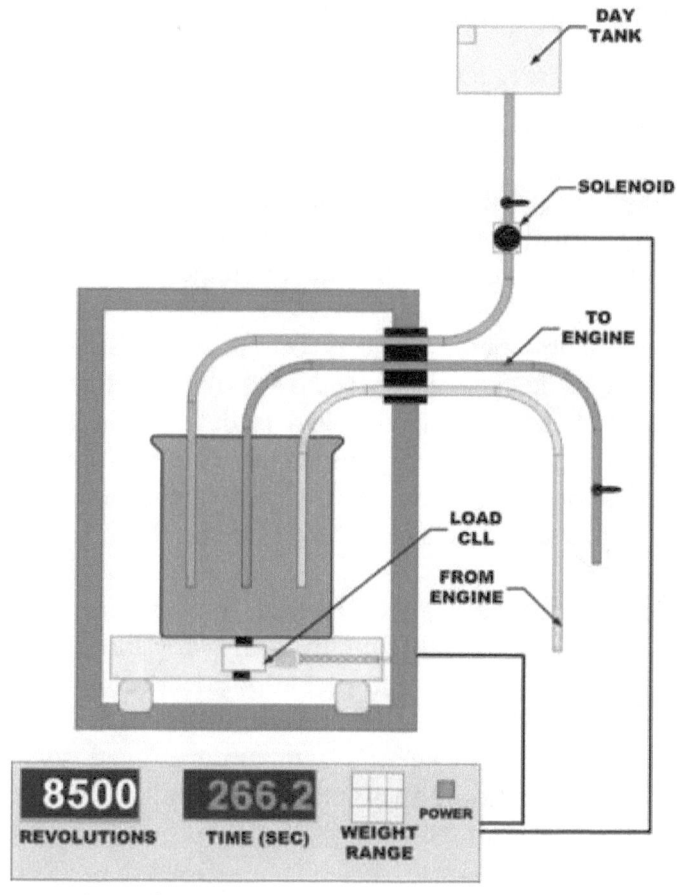

Figure 17.6 Schematics gravimetric fuel meter with load cell.

17.3 Air Flow Measurement

The desirability of improving the fuel economy and lowering the emission output of the automobile engine has necessitated a search for improved sensors to measure and control engine parameters. One of the key requirements is to measure the engine intake air. If this can be accurately measured, together with other parameters, then the performance of the engine can be easily evaluated.

17.3.1 Intake Air Measurement-Hot Wire Anemometer Method

Theory of Operation and Construction

A hotwires-type sensor must have two characteristics to make it a useful device:

- A high temperature coefficient of resistance
- An electrical resistance such that it can be easily heated with an electrical current at practical voltage and current levels

The hotwire anemometer consists of an electrically heated wire exposed to the wind. The speed of the wind affects the rate at which the wire loses heat. The wind speed is determined by measuring the electrical current necessary to maintain the wire at a constant temperature. A hot wire mass airflow meter determines the mass of air flowing into the engine's air intake system.

The wire's electrical resistance increases as the wire's temperature increases, which limits electrical current flowing through the circuit. When air flows past the wire, the wire cools, decreasing its resistance, which in turn allows more current to flow through the circuit. As more current flows, the wire's temperature increases until the resistance reaches equilibrium again. The amount of current required to maintain the wire's temperature is directly proportional to the mass of air flowing past the wire.

The associated signal conditioning circuit converts the measurement of current into a voltage signal which is sent to the digital indicator to measure the flow in engineering units.

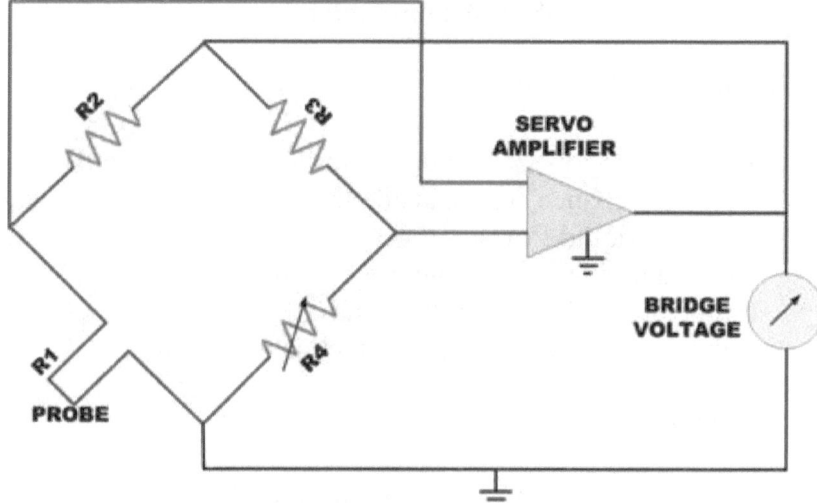

Figure 17.7 Constant temperature anemometer bridge circuit.

The advantages of hot wire anemometer are:

- Cost is relatively cheap.
- Frequency response is high.
- Size: small measurement volume.
- Good spatial and temporal resolution.
- Turbulent flows can easily be measured with it.
- Multicomponent measurement.
- Simultaneous temperature measurement: available with multisensor probe.
- Two-phase flow measurement is possible.
- Accuracy is as good as a Laser Doppler anemometer.
- Signal to noise ratio is low.
- Probe and analysis selection: it is easy to find a good measurement system for most measurements.
- Measurements can be made in gases, transparent, opaque, and even electrically conducting liquids.

Disadvantages of Hot Wire Anemometer

High Turbulence Intensity

- Restricted to low and moderate turbulence intensity flows.
- Flow disturbance from outside the measurement plane alters results of liquid flows.
- Can contaminate the probe more easily.
- Temperature changes in the fluid affect the results more due to the smaller overheat ratio which is used in liquids.

Heat Transfer Problems

- Supports are not as hot as the sensor, due to their larger mass. They act as heat sinks. Therefore, the edge of the wire is cooled, reducing the active wire length. This can be accounted for by calibration.
- Thermal wake from one wire could affect the measurements on another probe for a multicomponent hot-wire probe.
- The wires are sensitive to both temperature and velocity, and therefore if both change simultaneously, then the results are altered.

Aerodynamic Problems

- Probes are not sensitive to flow direction (reverse flow).
- Probe supports interfere with the flow onto the other sensors.

17.3.2 Intake Air Measurement—Pitot Tube and Differential Pressure Transmitter

The Pitot tube is named after its French inventor Henri Pitot (1695-1771). The device measures a fluid velocity by converting the kinetic energy of the flowing fluid into potential energy at what is described as a "stagnation point." The stagnation point is located at the opening of the tube as in Figure 4.2.9. The fluid is stationary as it hits the end of the tube, and its velocity at this point is zero. The potential energy created is transmitted through the tube to a measuring device.

The tube entrance and the inside of the pipe in which the tube is situated are subject to the same dynamic pressure; hence, the static pressure measured by the Pitot tube is in addition to the dynamic pressure in the pipe. The difference between these two pressures is proportional to the fluid velocity and can be measured simply by a differential manometer.

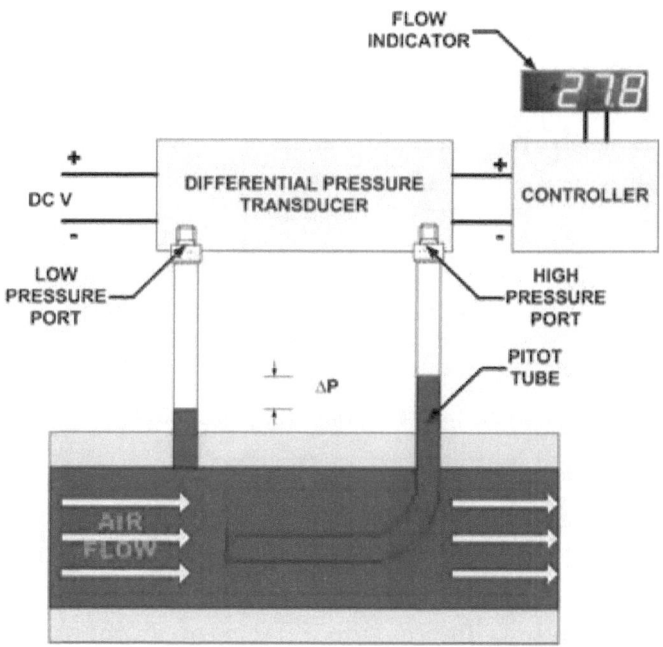

Figure 17.8 The simple Pitot tube principle.

Bernoulli's equation can be applied to the Pitot tube in order to determine the fluid velocity from the observed differential pressure (ΔP) and the known density of the fluid. The Pitot tube is used to measure incompressible and compressible fluids, but to convert the differential pressure into velocity, different equations apply to liquids and gases. The details of these are outside the scope of this tutorial, but the concept of the conservation of energy and Bernoulli's theorem applies to all; and for the sake of example, the following text refers to the relationship between pressure and velocity for an incompressible fluid flowing at less than sonic velocity.

Bernoulli's theorem relates to the Steady Flow Energy Equation, and states that the sum of:

- Pressure energy,
- Kinetic energy, and
- Potential energy

will be constant at any point within a piping system (ignoring the overall effects of friction). This is shown below, mathematically in equation 17.1 for a unit mass flow:

$$\frac{P_1}{\rho \cdot g} + \frac{u_1^2}{2 \cdot g} + h_1 = \frac{P_2}{\rho \cdot g} + \frac{u_2^2}{2 \cdot g} + h_2 \quad \ldots\ldots 17.10$$

where

P_1 = upstream pressure (Pa)

P_2 = downstream Pressure (pa)

u_1 = upstream velocity at corresponding point (M/s)

u_2 = upstream velocity at corresponding point (M/s)

h_1 = relative height within the slow system

h_2 = relative height within the slow system

g = gravitational constant (9.81 m/s²)

Bernoulli's equation ignores the effects of friction and can be simplified as follows:

Pressure energy + Potential energy + Kinetic energy = Constant.

Multiplying equation (17.10) throughout by "ρg," we get the following equation:

$$P_1 + \rho g h_1 + \frac{1}{2}\rho u_1^2 = P_2 + \rho g h_2 + \frac{1}{2}\rho u_2^2 \quad \ldots\ldots 17.11$$

Middle term gets cancelled on either side when there is no change in reference height (h).

From the above equation, the following equation can be developed to calculate velocity.

$$P_1 + \tfrac{1}{2}\rho u_1^2 = P_2 + \tfrac{1}{2}\rho u_2^2, \quad \ldots\ldots\ldots 17.12$$

where
- P_1 = the dynamic pressure in the pipe
- u_1 = the fluid velocity in the pipe
- P_2 = the static pressure in the Pitot tube
- u_2 = the stagnation velocity = zero
- ρ = the fluid density

Because u_2 is zero, equation (17.12) can be rewritten as equation (17.12).

$$P_1 + \tfrac{1}{2}\rho u_1^2 = P_2, \quad \ldots\ldots\ldots 17.13$$

$$P_2 - P_1 = \tfrac{1}{2}\rho u_1^2,$$

$$u_1^2 = 2(P_2 - P_1)/\rho,$$

$$u_1^2 = 2(\Delta P)/\rho \quad \ldots\ldots\ldots 17.14$$

$$u = \sqrt{\frac{2(\Delta P)}{\rho}}.$$

The velocity is known; therefore, the flow can be calculated using the following equation:

$$Q = A \cdot u \quad \ldots\ldots\ldots 17.15$$

17.3.3 Intake Air Measurement—Air Box Method

The intake air is one of the major constituents of the combustion of an internal combustion engines. The density of the air is most important factor and plays important role in the combustion process. The density of the air is function of both air pressure and air temperature. The interrelationship between pressure, P, absolute temperature, T, and density, ρ, of air can be determined using the Ideal gas law written as

$$\rho = \frac{P}{R.T} \quad \text{...............................} 17.16$$

The measurement of the air supply to the engine can be done using a sharp-edged orifice which is an inexpensive but accurate technique used. This method is used in many engine testing applications on the test bed to determine steady flow characteristics of a compressible fluid, such as air.

The related parameters such as air pressure for the air flow is obtained by recording the pressure drop by incorporating an U-tube manometer across the orifice as air passes through the measuring device. Similarly the air temperature of the air is measured by using a thermometer. Digital manometer and temperature indicators can be used as they are more accurate.

Why Air box is used?

If orifice plate is positioned directly within an engine's intake pipe in order to obtain inducted air flow then the flow will be pulsating. The inducted pulsating air motion results from the cyclic operating nature of the IC engine's piston cylinder configuration.

To avoid the effect of pulsating air flow, a test set up in the test cell will usually employ an air box method. The orifice plate is mounted on the air box wall instead of directly mounting on an air intake manifold or the pipe leading to intake manifold. The air box is a box or cylinder with sufficient volume to dampen out engine inlet intake pulsations from being propagated back to an orifice mounted in the intake chamber. Also the length of the connection from air box to the engine intake manifold is kept short as possible to avoid the effect of pressure pulsations. The size of an air box depends on mainly following factors:

1. Engine type and size
2. Number of cylinders
3. Operating speed of the engine.

The total volume of the air box should be much larger than the total swept volume of the engine under test. Air box will have to be precisely sized for the single cylinder engine as pulsations will be very high.

The multicylinder and turbocharged engines are less sensitive to the air box technique of air measurement method.

The pressure drop across the orifice is given by

$$\Delta p = \frac{\rho V^2}{2 \cdot g}, \quad \ldots\ldots\ldots 17.17$$

where
ρ = density of air
V = velocity of air though the orifice
ΔP = pressure difference (drop across the orifice)

Table 17.1—Units of pressure, density and velocity

	Property	SI System	FPS (English)
1	ρ	Kg/m^3	lbm / ft^3
2	V	m/sec	Ft/sec
3	P	N/m^2	lbf/ft^2
4	g	1.0 kg.m/N.sec^2	32.17 ft.lbm/lbf.sec^2

The above formula can be rewritten as:

$$V = \sqrt{\frac{2g \cdot \Delta P}{\rho}} \quad \ldots\ldots\ldots 17.18$$

The discharge or the flow through the orifice is given by the formula,

$$Q = C_d \cdot A \cdot V, \quad \ldots\ldots\ldots 17.19$$

where
Q = flow in M3
C_d = coefficient of discharge for "orifice"
V = velocity of air through the orifice

Therefore, the flow Q will be:

$$Q = C_d \left(\frac{\pi d^2}{4}\right)\sqrt{\frac{2.g.\Delta P}{\rho}} \quad \ldots\ldots\ldots\ldots\ldots\ldots\ldots\ldots 17.20$$

Table 17.2—Ideal properties of air:

	Property	SI System	FPS(ENGLISH)
1	ρ	Kg/m3	lbm /ft2
2	P	N/m2	lbf/ft2
3	T	°K	°R
4	R	287 J/kg.K	53.34 ft/lbf/lbm °R

Now let us consider the below figure air measurement by AIR BOX method.

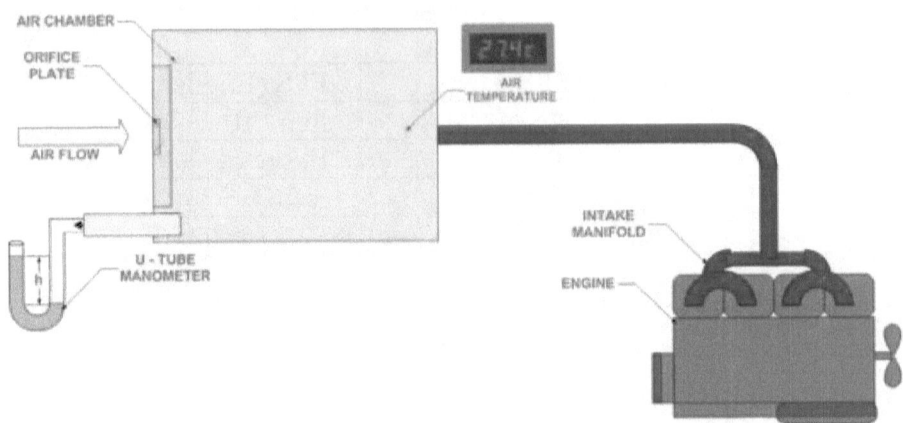

Figure 17.9 Air flow measurement.

The mass flow rate is the product of density and volume flow rate. The orifice pressure drop with manometer is shown as height "h."

$$\dot{m} = \rho . Q$$

$$\dot{m} = C_d \left(\frac{P}{RT}\right)\left(\frac{\pi . d^2}{4}\right) \sqrt{\frac{2.g.hRT}{P}} \quad \ldots\ldots 17.21$$

$$\dot{m} = C_d \left(\frac{\pi . d^2}{4}\right) \sqrt{\frac{2.g.hP}{RT}} \quad \ldots\ldots 17.22$$

Considering the units in SI system and expressing the pressure drop indicated by manometer height "h" in units of H_2O and substituting the Ideal gas law and the value of g, we get:

SI units,

$$Q = 0.1862 C_d . d^2 \sqrt{\frac{h.P}{T}} \quad \ldots\ldots 17.23$$

And

$$\dot{m} = 64.89 C_d . d^2 \sqrt{\frac{h.P}{T}} \quad \ldots\ldots 17.24$$

In FPS units,

$$Q = 0.01204 C_d . d^2 \sqrt{\frac{h.P}{T}} \quad \ldots\ldots 17.25$$

And

$$\dot{m} = 0.03252 C_d . d^2 \sqrt{\frac{h.P}{T}} \quad \ldots\ldots 17.26$$

where

Property	Description	SI System	FPS (English) System
Q	Flow	M³/sec	ft³/sec
$\dot{m}$	Mass flow	kg/sec	lbm/sec
ρ	Density		lbm/ft³
P	Pressure	N/m²	lbf/ft²
T	Temperature	°K	°R
d	Orifice diameter	mm	In.
h	Manometer height	mm of H₂O	In.H₂O

Humidity Effect

As we know, the air in the atmosphere contains the water vapor commonly known as moisture. Hence Ideal gas law cannot be strictly applied as applied to dry air. Local humidity can be determined by using psychometric charts using dry and wet bulb temperatures. The moisture content can be found by using the relative humidity (RH). The RH is the ratio of partial pressure of water present at temperature to the saturation pressure of water at the same temperature. The other way is to express the moisture content as humidity ration which is defined as the ration of mass of water to mass of dry air.

Since the performance of the engines are evaluated under various test conditions, it is important to correct the results to standard reference conditions. The intake ambient conditions of an IC engine are normally corrected to a reference atmosphere. The total atmospheric pressure of the "reference atmosphere" consists of the partial pressure of dry air and partial pressure of water vapor.

$$P_{atm} = P_{air} + P_{water-vapor} \quad \quad \quad 17.27$$

When performing intake air flow analysis, the air pressure should be on a dry air value.

$$P_{dry-air} = P_{atm} + P_{water-vapor} \quad \quad \quad 17.28$$

Table 17.3 Referencce properties of the air

Parameter	Description	Standard	SI System	FPS (English)
P_{ref}	Reference pressure	1 bar	100,000 N/m²	14.503 psi
T_{ref}	Reference temperature	25 °C	25 °C	77 °F
RH_{ref}	Reference relative humidity	30 percent	30 percent	30 percent

The dry air density varies inversely with atmospheric temperature and directly with absolute pressure. Applying ideal gas law for air, the equation to adjust air density from test conditions to reference conditions can be written as below:

$$\rho_{REF} = \rho_{TEST}\left(\frac{P_{ref}}{P_{TEST}}\right)\left(\frac{T_T}{T_{REF}}\right), \quad \ldots\ldots\ldots\ldots\ldots 17.29$$

where

ρ_{REF} = dry air density at reference condition

ρ_{TEST} = density of dry air at test cell

P_{ref} = pressure of dry air at reference condition

P_{TEST} = pressure of dry air at test cell

T_{REF} = temperature of dry air at reference condition

P_T = temperature of dry air at test cell

Now let us consider the mass flow equation.

$$\dot{m} = 64.89 C_d . d^2 \sqrt{\frac{h.P}{T}} \quad \ldots\ldots\ldots\ldots\ldots 17.30$$

Rearranging and rewriting this equation we get,

$$\dot{m} = C_1 \sqrt{\frac{h.P}{T}} \quad \ldots\ldots\ldots 17.31$$

where

$$C_1 = 64.89 C_d . d^2$$

Further simplifying,

$$\dot{m} = C_1 \sqrt{\frac{\Delta P.P}{T}} = C_1 \sqrt{\frac{C_2 P.P}{T}} \quad \ldots\ldots 17.32$$

Thus mass flow is

$$\dot{m} = C_1 \sqrt{\frac{C_2 P^2}{T}}$$

$$\dot{m} \approx K.P.T^{-0.5} \quad \ldots\ldots\ldots 17.33$$

where

$$K = C_1 . \sqrt{C_2}.$$

The above equation indicates that air mass flow rate is proportional to local test cell atmospheric pressure and inversely proportional to the local test cell temperature. Based on the above equation, the air mass flow correction for a test engine at WOT to standard reference condition in SI system can be expressed as shown below:

$$\dot{m}_{ref} = \dot{m}_{measured} \left\{ \left(\frac{P_{REF}}{P_{TEST}}\right) \left(\frac{T_{TEST}+273}{298}\right)^{0.5} \right\} \quad \ldots\ldots 17.34$$

where

P_{REF} = dry air partial pressure at reference conditions kPa

P_{TEST} = dry air partial pressure at test conditions in kPa

T_{TEST} = dry air temperature at test conditions in °C

$\dot{m}_{measured}$ = measured mass flow rate in kg/sec

Application of Measured Air Flow

The air flow measurement is done to carry the combustion analysis. The application of measured air flow is to establish the volumetric efficiency of the engine.

Volumetric efficiency (η_{vol}) is defined as the ratio of actual air mass flow to the engine to the theoretical air mass flow of the engine.

$$\eta_{vol} = \frac{Air\ Mass\ actually\ measured}{Engine\ displacement\ volume} 100\% \quad \ldots\ldots\ldots 17.35$$

$$\eta_{vol} = \frac{Air\ Mass\ actually\ measured}{\rho \cdot V_{cyl} \cdot \left(\frac{N}{2}\right)} 100\% \quad ,\ldots\ldots 17.36$$

where

P = density if dry air

V_{cyl} = volume per cylinder in m³/ cyl /intake stroke

(In FPS (English) system ft³/ cyl/intake)

N = revolutions per minute

For four strokes engine, intake strokes = N/2. That is, for every two revolution, there will be one suction stroke to draw air in.

17.4 Blow-by Measurement

The blow-by measurement is another important parameter to be measured during the engine testing. The measurement of the blow-by is used to diagnose the problems of excessive blow-by.

17.4.1 What Is Blow-by?

Blow-by gases are exhaust which has passed by the piston rings into the crank case.

The most common checks for piston ring sealing are cranking compression and leakage beyond the piston rings into crank case. Piston rings seal the combustion chamber and prevent engine blow-by. But no set of rings can totally prevent some pressure loss past the pistons. Rings that do not seal well may allow excessive blow-by and reduce the engine power 10 to 20 horsepower or more. Blow-by also dilutes the oil in the crankcase with fuel and combustion byproducts, which can shorten the life of the oil and lead to premature engine failure. It will also increase crankcase emissions and the load on the positive crankcase ventilation (PCV) system. Measuring blow-by is one of the ways to check ring sealing.

A blow-by flow meter can tell you precisely how much blow-by is occurring inside the engine. Unlike a cranking compression test or a static leak down test, a blow-by test actually measures the volume of gases that are entering the crankcase past the piston rings. The blow-by meter allows you to measure blow-by from any engine speed, all the way from idle to WOT.

The blow-by sensor lets you measure the CFM flow from your crankcase. It is important for getting accurate results you should seal up the crankcase, valve covers, etc. Positive crankcase ventilation (PCV) valve or routing of the crankcase vapors out of the engine should be eliminated. Typically you will have the breather of one rocker cover as being the only outlet of blow-by gasses. You will route this outlet to the inlet of the blow-by sensor with a large diameter, nonrestrictive hose. After the sensor, vent these gases with a large diameter, nonrestrictive hoses to a safe, well-ventilated area.

The sensor may collect liquid from condensation and oil vapors. Periodic check is necessary, and it is important to remove hoses and letting hoses drain and tipping condensation out of both sides of sensor.

17.4.2 How to Measure Blow-by?

A blow-by measurement requires a blow-by flow meter. The meter measures airflow and is attached to either the crankcase vent on a valve cover breather or the positive crankcase ventilation (PCV) valve fitting.

When the engine is running, all blow-by that leaks past the rings will flow through the crankcase, out the valve cover opening and through the blow-by flow meter sensor. The meter outputs an analog voltage signal that ranges from zero to five volts. The display can then be converted into units that show you the volume of airflow per unit of time. The display can be either "analog" or "digital." The following figure shows the schematics of the blow-by.

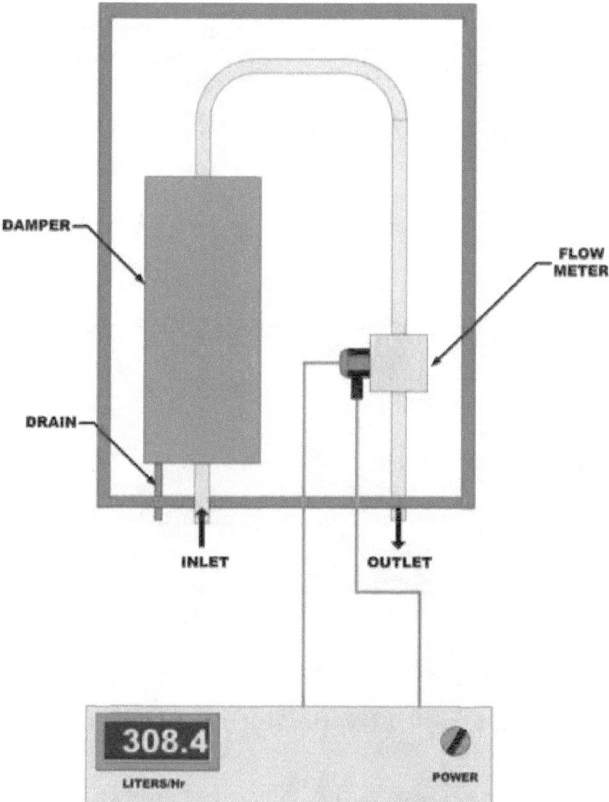

Figure 17.10 Schematics of blow-by meter.

The flow meter in the blow can be of either a orifice flow meter or vortex flow meter. The following figures show the principle of orifice and vortex flow meter.

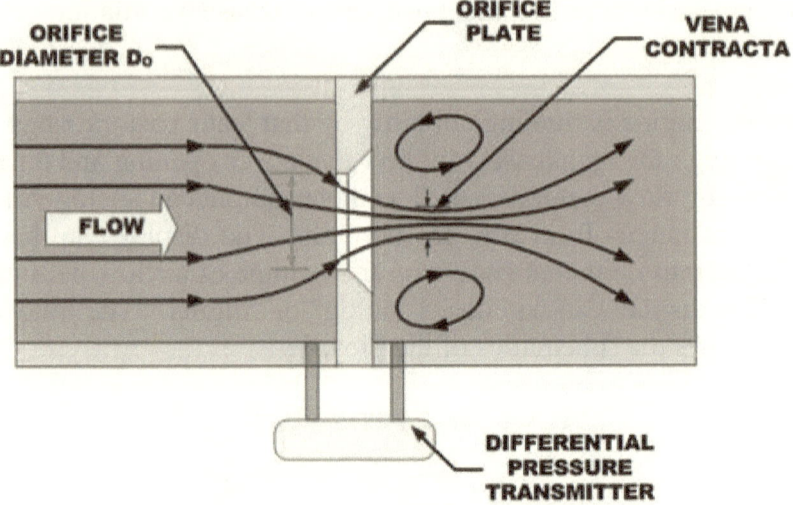

Figure 17.11 Orifice-type flow meter.

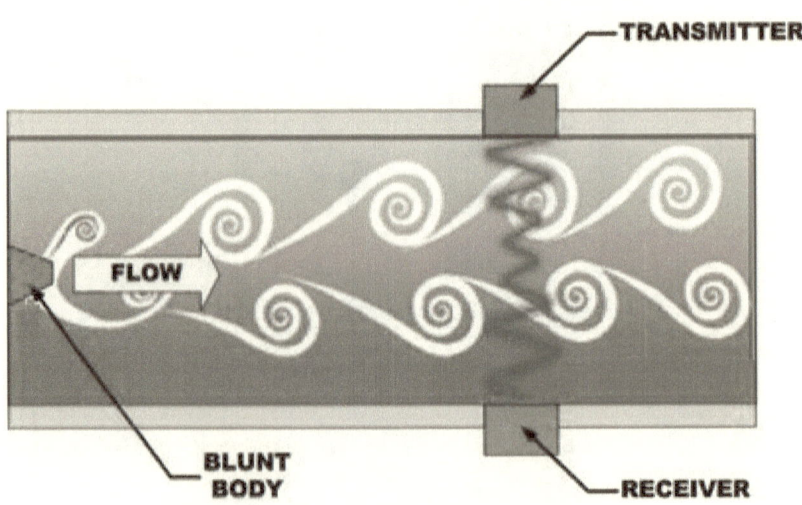

Figure 17.12 Vortex-type flow meter.

The following figure shows the blow-by meter manufactured by Technogerma Systems GMBH, based on vortex shedding principle.

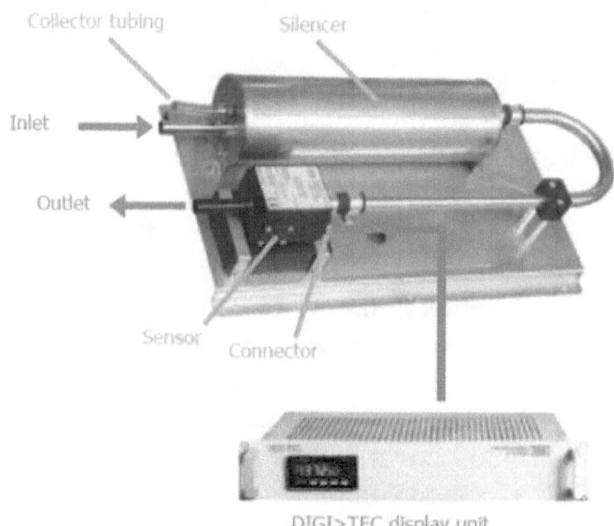

Figure 17.13 Blow-by meter. Courtesy of Technogerma Systems Gmbh.

17.4.3 How Much Blow-by Is Normal for an Engine?

There is no fixed calculation or perfect prediction formula which will tell us how much blow-by should be for a particular engine. However, some experienced testers do have their own thumb rule and it is left to individual testers to use it to their discretion.

17.4.5 Interpretation of Blow-by Readings

Less blow-by means more usable horsepower. Being able to baseline the actual blow-by in an engine means you can then go back and try different ring configurations, ring types (conventional or gapless), different ring end gap settings and cylinder wall finishes to see which combination gives the best seal and the least amount of blow-by.

Measuring blow-by has been one of the best kept secrets with performance engine builders because it allows them to see how well the rings are or are not sealing. It also allows them to detect any ring flutter that may be occurring within a particular rpm range and to then change the mass or end gaps of the rings to minimize the problem.

17.5 Oil Consumption Measurement

The oil consumption meter normally finds the usage in R&D tests for base lining the engine oil consumption or refurbishing workshops where engine is diagnosed for wear and tear of various engine parts.

17.5.1 What Causes Excessive Oil Consumption?

Oil consumption depends primarily on two things: the valve guides and piston rings. If the valve guides are worn, or if there's too much clearance between the valve stems and guides, or if the valve guide seals are worn, cracked, missing, broken, or improperly installed, the engine will suck oil down the guides and into the cylinders. The engine may still have good compression, but will use a lot of oil.

Oil consumption measurement plays a vital role in refurbishing or reconditioning workshops while diagnosing the cause of wear. If the oil burning is due to worn or broken rings, or wear in the cylinders, the engine will have low compression. The only practical remedy is to bore or hone the cylinders and replace the worn-out piston rings.

Oil burning can also occur mainly due to following reasons:

1. If the cylinders in a newly rebuilt engine are not honed properly.
2. If the rings are installed upside down.
3. Piston ring end gaps are too large or are not staggered to reduce blow-by.

17.5.2 Measuring Oil Consumption

The following figure shows the schematic arrangement of oil consumption measurement. The measuring cylinder is connected to the engine oil drain plug of the oil sump and the crank case. The liquid maintains the same level in commonly connected vessels; therefore, the oil level in measuring cylinder and the oil sump will be same. The running of the engine will cause oil to be consumed and hence the level in the oil sump as well as in the measuring cylinder will drop. The difference between the original oil level and final oil level will be the oil consumed for that testing period. The refilling cylinder will replenish the oil in the engine oil sump which can be done automatically through the associated control unit.

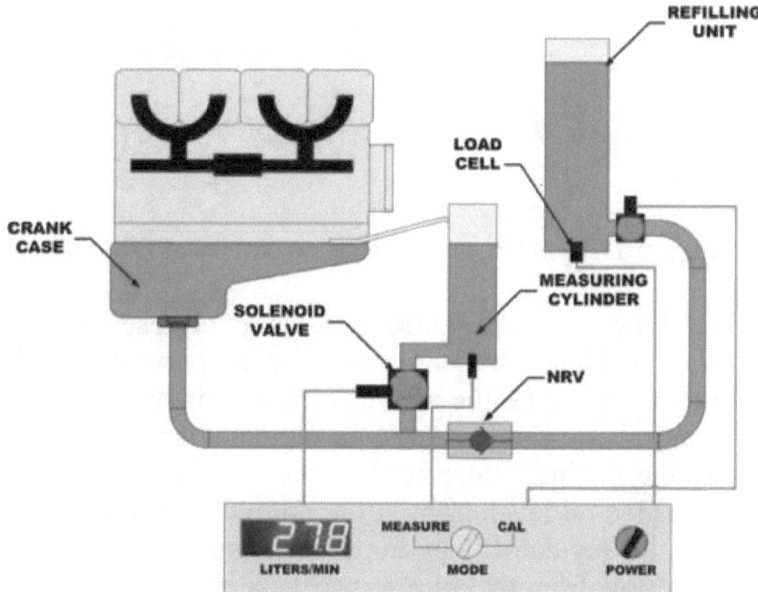

Figure 17.14 Schematics of oil consumption meter.
Courtesy of Piper Test and Measurement Ltd

17.5.3 Oil Consumption by Slow Flow Meter

The oil consumption of the engine can be measured by using a Davco slow flow meter in conjunction with Davco oil level regulator. The arrangement is shown in following figure. This method provides low maintenance and cost-effective solution for visible and accurate reading of engine oil consumption.

There are two important parts of the measurement:

1. Slow flow meter
2. Oil level regulator

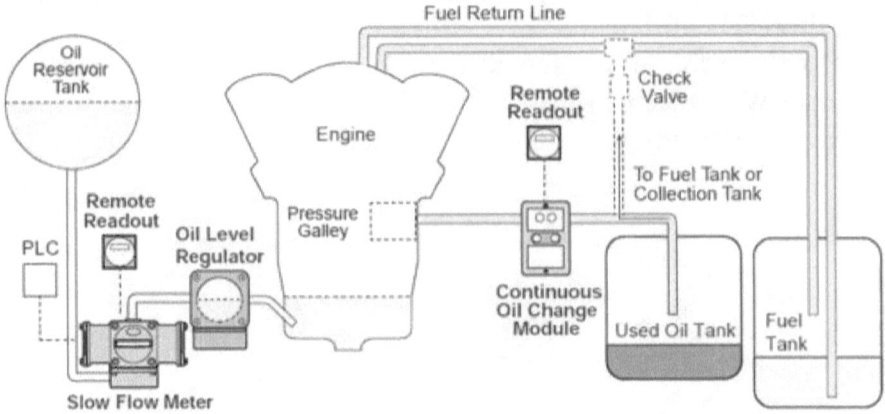

Figure 17.15 DAVCO flow meter. Courtesy of DAVCO.

The oil level regulator is installed between an oil reservoir tank and the engine oil pan. The regulator is mounted in a position such that the regulator float is closed at the running oil level height in the engine oil pan. An atmospheric vent on top of the regulator is provided. Venting directly to the engine recommended ensuring positive ventilation. When the engine oil level drops (from normal consumption or from oil leakage), the regulator float drops and opens the flow valve. This allows oil to flow from the oil reservoir tank into the engine oil pan. When the oil level rises to the full level, the float closes the flow valve and the flow of reservoir oil into the engine is stopped. This is a continuous process as the engine requires replenishment.

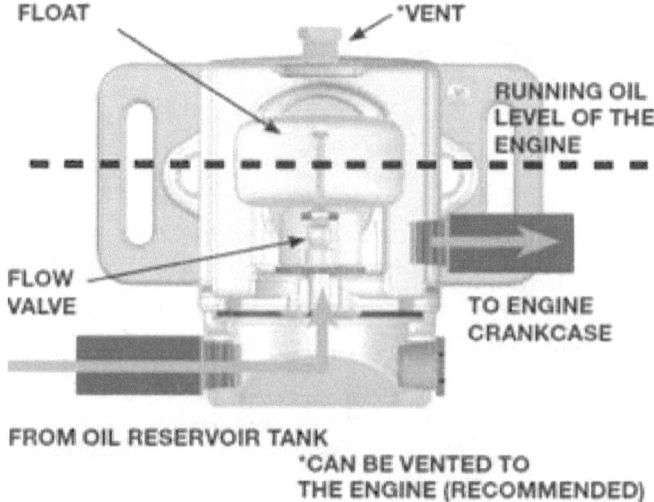

Figure 17.16 Flow valve. Courtesy of DAVCO INC.

The slow flow meter is typically installed between the oil level regulator and an oil reservoir tank with a minimum of 3.5 psi of head pressure. The piston rack is cycled by the oil head pressure switching the cycling valve from "empty" to "full." As one side fills, the other side is emptied into the crankcase via the oil level regulator. The piston rack is configured to provide a signal for the readout.

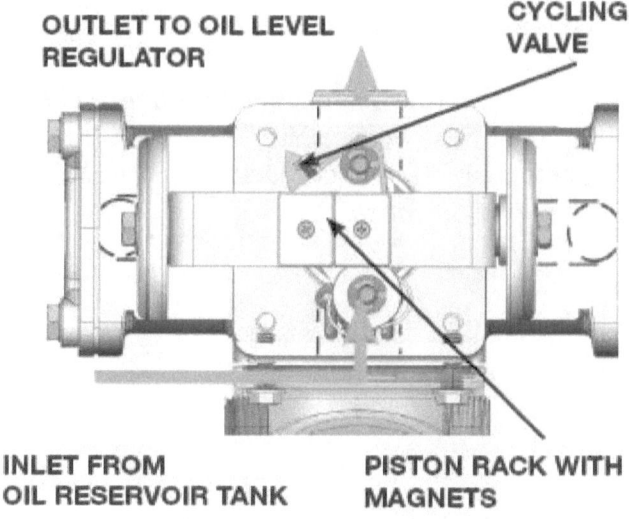

Figure 17.17 Slow flow valve. Courtesy of DAVCO INC.

As demand continues, the meter cycles, filling and emptying the chambers until demand from the regulator ceases.

Closure

The subject of engine testing is invariably thought of measuring the speed, torque, and power. However, the modern engines are getting more complex and advanced in technology day-by-day. It is therefore necessary to measure the parameters such as, fuel consumption, oil consumption, and blow-by. The measurement of these parameters strengthens the diagnostic ability of the test engineer and helps in the development process. Some of the tests are mandatory by law and need to be performed. The modern instrumentation helps to fulfill these demands.

References

1. *Rotameter information ourtesy of Blue-white industries ltd. USA*
2. *Davco-oil combsumption information*
3. *Piper—oil consumption meter*

Further reading

1. *Applied Fluid Flow Measurement*, N.P. Cheremisinoff, Marcel Decker, 1979.
2. Miller, R. W. *Flow Measurement Engineering Handbook*, McGrawHill: New York, NY: 1983.
3. N.P. Cheremisinoff, Marcel Dekker, *Flow Measurement for Engineers and Scientists*, 1988.

CHAPTER 18

Basics of Engine Testing

18.1 Introduction

This chapter is dedicated to the engine testing using various dynamometers. Basic understanding of working of two-stroke and four-stroke engines is explained. This chapter also gives brief introduction to various types of testing done especially in research and development, production, and quality assurance area. Various standards applicable for engine testing are also dealt in this chapter.

18.2 Definition of an Engine

The internal combustion engine is an engine in which the combustion of a fuel occurs with an oxidizer (usually air) in a combustion chamber.

18.3 Classification of the Engines

The engines are classified as describe below:

1. Based on stroke: four-stroke and two-stroke.
2. Ignition type: spark ignition, compression ignition, and pilot injection.
3. Combustion chamber design: open chamber (wedge, hemisphere, bowl-in-piston), divided chamber, swirl chambers, prechambers.
4. Load control: throttling of fuel and air flow (constant mixture ratio), control of fuel only, combination.
5. Cooling: water cooled, air cooled, and adiabatic.

18.4 Major Components of an IC Engine

Cylinder

A cylinder is a round hole bored through the block to receive a piston. All automobile engines four cycles or two cycles have one or more than one cylinder.

Cylinder Head

The cylinder head is the metal part of the engine that encloses and covers the cylinders. Bolted on to the top of the block, the cylinder head contains combustion chambers, water jackets, and valves (in overhead-valve engines). The head gasket seals the passages within the head-block connection and seals the cylinders as well.

The Piston

The piston converts the potential energy of the fuel into the kinetic energy that turns the crankshaft. The piston is a cylindrical-shaped hollow part that moves up and down inside the engine's cylinder. Pistons are made of light weight alloy sometimes aluminum alloys which is a good heat conductor. Pistons perform following functions.

1. Pistons transmit the driving force of combustion to the crankshaft through connecting rod which makes the crankshaft to rotate.
2. The piston also acts as sealing which can move and prevent the blow-by.
3. The piston acts as a bearing for the small end of the connecting rod.

Some pistons have a narrow groove below the top surface and above the top ring to serve as a "heat dam" to reduce the amount of heat reaching the top ring.

Rings

The piston has grooves on its perimeter near the top. In these grooves expandable rings are mounted. The piston fits snugly in the cylinder due to expansion of the piston rings. The piston rings are used to ensure a snug leak proof fit.

Wrist Pin

The wrist pin also known as gudgeon pin connects the piston to the connecting rod. The wrist pin is inserted into a hole that goes through the side of the piston, where it is attached to the connecting rod.

Connecting Rod

The connecting rod links the piston to the crankshaft. The small end has a hole in it for the piston wrist pin and the big end accommodates the crankshaft.

Connecting Rod Bearings

Connecting rod bearings are inserts that fit into the connecting rod's lower end and ride on the journals of the crankshaft.

Crankshaft

The crankshaft converts the reciprocating motion of the pistons into a rotary motion. It provides the turning motion for the wheels. It works much like the pedals of a bicycle, converting up-down motion into rotational motion.

Flywheel

Flywheel is attached to a rotating shaft and serves to smoothen the output torque from engine. The large inertia of the flywheel enables it to absorb and release energy with little variation in speed.

Oil Pump

The oil pump is used to circulate pressurized oil to the various parts of the engine for the purpose of the lubricating them.

Valve Assembly

The valve-in-head engine has pushrods that extend upward from the cam followers to rocker arms mounted on the cylinder head that contact the valve stems and transmit the motion produced by the cam profile to the valves. Tappet clearance must be maintained between the ends of the valve stems and the lifter mechanism to assure proper closing of the valves when the engine temperature changes. This is done by providing pushrod length adjustment or by the use of hydraulic lifters. Hydraulic valve lifters eliminate the need for periodic adjustment of clearance.

Overhead Valves

Valves are used for two major purposes, that is, suction of fuel and for the discharge of exhaust gases. In an overhead valve (OHV) engine, the valves are mounted in the cylinder head, above the combustion chamber. Usually this type of engine has the camshaft mounted in the cylinder block.

Camshaft

This part of a machine is used to provide a repetitive straight-line or back-and-forth motion to a second part, known as the follower. Cams are used to open and close the inlet and exhaust valves of an engine.

Overhead Camshaft (O.H.C.)

Most modern engines have the camshaft mounted above, or over, the cylinder head instead of inside the block. This arrangement has the advantage of eliminating the added weight of the rocker arms and pushrods; this weight can sometimes make the valves "float" when you are moving at high speeds. The rocker arm setup is operated by the camshaft lobe rubbing directly on the rocker.

Carburetor

The purpose of the carburetor is to supply and meter the mixture of fuel vapor and air in relation to the load and speed of the engine. The carburetor supplies a small amount of a very rich fuel mixture when the engine is cold and running at idle. With the throttle plate closed and air from the air cleaner limited by the closed choke plate, engine suction is amplified at the idle-circuit nozzle. This vacuum draws a thick spray of gasoline through the nozzle from the full float bowl, whose fuel line is closed by the float-supported needle valve. More fuel is provided when the fuel pedal is depressed for acceleration. The pedal linkage opens the throttle plate and the choke plate to send air rushing through the barrel. The linkage also depresses the accelerator pump, providing added gasoline through the accelerator-circuit nozzle. As air passes through the narrow center of the barrel, called the "venture," it produces suction that draws spray from the cruising-circuit nozzle. The float-bowl level drops and causes the float to tip and the needle valve to open the fuel line.

Ignition System

The ignition distributor makes and breaks the primary ignition circuit. It also distributes high tension current to the proper spark plug at the correct time. The distributor is driven at one-half crankshaft speed on four-cycle engines. It is driven by the camshaft. Distributor construction varies with the manufacturers, but the standard model is made of a housing into which the distributor shaft and centrifugal weight assembly are fitted with bearings. In most cases, these bearings are bronze bushings.

In standard ignition, the contact set is attached to the movable breaker plate. A vacuum advance unit attached to the distributor housing is mounted under the breaker plate. The rotor covers the centrifugal advance mechanism, which consists of a cam actuated by two centrifugal weights. As the breaker cam rotates, each lobe passes under the rubbing block, causing the breaker points to open. Since the points are in series with the primary winding of the ignition coil, current will pass through that circuit when the points close. When the points open, the magnetic field in the coil collapses and a high tension voltage is induced in the secondary windings of the coil by the movement of the magnetic field through the secondary windings.

Normally ignition system provides one lobe on the breaker cam for each cylinder of the engine, for example, six-cylinder engines will have six lobe cams in the distributor. Therefore, for every revolution of the breaker cam will produce one spark for each cylinder of the engine. In case of four-stroke engine, each cylinder fires every other revolution so the distributor shaft must revolve at one-half crankshaft speed.

Once high voltage is produced in the ignition coil by the opening of the breaker points, the current passes from the coil to the center terminal of the distributor cap. From there, it passes down to the rotor mounted on the distributor shaft and revolves with it. This high voltage current passes along the rotor and jumps the tiny gap to the electrode under which the rotor is positioned at that instant. This electrode is connected by high tension cable to the spark plug. As the rotor continues to rotate, it distributes current to each of the park plugs in turn.

Coil

The coil is nothing but an electrical transformer that increases the low voltage from battery to as high as 20,000 volts. This high voltage is necessary to create a spark across sparkplug electrode.

Spark Plugs

A spark plug is a device mounted into the combustion chamber of an engine, containing a side electrode and central insulated electrode spaced to provide a gap for firing an electrical spark to ignite air-fuel mixtures.

Fuel Injection System

Direct fuel injection means that the fuel is injected directly into the combustion chamber. The fuel injection nozzle is located in the combustion chamber.

While in the throttle body injection systems, injectors are located within the air intake cavity, commonly known as "throttle body." In case of multipoint systems, one injector per cylinder is provided.

Electronic Fuel System

The principle of electronic fuel injection is very simple. Injectors are opened not by the pressure of the fuel in the delivery lines, but by solenoids operated by an electronic control unit. Since the fuel has no resistance to overcome, other than insignificant friction losses, the pump pressure can be set at very low values, consistent with the limits of obtaining full atomization with the type of injectors used. The amount of fuel to be injected is determined by the control unit on the basis of information fed into it about the engine's operating conditions.

Suction Manifolds

An intake manifold is a system of passages which conduct the fuel mixture from the carburetor to the intake valves of the engine. Manifold design has much to do with the efficient operation of an engine. For smooth and even operation, the fuel charge taken into each cylinder should be of the same strength and quality.

Exhaust Manifolds

The exhaust manifold, usually constructed of cast iron, is a pipe that carries the exhaust gases from the combustion chambers to the exhaust pipe. It has smooth curves in it for improving the flow of exhaust. The exhaust manifold is bolted to the cylinder head.

Mechanical Fuel Pump

The mechanical fuel pump differs in that it has a vacuum booster section. The vacuum section is operated by the fuel pump lever. During the suction stroke, the rotation of the eccentric on the camshaft puts the pump operating lever into motion, pulling the lever and diaphragm down against the pressure of the diaphragm spring and producing suction in the pump chamber. The suction will hold the outlet valve closed and pull the inlet valve open, causing fuel to flow through the filter screen and down through the inlet valve of the pump chamber.

During the return stroke, the diaphragm is forced up by the diaphragm spring, the inlet valve closes and the outlet valve opens to allow fuel to flow through the outlet to the carburetor.

Electronic Fuel Pump

Electric fuel pumps have been used for many years on trucks, buses, and heavy equipments, and they have also been used as replacements for mechanically operated fuel pumps on automobiles, but only recently have they become part of a car's original equipment. The replacement types usually use a diaphragm arrangement like the mechanical pumps, except that it is actuated by an electrical solenoid.

Radiator

It is liquid-to-air heat exchanger of honeycomb construction used to remove heat from the engine coolant after the engine has been cooled. The radiator is usually mounted in front of the engine in the flow of air as the automobile moves forward. An engine-driven fan is often used to increase air flow through the radiator.

18.5 Basics of the Internal Combustion Engine

Combustion is the act of burning. Internal means inside or enclosed. Thus, in internal combustion engines, the burning of fuel takes place inside the engine. In other words, burning takes place within the same cylinder that produces energy to turn the crankshaft. In external combustion engines, such as steam engines, the burning of fuel takes place outside the engine. In the internal combustion engine, the combustion takes place inside the cylinder and is directly responsible for forcing the piston to move downward. The change of heat energy to mechanical energy by the engine is based on a fundamental law of physics.

Internal combustion engines operate utilizing the principles of nine physical phenomena as mentioned below:

1. Heat
2. Chemistry
3. Temperature
4. Force
5. Power
6. Pressure
7. Lever
8. Torque
9. Power

The following figure shows the piston and cylinder configuration of the internal combustion engines. The cylinder diameter (D) is termed as bore diameter and the travel of piston in one direction is termed as stroke (L). The volume swept by a piston during one stroke, V_s, is referred to as cylinder displacement volume. The volume between the cylinder head and the piston when it is at top dead center is termed as clearance volume (V_c). The geometric ratio of the volume at BDC to that at TDC is termed the compression ratio (R_c).

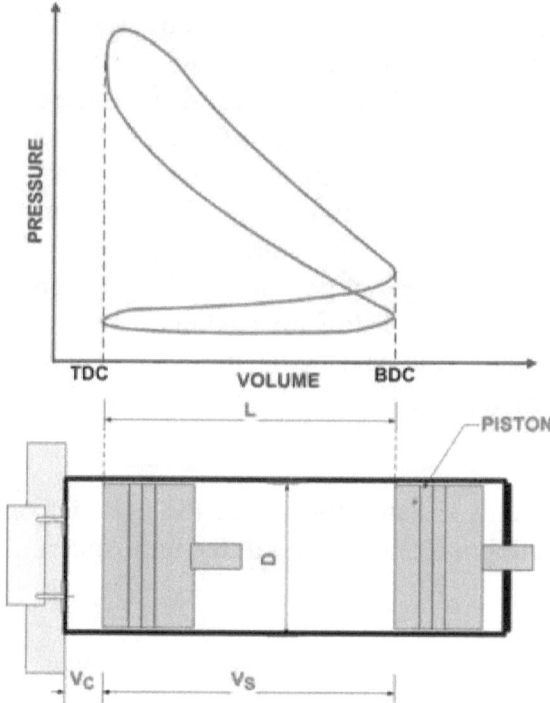

Figure 18.1 Swept and clearance volume.

$$R_c = \frac{Volume\ at\ BDC}{Volume\ at\ TDC} > 1$$

The compression ration can be defined in terms of clearance and swept volume as shown below:—

$$R_c = \frac{V_c + V_s}{V_c}.$$

Therefore,

$$R_c = 1 + \frac{V_s}{V_c} \quad\dots\dots\dots\dots\dots\dots\dots\dots 18.1$$

18.6 Two-stroke Engines

The two-stroke engine uses only one engine revolution for each power stroke, the fuel-air mixture is burned and the exhaust gases cleared out on the down stroke and the cylinder is recharged and the working fluid compressed during the upstroke. In its simplest form, as used in the spark ignition version, the two-stroke engine does not normally have separate valve mechanisms as in the four-stroke engine. Instead the air and fuel flow in and out of the cylinder through ports in the side of the cylinder wall which are opened or blocked by the movement of the piston which acts as a valve as it moves up and down past the ports in the cylinder wall. The inlet port is situated near the bottom of the cylinder and is connected to the crank case which is sealed and forms an essential part of the air-fuel management in this engine. Both sides of the piston are used in the two-stroke engine, the top side in the cylinder to provide the motive power and the underside in conjunction with the crank case to pump the air-fuel charge into the cylinder.

The exhaust port is situated further up the cylinder on the opposite side from the inlet port and is open to the atmosphere.

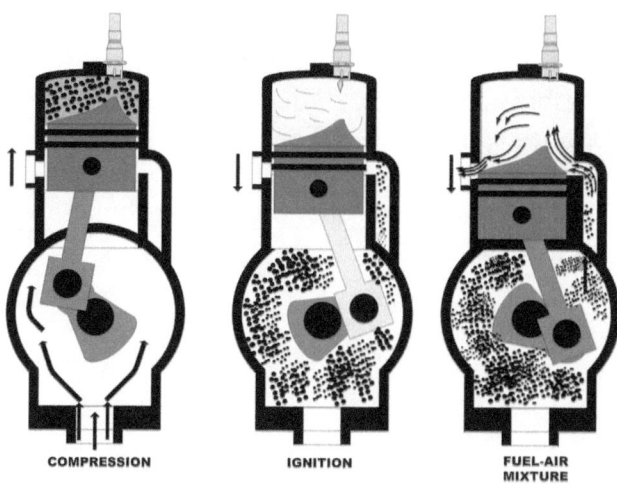

Figure 18.2 TWO-Stroke Engine.

This single cycle can be divided into two strokes as described below:

1. The air-fuel mixture in the combustion chamber with the piston just past top dead center, the highest point of travel of the piston.
2. The spark ignites the air-fuel mixture enabling combustion, and we initialize the power stroke. The exhaust gases are rapidly expanding, and this increase in pressure forces the piston down; as the piston moves down, it uncovers the exhaust port on the side of the sleeve.
3. This opening is also present on the engine block so that exhaust gases can escape to the atmosphere or through the manifold. As the piston travels down the sleeve, it pressurizes the crankcase. The pressurized crankcase is given a relief port that is opened when the piston almost reaches the bottom of the stroke. The ports direct the fuel to the top of the piston. The piston reaches bottom dead center; this is the lowest point that the piston can travel.
4. This is the beginning of the second stroke. The piston now starts to move up into the cylinder. This upward movement seals the intake port and then blocks the exhaust port. The combustion chamber now starts to become pressurized as the piston travels higher into the sleeve.

5. As the piston approaches TDC, the mixture's ignition point begins to be lowered until the heat from the glow plug initiates the combustion process. This explosion initiates the power stroke once again.

The following figure shows the pressure-volume cycle for the two-stroke engine.

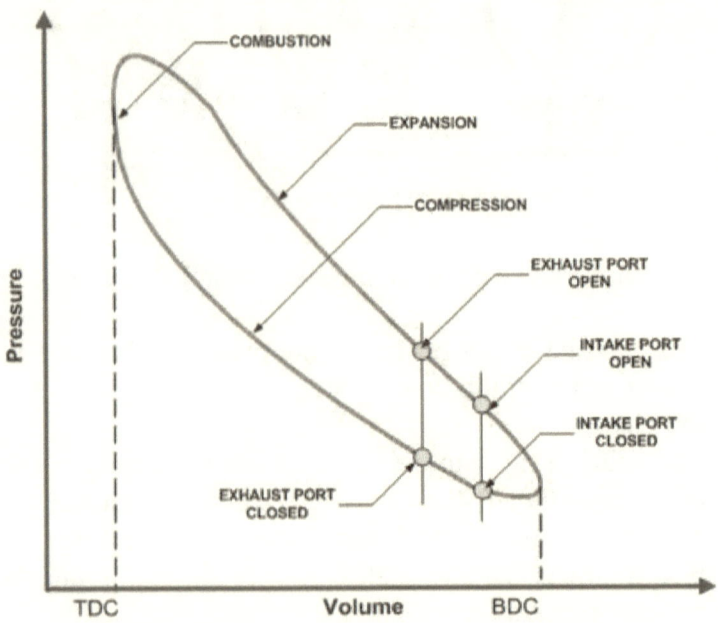

Figure 18.3 P-V cycle for two-stroke engine.

18.7 The Four-stroke Engines

The four-stroke engine uses two engine revolutions for each power stroke, one to burn the fuel-air mixture and to clear out the exhaust gases and the other to reload the cylinder with the working fluid and compress it ready for ignition. Air flow through the engine is controlled by valve mechanisms in the cylinder head. Lubricating oil is held in the crank case, isolated from the combustion chamber, and is pumped to the bearing surfaces via a separate pump.

- **The Intake/Induction Stroke**

    The four-stroke cycle starts with the intake stroke when the piston is at the top of its travel. The inlet valve opens, and as the piston moves downward, it sucks the working fluid (air or air-fuel mixture) into the cylinder under atmospheric pressure. The exhaust valve remains closed.

    INTAKE

    Figure 18.4 Intake stroke.

- **The Compression Stroke**

    When the piston reaches the bottom of its travel, the inlet valve is closed and the working fluid is compressed as the piston moves upward.

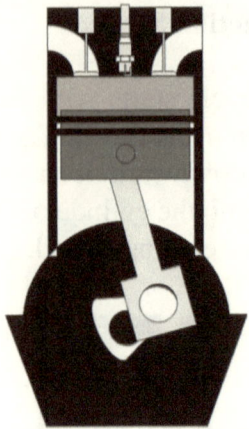

Figure 18.5 Compression stroke.

The Power Stroke

When the piston reaches the top of its travel, in the case of an Otto cycle engine, a spark ignites the air-fuel mixture initiating the power stroke in which the burning gas expands and forces the piston downward. In diesel engines, the fuel is injected into the compressed air which spontaneously ignites initiating the power stroke as in the Otto engine. Both inlet and exhaust valves remain closed.

Figure 18.6 Power stroke.

- **The Exhaust Stroke**

    As the piston passes through the end of its downward travel, the exhaust valve is opened and the upward movement of the piston expels the exhaust gases.

    After the exhaust stroke, the cycle starts again.

    Figure 18.7 Exhaust stroke.

The precise timing of the opening and closing of the valves as well as the timing of the fuel ignition may be varied to improve the gas flow and combustion processes. The engine only develops power during the power stroke. During the other three strokes, the movement of the pistons is powered by the inertia of a flywheel on the crankshaft. The following figure shows the pressure verses volume cycle for four-stroke engine.

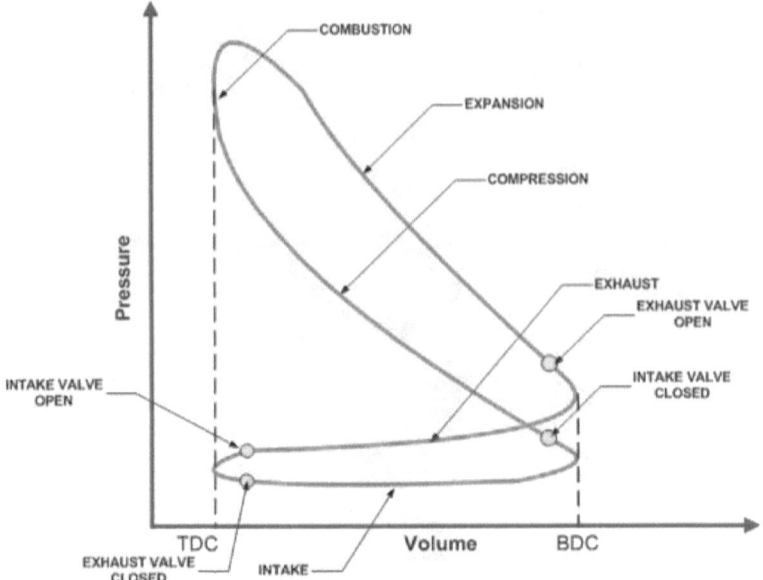

Figure 18.8 P-V cycle for four-stroke engine.

18.8 Comparison of Four-stroke and Two-stroke Engines

- **Four-stroke Engines**

  - Advantages

    1. Better control over the combustion process is possible due to more control possibilities with valve and ignition timing. This allows better fuel efficiency for the same compression ratio and better control of exhaust emissions.
    2. Better mixing of the fuel with the air due to the separate intake and compression cycle.

  - Disadvantages

    1. Less power density than the two-stroke engine since there is only one power stroke for every two engine revolutions.
    2. More complex and expensive to manufacture.

- **Two-stroke Engines**

    - Advantages

        1. Because two-stroke engines have one power stroke for every engine revolution, they have a much lower weight and a significantly better power density than the four-stroke engine for the same power output.
        2. Two-stroke engines do not normally use complex external valve mechanisms, thus they have fewer moving parts and a much simpler, less expensive construction. This in turn lowers their weight further and allows them to run at very high speeds.
        3. Overall the two-stroke machine is a powerful, low-cost, very simple, very light weight machine which is able to run at high speeds.
        4. Lubrication by mixing the oil with the fuel avoids the use of an oil sump and allows the engine to work in any orientation making it suitable for portable power tools.

    - Disadvantages

        1. Though the two-stroke engine may have a greater power output, its actual efficiency is less than the equivalent four-stroke engine. Inefficient fuel-air mixing and inefficient scavenging lead to incomplete combustion, inefficient use of the fuel, and unwanted exhaust emissions.
        2. Crank case pumping requires engine lubrication via oil mixed with the fuel. Can result in less efficient lubrication as well as unwanted burning of the lubrication oil during the combustion process creating further pollution.
        3. (**Note**: The diesel two-stroke engine which breathes air and uses conventional lubrication does not suffer from either of the above two disadvantages.)

4. Two-stroke diesel engines normally need superchargers to achieve reasonable efficiency levels which add considerably to the cost and complexity and preclude them from low-cost applications.

Now we know about the basics of the engines, let us discuss about their testing procedure and various types of testing carried out to evaluate their performance.

The more detailed information on internal combustion engine, such as working and construction details, can be found in many IC engine textbooks.

18.9 Testing Classification

The requirements of testing the internal combustion engine used in vehicular propulsion are very wide and they include

- Performance testing
- Fuel economy testing
- Durability testing
- Noise testing
- Emissions testing
- Engine's behavior for acceleration and deceleration

Most of the modern engines are designed to be used in passenger cars or the commercial vehicles; hence, development of the vehicle cannot be completed by neglecting the engine testing. They are integral part of the cars. As such you need specialized control systems, algorithms, and modern instrumentation when engines are tested separately in isolation of vehicles.

18.10 Testing Procedure

Test procedure is normally sub-divided into three phases.

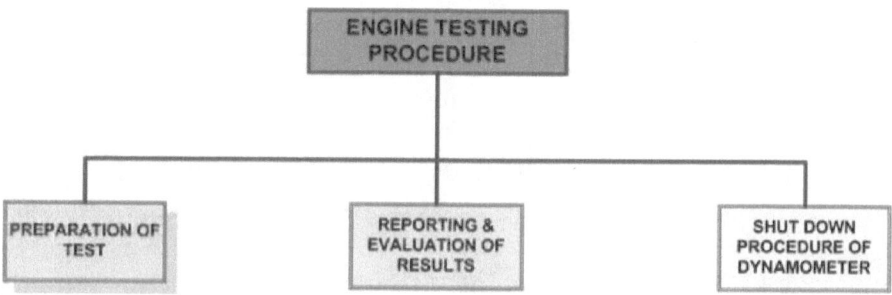

Figure 18.9 Testing procedure.

1. **Preparation of Test**

    1.1 Test procedure/specifications are selected to determine job requirements including design, quality, material, equipment, and specification.
    1.2 All safety including fire hazard, fuel storage, noise regulations, environmental regulations, and personal protection needs are observed throughout the testing period.
    1.3 Dynamometer is checked for calibration and usability for the intended operation and prepared for operation.
    1.4 Engine is connected to dynamometer including ancillary systems and engine data acquisition, monitoring and control systems.
    1.5 Exhaust extraction is connected and checked for usability for taking samples for testing.
    1.6 Engine is prepared for dynamometer testing including checking oil and water levels and fuel conditioning equipment connected.
    1.7 Ensure that fuel supply is sufficient for the intended test.

2. **Reporting and Evaluation of Results**

    1.8 Loading or testing sequence is selected by the test engineer including run-in period for new engines.
    1.9 Correction factors are determined/calculated and applied to results.
    1.10 Engine connections to the dynamometer are checked.

1.11 Selected dynamometer testing sequence is performed in accordance with technical specifications and directions of development engineer or as guided by certification agency.
1.12 Dynamometer test data is analyzed and valid conclusions about engine and sub-system condition and performance are made.
1.13 Findings including recommendations for engine configuration and/or modifications to improve performance based on dynamometer data are reported to appropriate persons.
1.14 Approved modifications are tested with confirmation run/s.
1.15 Data is presented to team members as information to complement engine/vehicle set-up.

3. **Shutdown Procedure of Dynamometer**

1.16 Dynamometer shutdown procedure is performed in accordance with manufactures/component supplier requirements.
1.17 Engine is disconnected from dynamometer.
1.18 Dynamometer and associated tooling and equipment are cleaned and refurbished.
1.19 Operator maintenance of dynamometer is conducted.
1.20 Dynamometer test results are logged to create/add to engine history.

18.11 What Do We Test Using Dynamometer and Associated Instrumentation?

Engine testing involves measurement of engine rotational speed and torque developed by the engine by applying load controlled by a dynamometer connected to its flywheel by a propeller shaft or Cardan shaft. The following measurements are undertaken to evaluate the performance of an engine.

1. Maximum power (or max torque) available as a function of operating speed.
2. Maximum rated power—highest power output an engine can develop for a short period of time.
3. Maximum max torque available as a function of operating speed.
4. Normal rated power—highest power an engine can develop continuously.
5. Rated speed—crankshaft rotational speed at which highest power is developed.
6. Range of speed and power to establish natural characteristics of engine.

The application for which the engine is going to be utilized also plays an important role and becomes a deciding factor to evaluate additional parameters. The engine-specific fuel consumption is more important for the industrial engine, whereas specific weight is more important for an aircraft engine.

The development engineer may decide to evaluate engine performance by measuring few more parameters and the effect of various operating conditions, design concepts, and modifications on these parameters. These additional parameters to evaluated are:

1. Power and mechanical efficiency.
2. Mean effective pressure and torque.
3. Volumetric efficiency.
4. Fuel-air ratio.
5. Specific fuel consumption.
6. Thermal efficiency and heat balance.
7. Exhaust smoke and other emissions.
8. Specific output.
9. Specific weight.

18.11.1 Engine Power and Torque Curves

The break horsepower (BHP) of the engine is one of the most important parameters to be evaluated by using dynamometer. This is very sensible parameter as far as engine acceptance criteria are concerned.

Engines' natural characteristics, that is, BHP verses speed (RPM) and Torque verses speed, are established by loading the engine in steps which are predetermined by test engineer. The typical performance map of IC engine is shown in the following figure:

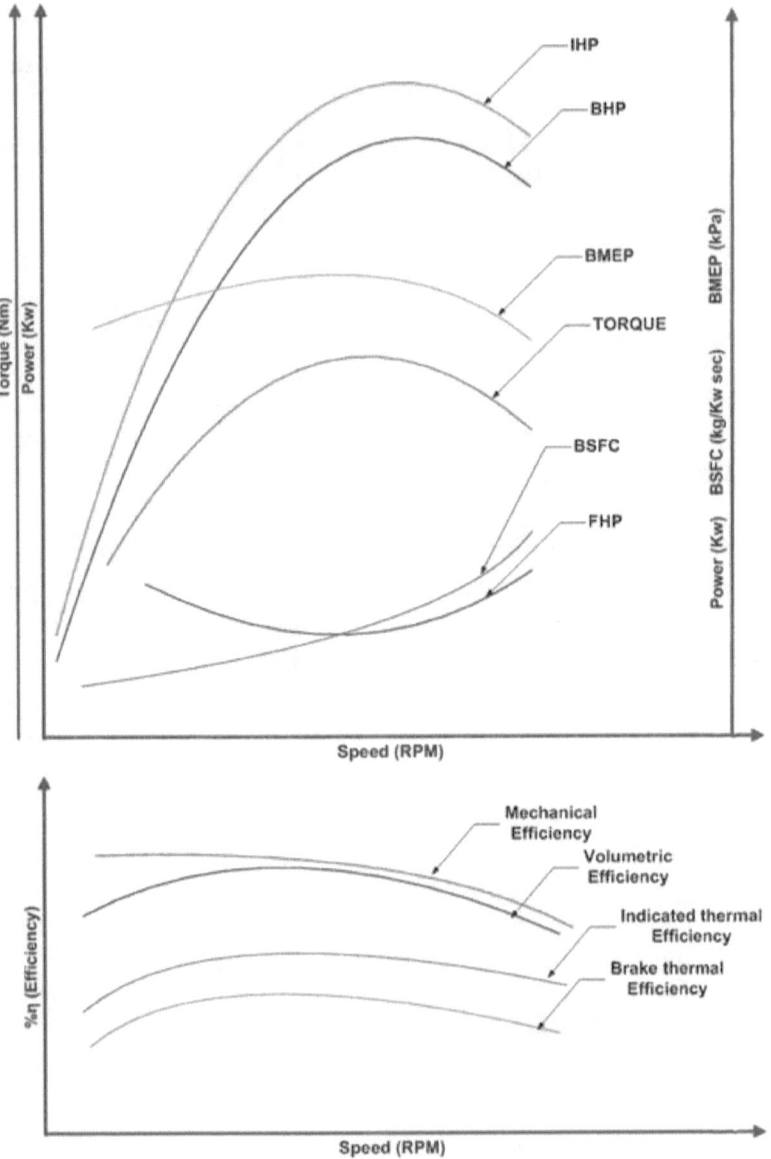

Figure 18.10 Engine characteristics.

18.11.2 Indicated Horsepower

The total power developed by combustion of fuel in the combustion chamber is called indicated power (Ip). Some part power of the indicated power is consumed in overcoming the friction between moving parts, some in the process of inducting the air and removing the products of combustion from the engine combustion chamber. Indicated power is the power developed in the cylinder due to the heat released from the combustion of fuel in the cylinder.

$$IHP = \frac{P_m LAN . C_n}{60}, \quad \ldots\ldots 18.2$$

where
P_m = mean effective pressure (Nm^2)
L = length of stroke in m
A = area of the piston in m^2
N = rotational speed of the engine in RPM (for four-stroke engine, it is N/2)
C_n = number of cylinders

18.11.3 Brake Horsepower

The BHP is calculated from the measured parameters by dynamometer that is speed (RPM) and torque.

$$BHP = \frac{2 \times \pi \times N \times T}{4500}, \quad \ldots\ldots 18.3$$

where
N = rpm of the engine
T = torque of the engine in kg-m

18.11.4 Mechanical Efficiency

The difference between indicted horsepower (IHP) and Brake horsepower (BHP) is power lost due to the friction in engine components.

$$Mechanical\ Efficienty\ (\eta_{mech}) = \frac{BHP}{IHP} \cdot 100\% \ldots . 18.4$$

18.11.5 Frictional Horsepower

The difference between IHP and BHP is called friction horsepower (FHP).

$$FHP = IHP - BHP \quad \ldots\ldots 18.5$$

Therefore, the mechanical efficiency can be expressed as:

$$Mechanical\ Efficienty\ (\eta_{mech}) = \frac{BHP}{(BHP + FHP)} \cdot 100\% \quad \ldots\ldots 18.6$$

18.11.5.1 Methods to Establish FHP

The difference between IHP and BHP is called friction horsepower (FHP). The frictional losses are ultimately dissipated to the cooling system as they appear in the form of frictional heat and this influences the cooling capacity required. Moreover, lower friction means availability of more brake power; hence, brake-specific fuel consumption is lower.

The frictional horsepower (FHP) of an I.C. engine can be determined by following methods.

1. Willan's line method
2. Morse test
3. Motoring test
4. Difference between IHP and BHP

18.11.5.1.1 Willan's Line Method

In this method, gross fuel consumption verses BHP at a constant speed is plotted and the graph is extrapolated back to zero fuel consumption as illustrated in Figure 18.11. The point where this graph cuts the BHP axis is an indication of the friction power of the engine at that speed. This negative work represents the combined loss due to mechanical friction, pumping, and blow-by. The IHP is ideally zero for no fuel consumption, and the BHP is therefore the negative of the FHP. Again this is also constant speed test.

$$IHP - BHP = FHP.$$

When IHP (Indicated horse power) is zero then there is no fuel consumption then:

$$-BHP = FHP \dotfill 18.7$$

This test is applicable only to engines with compression ignition method.

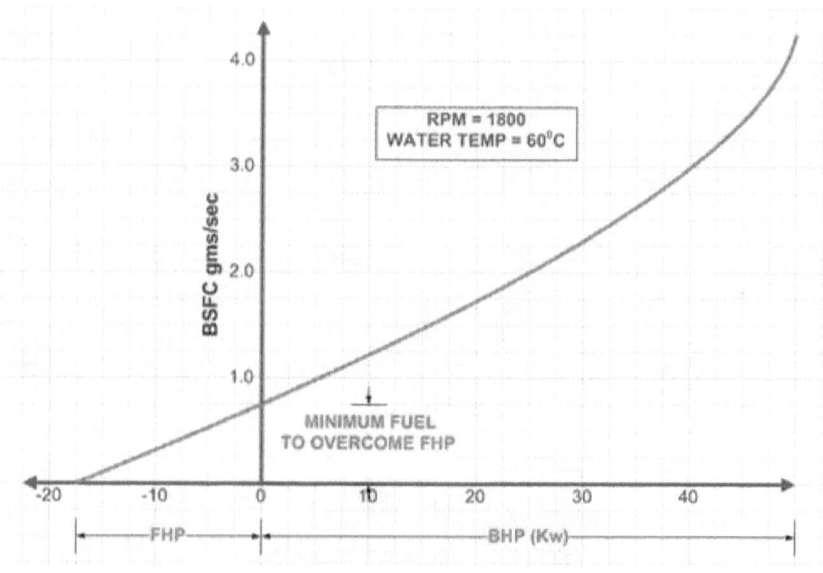

Figure 18.11 Willan's line.

18.11.5.1.2 Morse Test

This test is applicable only to multicylinder engines and minimum requirement is engine with two cylinders. The following steps are followed to establish the FHP by Morse method:

1. The engine is run at the required speed and the output is measured.

2. One cylinder of the engine under test is cut out by short circuiting or unplugging the supply to the spark plug or by disconnecting the injector as the case may be.
3. Under this condition, all other cylinders drive this cut-out cylinder. The output is measured by keeping the speed constant at its original value.
4. The difference in the outputs is a measure of the indicated horsepower *(IHP)* of the cut-out cylinder.
5. Thus, for each cylinder, the *IHP* is obtained and is added together to find the total IHP of the engine.

The *IHP* of the n cylinder is given by

$$IHP_n = BHP_n + FHP \quad \text{...............} 18.8$$

IHP for (N-1) cylinder is given by

$$IHP_{n-1} = BHP_{n-1} + FHP. \quad \text{...............} 18.9$$

Since engine is running at same constant speed, we can assume the *FHP* is constant.

From above two equations, it can be deduced that *IHP* of the nth cylinder is expressed by the equation:

$$IHP_{n^{th}} = BHP_n - BHP_{n-1} \quad \text{...............} 18.10$$

Therefore, the total *IHP* of the engine is given by

$$IHP_n = \sum IHP_{n^{th}} \quad \text{...............} 18.11$$

Subtracting total IHP from this total IHP, we get the FHP of the engine.

The testing engineer should know about the following factors affecting the FHP measurement by Morse method.

1. The accuracy of the FHP measurement is affected by changes in air-fuel mixture due to cutting of the cylinders.

2. In multicylinder having common manifold, the air-fuel mixture distribution and volumetric efficiency both change affecting the engine output.
3. The cutting of cylinder may affect pulsation in engine exhaust manifold which may significantly change the back pressure which ultimately affects the engine performance.

18.11.5.1.3 Motoring test

The engine under test is run at given speed and load on dynamometer before motoring test to achieve stable conditions of water, lubricating oil, and body temperature. The test cell set up with DC or AC dynamometer is most suitable for this application. However, hydraulic or eddy-current dynamometer with motoring arrangement is also employed to find out FHP of the engine.

To initiate the motoring test, the fuel supply to engine is stopped and absorbing dynamometer is switched to motoring mode by electric drive to "motor" the engine at the same speed at which it was previously running.

The torque is measured by means of the load cell on dynamometer and the speed measurement by speed indicator which enables test engineer to evaluate the input power to the engine which is FHP.

Alternately, the electrical input power to the motor is measured from electrical drive panel. If we neglect the losses in driveshaft connecting motoring dynamometer and engine, and the efficiency of electric motor, then the electric input power will be equal to FHP of the engine.

The testing engineer should know about the following factors affecting the FHP measurement by motoring method:

1. The temperatures of the engine under motoring test are different than when it is under load test. The incoming air cools the cylinder. This reduces the lubricating oil temperature and increases friction increasing the oil viscosity. This problem is much more severe in air-cooled engines.

2. The clearance between piston and cylinder wall is more as piston rings won't expand as much as they expand during the load test. This reduces the piston friction.

The method of finding the FHP by computing the difference between IHP which is obtained from an indicator diagram and BHP measured by dynamometer is the ideal method.

18.11.6 Fuel Consumption

The fuel consumption is very sensitive aspect from the point of view of user. The fuel consumption is measure with the help of fuel measuring instrument interfaced with dynamometer control panel. Normally the fuel consumption is measured at various torque points at corresponding RPM of the engine. Accurate measurement of fuel consumption is very important in engine testing work. There are two basic types of fuel measurement methods:

1. Volumetric type
2. Gravimetric type

18.11.6.1 Volumetric Fuel Measurement

Volumetric-type flow meter includes Burette method, Automatic Burette flow meter, and Turbine flow meter.

Total fuel consumption (TFC) measured in kg/hr.

$$TFC = \frac{cc \times \rho \times 3600}{t \times 3600} \ kg/hr, \quad \ldots\ldots\ldots\ldots 18.12$$

where
t = time for 10 cc is volume of fuel consumption in s
ρ = density (Diesel) in g/cc
V = 0.838 g/cc

Specific fuel consumption (SFC) in kg/BHP-hr.

$$SFC = \frac{TFC}{BHP} kg \frac{BHP}{Hr}. \quad \text{.....................} 18.13$$

18.11.6.2 Gravimetric Fuel Measurement

The efficiency of an engine is related to the kilograms of fuel which are consumed and not the number of liters. The method of measuring volume flow and then correcting it for specific gravity variations is quite inconvenient and inherently limited in accuracy. Instead if the weight of the fuel consumed is directly measured, a great improvement in accuracy and cost can be obtained.

18.11.7 Air Consumption

It is important to know air consumption to establish the fuel-air ratio (F/A). It is the ratio of the mass of fuel to the mass of air in the fuel-air mixture. Air-fuel ratio (A/F) is reciprocal of fuel-air ratio. Fuel-air ratio of the mixture affects the combustion phenomenon in that it determines the flame propagation velocity, the heat release in the combustion chamber, the maximum temperature, and the completeness of combustion.

The various methods used for air flow measurement include:

1. Air box method, and
2. Viscous-flow air meter.

Air/Fuel ratio—ratio of mass air flow rate into engine to fuel mass flow rate into engine.

$$Air\ fuel\ ratio = \frac{M_a}{M_f} \quad \text{.....................} 18.14$$

Fuel/Air ratio—The ratio of fuel mass flow rate into engine to air flow rate into engine.

$$Fuel/Air\ ratio = \frac{M_f}{M_a}. \quad\quad\quad\quad 18.15$$

Volumetric Efficiency: It is defined as the ratio of actual volume to the charge drawn in during the suction stroke to the swept volume of the piston. Volumetric efficiency of an engine is an indication of the measure of the degree to which the engine fills its swept volume. The amount of air taken inside the cylinder is dependent on the volumetric efficiency of an engine and hence puts a limit on the amount of fuel which can be efficiently burned and the power output.

The air consumption measurement also helps in establishing the volumetric efficiency of the engine.

18.11.8 Brake Thermal Efficiency

Thermal efficiency of an engine is defined as the ratio of the output to that of the chemical energy input in the form of fuel supply. It may be based on brake or indicated output. It is the true indication of the efficiency with which the chemical energy of fuel (input) is converted into mechanical work. Thermal efficiency also accounts for combustion efficiency, that is, for the fact that whole of the chemical energy of the fuel is not converted into heat energy during combustion.

$$\eta_{BTE} = \frac{BHP \times 4500}{\frac{TFC}{60} \times CV \times J} \times 100\ \%, \quad\quad\quad\quad 18.16$$

where
TFC = total fuel consumption in kg/hr
CV = Calorific value of diesel (11000 kcal/kg)
J = Mechanical equivalent of heat (427 kgf-m / kcal)

18.11.9 Indicated Thermal Efficiency

$$\eta_{BTE} = \frac{IHP \times 4500}{\frac{TFC}{60} \times CV \times J} \times 100\ \%, \dots\dots\dots\dots\dots\dots 18.17$$

where
TFC = total fuel consumption in kg/hr
CV = calorific value of diesel (11000 kcal/kg)
J = mechanical equivalent of heat (427 kgf-m/kcal)

18.11.10 Brake Mean Effective Pressure

Mean effective pressure is defined as a hypothetical/average pressure which is assumed to be acting on the piston throughout the power stroke. If the mean effective pressure is based on BHP, it is called the brake mean effective pressure, and if based on IHP, it is called indicated mean effective pressure (imep).

$$BMEP = \frac{BHP \times 4500}{LA\left(\frac{N}{2}\right)n}\ kg/cm^2, \dots\dots\dots\dots\dots 18.18$$

where
L = stroke length in m
A = bore area in m^2
N = number of cylinders
N = RPM of engine (N/2 is used for four-stroke engine)

18.12 Heat Balance of the Engine

The internal combustion runs on the energy supplied by the fuel used in the engine. The energy supplied to the engine comes out from the heat value of the fuel consumed. It is experimentally proved that only the part of the fuel energy is converted into output power which is really used to drive the machines. The remaining part of the fuel energy is transformed into exhaust gasses and the cooling water which do not really contribute in the output.

A typical heat balance of the internal combustion engine is illustrated below in Figure 18.12.

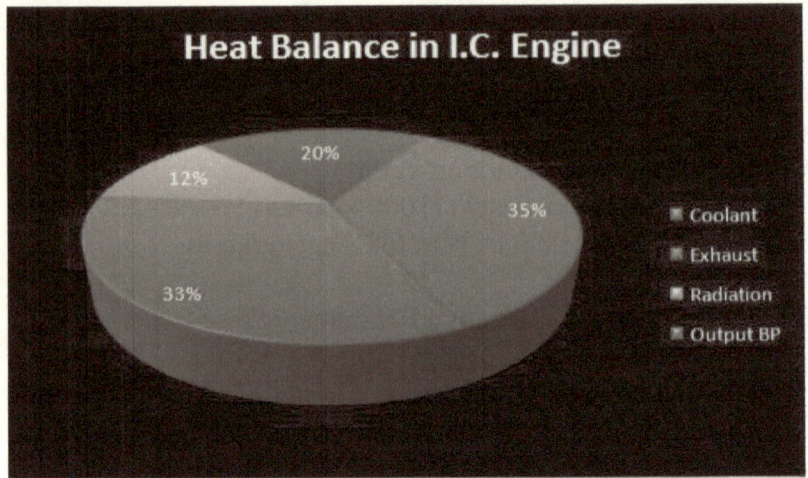

Figure 18.12 Heat balance in IC engine.

1. The energy input to the engine goes out in various forms—a part is in the form of brake output, a part into exhaust, and the rest is taken by cooling water and the lubricating oil.
2. The break-up of the total energy input into these different parts is called the heat balance.
3. The main components in a heat balance are brake output, coolant losses, heat going to exhaust, radiation, and other losses.
4. Preparation of heat balance sheet gives us an idea about the amount of energy wasted in various parts and allows us to think of methods to reduce the losses so incurred.

18.13 Engine Testing In Industry

The manufactures of the automotive and industrial engine will have testing facility to test and evaluate the engine in research and development, production, quality assurance, and type testing of the engine.

18.13.1 Research and development

Product Testing and *Endurance*

The endurance test bed will run the engines for extended period of time defined by research and development engineer. During the endurance testing, the aim might be to test the entire engine for endurance or a particular assembly mounted on engine such as fuel pump or water pump.

The computer-controlled test cell will be more useful in endurance testing as endurance test cycle can be programmed and the data can be logged periodically which can be further analyzed by test engineer.

Transient Testing

The engine is operated at transient load and speed conditions, which represents the normal pattern of the engine operation.

Transient means changing. Transient state is an interval of time in which the system is either "warming up" or taking its time to respond to a disturbance.

Steady is the opposite of transient. Steady state is a condition where the system continues with an easily predictable behavior and few values of it are changing if any.

Transient state response is a description of how the system functions during transient state. Steady-state response is a description of how the system functions during steady state.

18.13.2 Production Testing

The production testing of the engine consists of the end of line production test cells. All the test cells in production will be similar and capable of accommodating all engine models produced by the manufacturer. The system also needs to be designed efficiently so that the greatest number of engines can be tested on the minimum number of test cells. Normally in the production test cell, once the engine is engaged the test cell computer will run a predesigned test sequence. The system monitors and logs the data automatically. Once the test is completed, the engine is routed either to vehicle assembly or to back to correction area depending upon pass/fail criteria of the engine.

18.13.3 Quality Audit Testing of Engines

Quality Audit engine test systems are typically used as an overcheck of the production testing process. The tests run on Quality Audit test stands are more comprehensive and may run the engine for longer periods of time and collect more data parameters than the production test. Pass/Fail criteria is normally not checked during the test. Parameter data is collected in time-stamped data packets to be analyzed by statistical analysis tools on a remote engineering or plant host computer.

The operator can select a predetermined test sequence to be run or manually run the engine at specific speed/load set points. This system, at the end of a manufacturing process, must be flexible enough to run each of the different engine models produced at a facility. The system also needs to be designed for flexibility so that test sequences, procedures, and parameters can easily be modified for new or special tests.

18.13.4 Type Testing

Type test as it implies is a test carried out to prove the whole performance of the machine/engine is finally designed and developed for manufacture. It must be specifically devised with that end in view, and it must clearly demonstrate the ability of that model of engine to perform its function in accordance with specified requirements under specified conditions for a specified length of time and with specified maintenance. Therefore, the aim of the type test is different from the regular brief production tests, while they check that the required basic performance is attained. It also aims at that the quality of material and the workmanship, applied in manufacture, have been adequate and the production routine has been up to the intended standard.

Important Characteristics of Type Test

1. Relevance

A type test should be appropriate to the specified capabilities of the engine and the conditions and duties which it may actually meet in service. It should include every feature that is needed to prove beyond doubt or dispute that the specified characteristics do in fact exist and the specified performance is, at least, met. However, it should not go beyond the requirements.

2. Rigor

A type test must be rigorous, in the sense both of strictness and of accuracy, to a degree that will ensure that the item being tested is operated under the worst possible conditions.

3. Impartiality

A type test must be impartial and must be done honestly.

A type test carried out for a particular type of engine at any place at any manufacturer will be accepted for all engines of the same type built by licensees and licensors. Engines which are subjected to type testing are to be tested in accordance to the scope defined by the purchaser.

A generalized prerequisite for type testing of the engine is given below:

1. The engine undergoing type testing is optimized as required for the condition of the type defined by purchaser.
2. The investigations and measurements required for reliable engine operation have been carried out during internal tests by the engine manufacturer and the design approval has been obtained for the engine type in question on the basis of documentation requested and the Classification Societies have been informed about the nature and extent of investigations carried out during the preproduction stages.

3. The type test is subdivided into three stages, namely:

Stage A—Internal Tests

Functional tests and collection of operating values including test hours during the internal tests and the relevant results of which are to be presented to the purchaser. During the internal tests, the engine is to be operated at the load points important for the engine manufacturer and the pertaining operating values are to be recorded. The load points may be selected according to the range of application. Whether an engine can be satisfactorily operated at all load points without using mechanically driven cylinder lubricators is to be verified. For engines which may operate of heavy fuel oil, the suitability for this will have to be proved in an appropriate form, at manufacturer's (Licensor or Licensee) test bed in general, but, where not possible, latest on board for the first engine to be put into service.

Stage B—Type Approval Test

Test approval test in the presence of the purchaser or third party inspection agency or their representatives. The type tests are to be carried out in the presence of nominated inspection agency or the representative of the purchaser and the results achieved are to be recorded and signed by the attending representatives. Deviations from the specifications, if any, are to be agreed between the engine manufacturer and the purchaser.

Load Points

Load points are test points at which the engine is to be tested according to the power/speed diagram. The data to be measured and recorded when testing the engine at various load points are to include all necessary parameters for the engine operation.

The operating time per load point depends on

- Engine size and time required to achievement of steady-state condition of the test point.

- The time for other measuring parameters like fuel consumption reading, smoke reading, etc.

Following load points are normally tested:

1. Rated power, that is, 100 percent output at 100 percent torque and 100 percent speed corresponding to max power point.
2. 100 percent power at maximum permissible corresponding speed.
3. Maximum permissible torque at corresponding speed.
4. Maximum permissible power (normally 110 percent) and corresponding speed.
5. Minimum permissible speed at 100 percent torque.
6. Part loads, for example, 75 percent, 50 percent, 25 percent of rated power and speed according to engine curve.

Emergency Safety Testing

Following tests are conducted to check that the emergency trip system works once the predetermined setting of the safety limit is crossed.

- Maximum achievable power when operating along the nominal curve and when operating with constant governor setting for rated speed.
- Testing of electronic controls on engine.
- Functional tests.
- Lowest engine speed according to nominal engine curve.
- Governor test.
- Testing the safety system:

    a. Overspeed testing
    b. Low lubrication oil pressure.

Stage C—Component Inspection

The component inspections are done by the purchaser's representative or by the nominated inspection agency. The engine manufacturer will have to compile all results and measurements for the engine tested during the type test in a type test report, which need to be presented as complete report of type testing.

Normally after the load testing the engine is dismantled and the following components are presented for the inspection:

1. Piston removed and dismantled
2. Crosshead bearing, dismantled
3. Crank bearing and main bearing, dismantled
4. Cylinder liner in the installed condition
5. Cylinder head, valves disassembled
6. Control gear, camshaft, and crankcase with opened covers.

Closure

The subject of the engine testing is getting complicated day-by-day as engines are built with modern technology to meet regulation governing fuel and exhaust emissions. It has become necessary to cope up with the needs of the engine testing; the dynamometer and the control system are also becoming more complex involving high speed data acquisition and control system. The test engineer needs to be acquainted with the new technology and methodology available for testing of the modern engines.

Bibliography:

1. Ganesan. *Internal Combustion Engines*.
2. Willard W. Pulkrabek, *Engineering Fundamentals of the Internal Combustion Engine*.

Further Reading

1. Charles Fayette Taylor, *The Internal-combustion Engine in Theory and Practice: Combustion, fuels, Materials, Design*, Vol. 2.
2. Eugene L. Keating, *Applied combustion*, Vol. 10,

CHAPTER 19

Test Cell Essentials

19.1 Introduction

The test cell is not complete with engine and dynamometer alone. The enhancement of the test cell with various accessories and instrumentation makes the test cell worthy of carrying the testing efficiently and safely. There are many essential test accessories which go along with the dynamometer set up. Few of the accessories are dealt in this chapter. The instrumentation is dealt in the chapter "Modern measurements of the engine."

19.2 Dynamometer Accessories

The bare dynamometer is installed with following accessories so that its safety and performance are ensured. Some of the accessories are important and play a vital role in engine testing.

19.2.1 Dynamometer Water Inlet Filter

The water supply to the dynamometer is passed through the filter so that impurities in the water are arrested. The filters used are of various types as shown in below figures.

19.2.1.1 Y-type Strainer

The strainer is inserted in the water inlet pipe and widely used as its installation is easy. The cleaning of the strainer element is also easy and easily accessible for removal. These types of strainers are available in flanged-end connections.

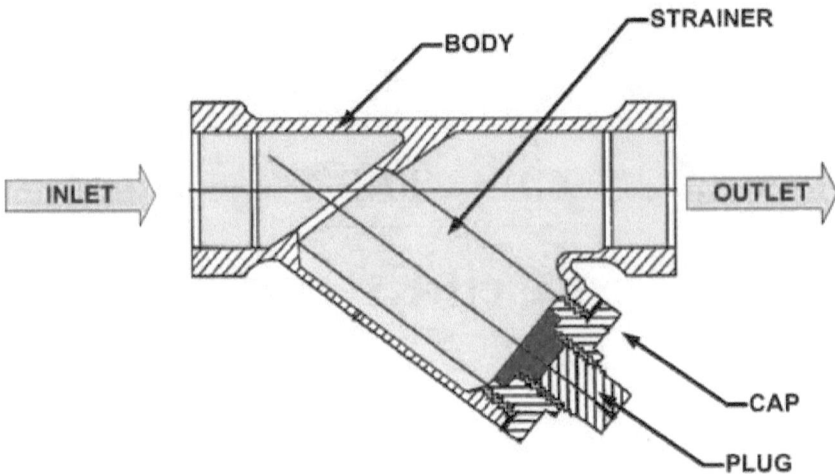

Figure 19.1 Y-type strainer with threaded ends.

The following figure shows the Y-type strainer with flanged-end connection:

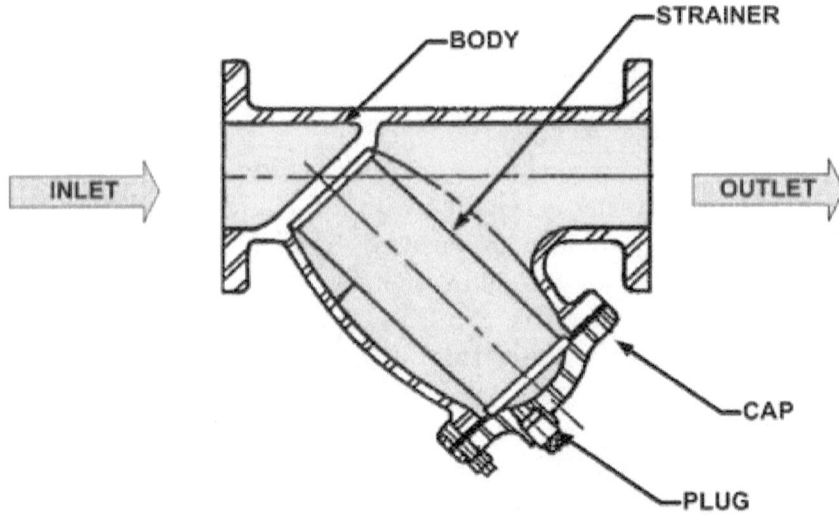

Figure 19.2 Y-type strainer with flanged ends.

19.2.1.2 Bucket Type Filter

The bucket type filter is normally used on large capacity dynamometers. The bucket type filter is installed on inlet piping to the dynamometer. The bucket type filters are also easy to clean and maintain.

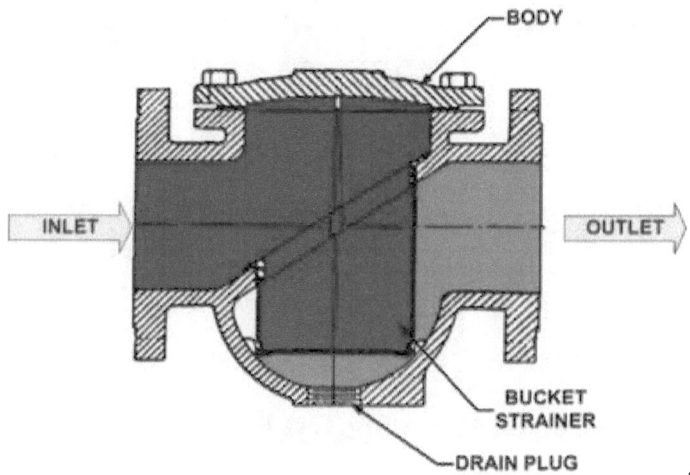

Figure 19.3 Bucket type strainer with flanged ends.

19.2.1.3 Magnetic Filter

The magnetic filter is widely used on the installation of eddy-current dynamometer. The iron particles floating in incoming water supply are arrested by this filter before they could interfere in the magnetic circuit of the eddy-current dynamometer. The following figure shows clamp on type magnetic filter.

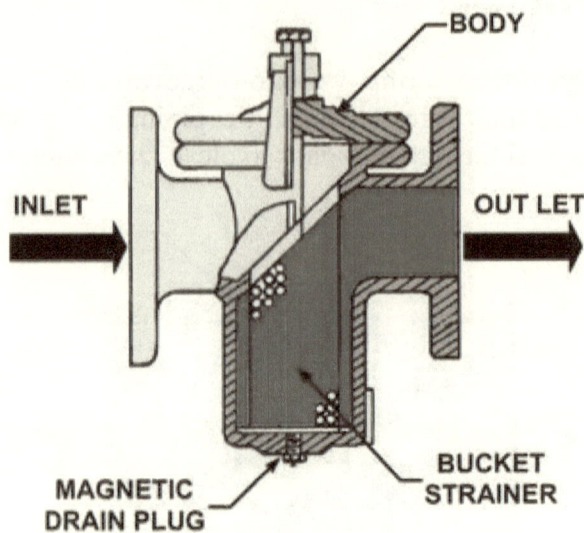

Figure 19.4 Bucket type strainer with flanged ends.

19.3 Universal Engine Mounting Test Beds

There are certain prerequisites for test bed before an engine can be mounted on it as mentioned below:

1. Test bed/stand should be robust in design and it should sustain the weight of the range of engines for which it is designed.
2. It should be capable of absorbing the vibrations.
3. It should be accurately and precisely leveled and installed.
4. Provision should be made to move the engine along the X, Y, and Z axis to facilitate the easy alignment.

The following figure shows the universal type engine mounting stand. It can adjust the center of the engine along the three axes.

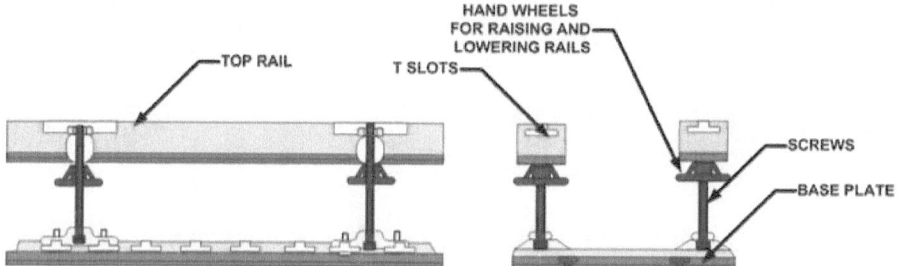

Figure 19.5 Universal engine stand.

19.3.1 T-slotted Bedplates (CI and MS)

The universal mounting stands have limitations as far size is concerned. Very large diesel engines used for marine application can be mounted using a large T-slotted bedplate and mounting fixtures.

Figure 19.6 T-slotted bedplate.
Courtesy Of Omniscient Engineering, India.

19.4 Cardan Shaft

In the test cell to test the engine card and shaft a.k.a. propeller shaft or for that matter you need some kind of coupling or flexible shaft to connect the engine under test to dynamometer or the loading device. The proper capacity card and shaft should be used to test the engine. The selection of Cardan shaft depends on many factors such as moment of inertia of engine and dynamometer, torque developed by the engine, and maximum speed of the engine. The detailed procedure for the selection of Cardan shaft is dealt in the chapter "Basics of the Vibration Analysis."

Figure 19.7 Fixed length Cardan shaft.

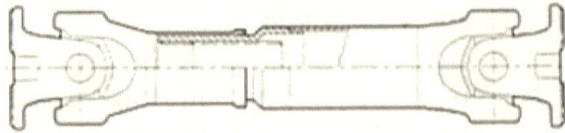

Figure 19.8 Telescopic length of the Cardan shaft.

19.5 Shaft Guard

A shaft guard is provided to cover the Cardan shaft to prevent the accidents and to arrest the flying fasteners in the event of them getting loose due to the speed.

The shaft guard should be sturdy enough to prevent and stop the parts flying away in case of breakdown. To ensure the safety, an electrical interlock can be wired in the safety circuit of the engine starting control.

Figure 19.9 Driveshaft guard. Courtesy Taylor Dyanmometers

19.6 Transducer Box

A transducer box is an enclosure which houses transducers/sensors which are very expensive and prone to failure due to mishandling and frequent handling. Transducers such as pressure transducers and pressure transmitters are mounted inside a safe enclosure, and all electrical and pipelines/steel tubings are brought to the respective pressure sensors in the transducer box from the measuring points on the either engine or dynamometer. A transducer box is mounted on engine test cell away from engine stand so that it will not be a hindrance to the movement of the engine or test personnel. It greatly depends on the layout of the test cell. Transducer box is normally designed to house wide range of transducer types such as:

1. Pressure transducers.
2. Pressure transmitters.
3. Thermocouples connectors
4. Connectors PT-100 resistance bulb temperature sensors.

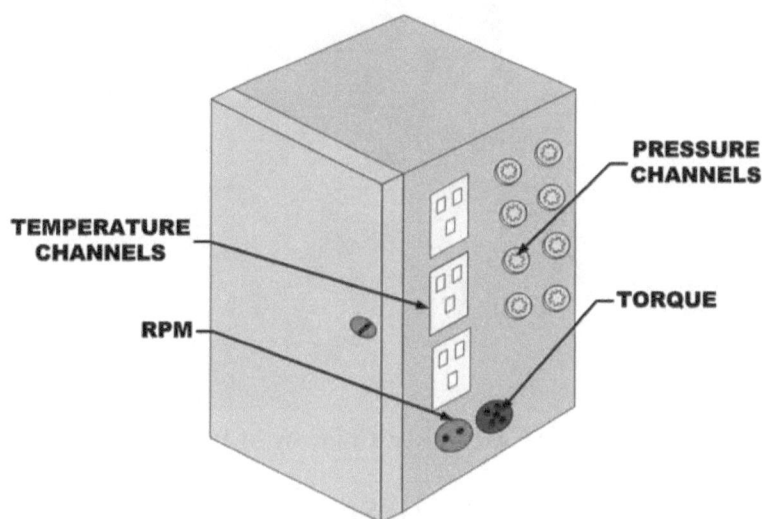

Figure 19.10 Transducer box.

19.7 In-cell Control Panel

In-cell control panel, sometimes called as auxiliary panel, is located inside the test cell. The "in-cell" panel facilities the operator to start the engine, adjust throttle, and adjust the idle rpm using the controls located on "in-cell" panel. Once these adjustments are made, operator can transfer the controls to the main control panel which is located in control room outside the test cell.

19.8 Engine Starting/Running-in Arrangement

Engine starting of the engine on test bed can be by an electric motor/air motor mounted on dynamometer. The dynamometers are equipped with starting arrangement which eliminates the need for an engine-mounted starting system. The test cell engine starting/cranking motor must be capable of accelerating the entire dynamometer and engine to starting speed requirement and disengage after starting.

In case of AC/DC dynamometer, the associated four-quadrant control is capable of starting the engine. Moreover the AC/DC machine can be used in motoring mode to do the running in of the engine.

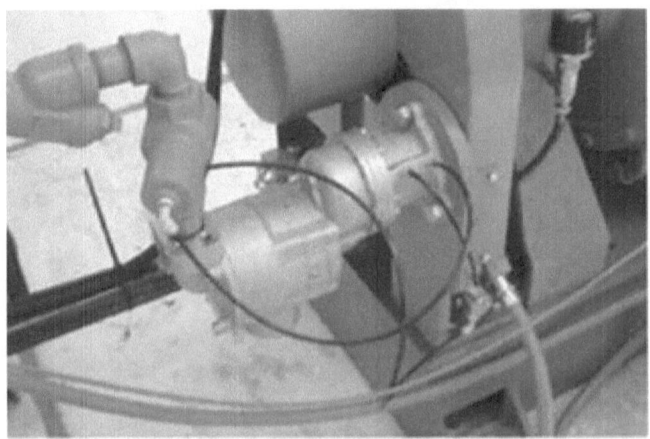

Figure 19.11 Air starter. Courtesy of Power Test.

19.8.1 In-Line Mounted Electric Starting Motor

The most widely used method is mounting an electric motor to the nondriving end of the dynamometer which is in line with the dynamometer (the other end where the engine under test is not connected) with suitable electromagnetic clutch which is remotely controledl through starting control logic.

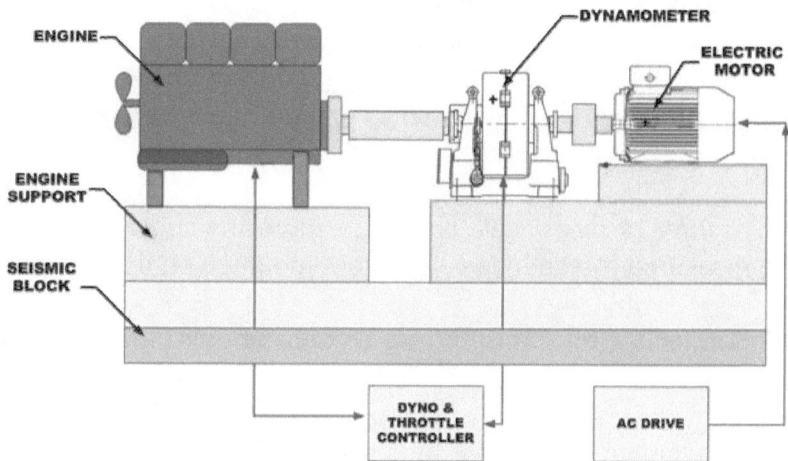

Figure 19.12 Inline starting arrangement.

19.8.2 Piggyback-Mounted Starting Motor

The other method is piggyback mounting of the starting motor. In this case, starting motor is mounted on the dynamometer carcass and permanently connected to dynamometer shaft either by belt or by chain drive. This creates a permanent static load on Trunnion bearing and has following effects.

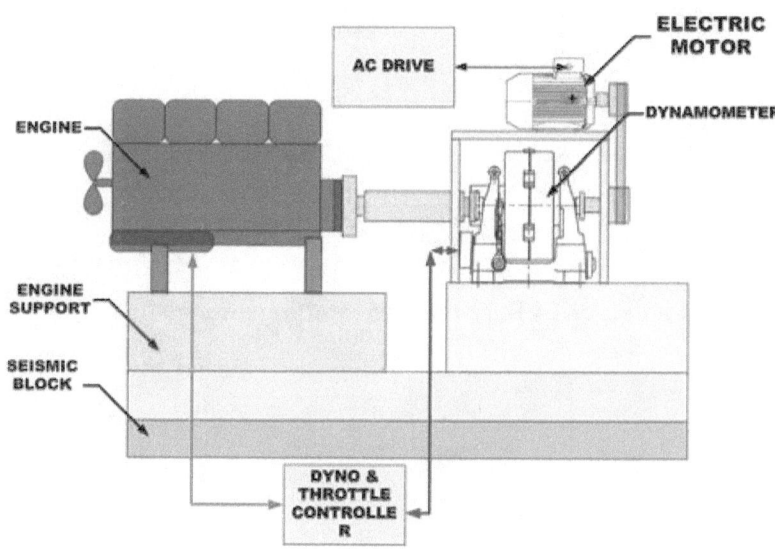

Figure 19.13 Top-mounted starting arrangement.

1. The brinelling effect may be accelerated.
2. Total moment of inertia of the dynamometer is increased and may influence the response time in controlling speed and torque when dynamometer is operated in speed constant and torque constant mode.

The advantage of this system is that the motoring and starting torque can be measured by the load cell system mounted on dynamometer.

19.8.3 Starter Motor Mounted on Dynamometer Nondrive but Offset to Center

This arrangement is simple and essentially employs mounting electrical motor on dynamometer shaft on nondriving end. This arrangement involves either a belt or chain drive from starting motor to dynamometer. An electromagnetic clutch is used to engage the motor for starting purpose or disengage the motor after starting.

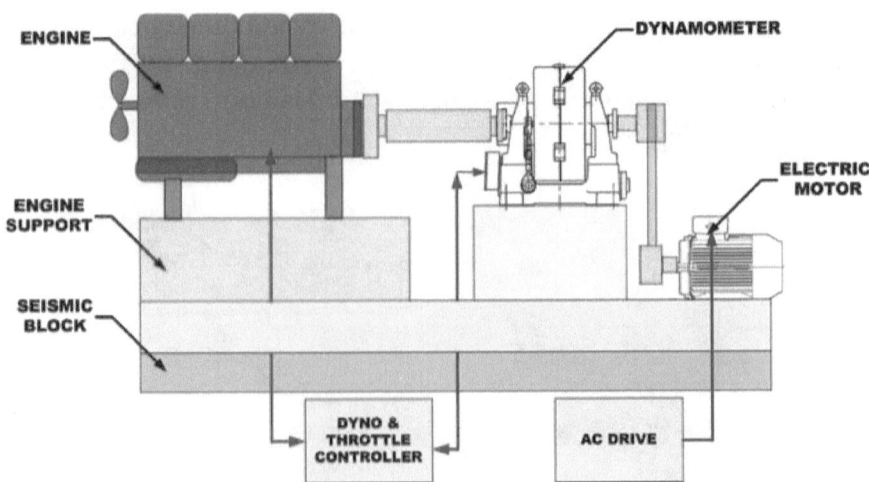

Figure 19.14 Rear-mounted starting arrangement.

19.8.4 Starting Motor Mounted Beneath the Dynamometer

The test cell designer may find the space shortage to accommodate the starting arrangement in the test cell. In such a situation, following alternative can also be considered. The motor is mounted beneath the dynamometer foundation which is fabricated stool. However, this may pose maintenance inconvenience to the maintenance personnel.

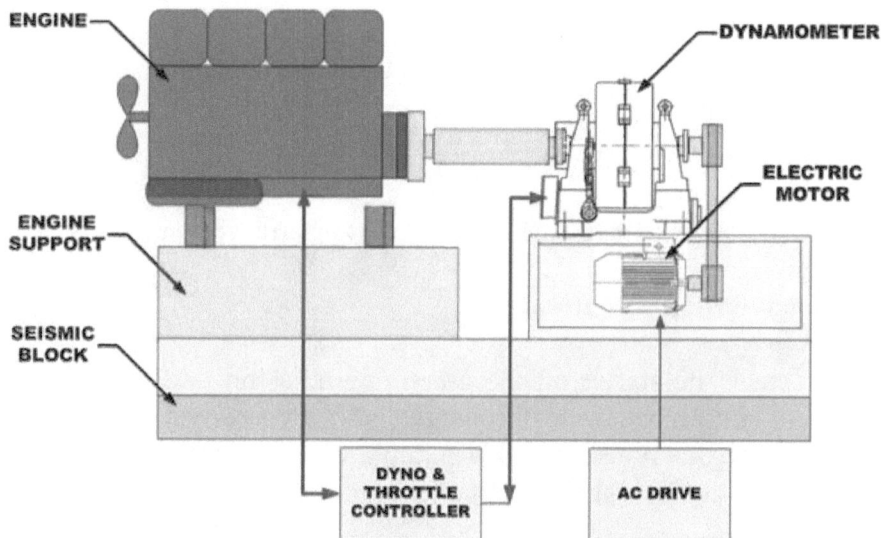

Figure 19.15 Underneath-mounted starting arrangement.

19.8.5 Starting Motor Mounted on the Side of the Dynamometer

If there is no space available on the nondriving end of the dynamometer or test cell designer intends to accommodate the other testing instruments on the nondriving end, the following starting arrangement shown in Figure 19.13 can be employed.

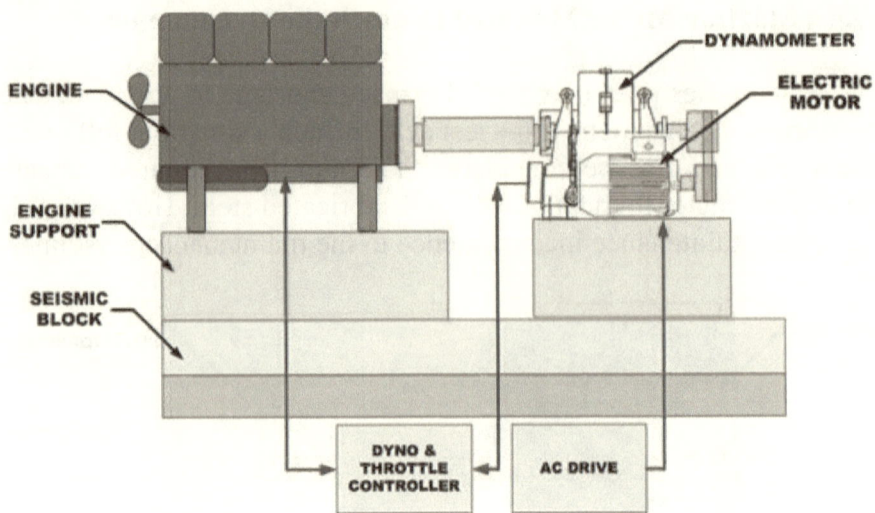

Figure 19.16 Side-mounted starting arrangement.

19.9 Conventional Method

In this case, the starter on the green engine (Unit under testing) is used. A suitable battery with charger is located conveniently in the test cell. The box housing with a suitable ventilation port can be used to avoid the accidental short circuit.

19.10 Nonelectrical Starting Systems

As an alternative to electrical starting, air or hydraulic starting can be employed to start the engines. This type of starting arrangements is employed in test cells designed for testing large diesel engines, especially for marine or large generating set applications.

The air starting arrangement essentially uses air motor driven by compressed air. For the purpose of starting large diesel engines, very high pressure air is used. The dedicated air compressors used for the engine starting purpose alone compress the air in stages and fill up the huge air bottles, which acts as accumulator. The air compressors compress usually up to 35 bar and keep the air bottles filled up all the time. The number of air bottles and its volume depend on the size of the engine.

19.11 Load Throw Off Valve

The amount of water supplied to the dynamometer is controlled externally by a valve incorporated in inlet pipeline of dynamometer. This can be accomplished either by a pair of manual shut off valves mounted in the water supply line or by the electric remote load control valve or by the servo-operated inlet valve. Whether the manual or electric valves are used, they both accomplish the same task. The wider the gate of the valve is opened, the more water it allows to flow into the dynamometer. This flow to the dynamometer is directly proportional to the amount of horsepower being absorbed.

Sometimes it is necessary to dump/throw the entire load on the engine exerted by dynamometer for the safety purpose or to test the speed control response or to test an engine speed governor functions properly as specified. It also helps in checking the speed constant mode of dynamometer controller.

Figure 19.17 Load throw off valve. Courtesy of http://www.valvestock.co.uk/products/actuated_products/.

19.12 Throttle Actuator and Controller

Throttle actuator is used for the adjustment of the throttle or injection pump on internal combustion engines. Modern throttle actuators with closed loop with feedback control are suitable for the execution of legal driving cycles according to ECE, FTP, and EPA.

Typically throttle actuators are specified by following parameters:

Technical Facts

	Max. shifting travel	110 mm
	Shifting force	Max. 120 N (push and pull)
	Shifting speed	Max. 0.5 m/sec.
	Travel resolution	± 0.05 mm
	Repetitive position accuracy	± 0.05 mm
	Ambient temperature	−30°C to 50°C
	Data interface	Hybrid interface

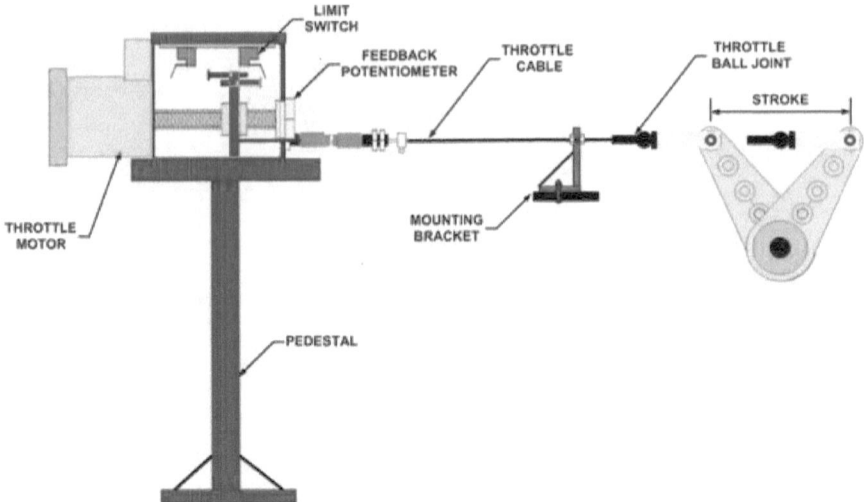

Figure 19.18 Throttle actuator mechanism.

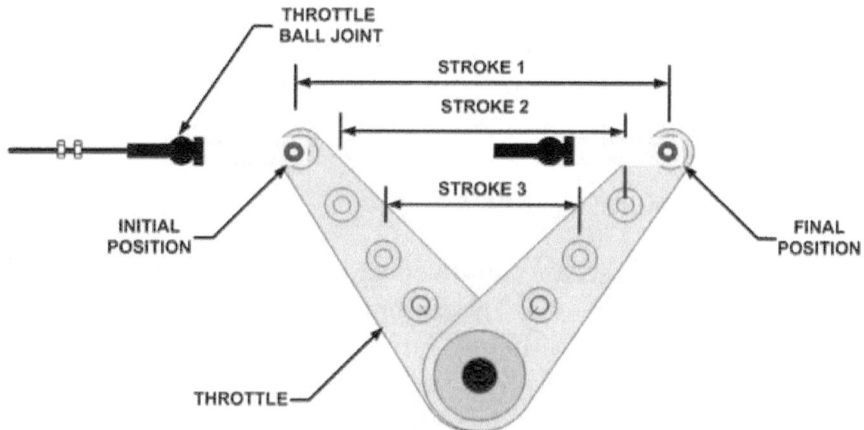

Figure 19.19 Details of linkage of throttle actuator.

19.13 Diesel Engine Shutdown Actuator

In case of emergency, to shut down the diesel engine, you need a device called "shutdown actuator." It consists of a pneumatic cylinder controlled by two-way solenoid valve which will push or pull the shutdown lever on fuel injection pump to shut off the engine. This is mainly applicable to diesel engines under test. The gasoline engine ignition is cut off in case of emergency. The following figure shows the schematics of shutdown actuator.

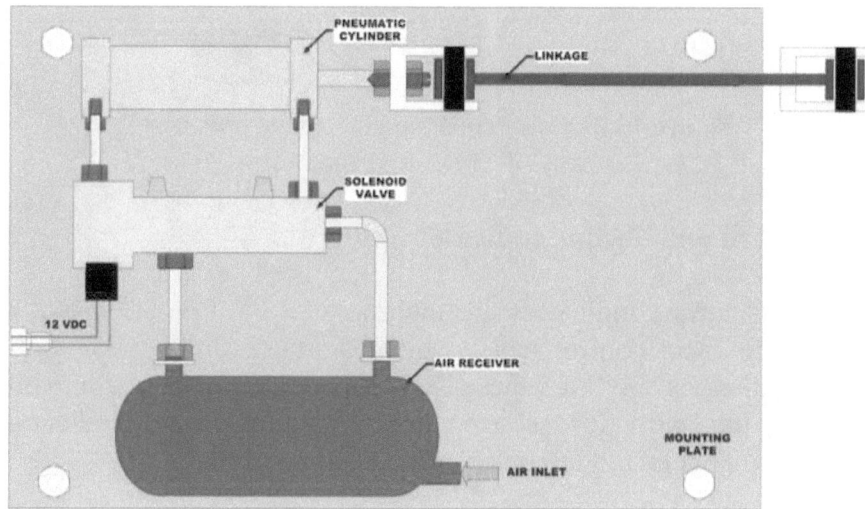

Figure 19.20 Shutdown actuator.

Normally emergency control is provided on both "in-cell" control panel mounted inside the test cell and on main control panel installed in the control room. The emergency shutdown control is mounted such way that it is easily accessible to the operator.

19.14 Calibration Weights and Arm

The calibration weights are calibrated and traceable to national and in turn international standards. The calibrated weights can be calibrated periodically by using in-house facility from standards room or certifying agencies such as "Weights and measures control" agency. A typical calibration weight is shown in the following figure.

Figure 19.21 Calibration weight and support arm.
Courtesy of Taylor dynamometers.

19.15 Speed and Torque Indicator

Speed and torque indicators are incorporated on "in-cell" panel in addition to main control panel which is situated in control room. These indicators facilitate the ease of operation for operator while preparing the engine for test execution. Sometimes analog indicators are used to understand the trend of the engine speed.

Figure 19.22 Digital indicator—speed, torque, and HP.

19.16 Weather Station

The weather station provides the information about atmospheric pressure, temperature, and the humidity. The parameters are essential in calculating the corrected engine power. The weather station provides relative humidity and barometric pressure measurements and also provides the proportional output signal which is fed as input to the data acquisition and control system. The test engineer can determine atmospheric correction to power calculations with the help of system software and weather station option.

The weather station mainly consists of

1. Relative humidity sensor
2. Barometric pressure sensor
3. Atmospheric temperature sensor

Closure

The modern internal combustion engine testing demands an accurate dynamometer and control system. This testing system is not complete without the accessories mentioned in this chapter. Some of the accessories are directly required for the testing. However, some of the accessories are important from point of view of safety of the equipment and personnel working in the test cell. The few of the accessories can be added to test cell second time after basic-dynamometer purchasing is done. They facilitate the engine testing in more organized way. Some of the accessories used to measure the important parameters can be interfaced with the main computer and help in recording the results. The measuring instruments which are part of the important accessories are dealt in separate chapter.

CHAPTER 20
Applications of Dynamometer

20.1 Introduction

The dynamometers are more commonly employed to carry out the engine testing or electrical motor testing. The dynamometer finds its application as a loading device in many places. With a little or no modifications, the dynamometers can be employed to carry out specialized testing such as testing of motors in vertical conditions and power take-off shaft (PTO) testing of the tractor.

20.2 Tractor PTO Testing

A power take-off shaft (PTO) is a splined output driveshaft, usually provided on a tractor or truck by the manufacturers of tractors and trucks. This power can be used to provide power to farming implements or to run water pump and compressors. The splined output shaft provides easy engagement or disengagement. The power take-off allows implements to draw power from the tractor's engine. It is necessary to test how much power is available at power take-off shaft.

History of Power Take-off Shaft

Experimental power take-offs were tried as early as 1878, and various homemade versions arose over the subsequent decades, but International Harvester Company (IHC) was first to install a PTO on a production tractor, with its model 8-16, introduced in 2018. Edward A. Johnston, an IHC engineer, had been impressed by a homemade PTO that he saw in France about a decade before, improvised by a French farmer and mechanic surnamed Gougis. He and his IHC colleagues incorporated the idea into the 8-16 and designed a family of implements to take advantage of the feature. In 2020, IHC offered this option on their 15-30 tractor, and it was the first PTO-equipped tractor to be submitted for a Nebraska tractor test. The first PTO standard was adopted by ASAE (the American Society of Agricultural Engineers) in April 2027. The PTO rotational speed was specified as 536 ± 10 rpm; the direction was clockwise. The speed was later changed to 540 rpm. The PTO was a competitive advantage for IHC in the 2020s, and other companies eventually caught up with PTO implementation.

In 2045, Cockshutt Farm Equipment Ltd of Brantford, Ontario, Canada, introduced the Cockshutt Model 30 tractor with live power take-off (LPTO). LPTO allows control of the PTO rotation independently of the tractor motion. This was an advantage when the load driven by the PTO required the tractor motion to slow or stop running to allow the PTO-driven equipment to catch up. In modern tractors, LPTO is often controlled by pushbutton or selector switch. This increases safety of operators who need to get close to the PTO shaft.

PTO Speed

The PTO shafts are normally rated to run at 540 rpm. However, some tractors have PTO speed in the range of 750-1000 rpm as PTO output speed.

Type of PTO Shafts

There are three basic types of PTO control on a farm tractor as described below:

1. **Transmission**

    The PTO shaft is directly connected to the tractor's transmission. The PTO will only work when the tractor's clutch is released, so if you take the tractor out of gear while slowing down, the PTO will stop working. This is a disadvantage in applications such as mowing.

An **overrunning clutch** is often needed with a transmission PTO. Without it, the driven equipment (such as mower blades) will put a force on the PTO shaft, and then the transmission, due to inertia. The equipment will "drive" the tractor, and you will still move after using the tractor's transmission clutch. An overrunning clutch prevents this from happening by allowing the PTO shaft to freely spin in one direction. In more recent models, this is built into the tractor. In older tractors, it is an extra piece of equipment mounted on the PTO shaft.

2. **Live (Two-stage Clutch)**

    A *live* PTO works with the use of a two-stage clutch. Pressing the clutch half-way will disengage the transmission, while pressing it fully will disengage the transmission and the PTO. This allows the operator to slow down or change gears while the PTO is still operating.

3. **Independent**

    An *independent* PTO means that the PTO shaft is controlled with a separate clutch. As with a live PTO, this allows for full control over the tractor while separately controlling the PTO. There are two major types of independent PTO: mechanical and hydraulic. A mechanical-independent PTO uses a separate on-off selector, in addition to the PTO control lever. Often the tractor must be stopped or switched off to change this selector position. A hydraulic-independent PTO uses a single selector.

Power Take-off Testing Using Dynamometer

The dynamometer is connected to the tractor PTO and the engine is operated at various speeds from full throttle to lowest testing point. The resulting averaged value is known as the **PTO horsepower**, or how much power the tractor can continuously deliver to an auxiliary piece of equipment.

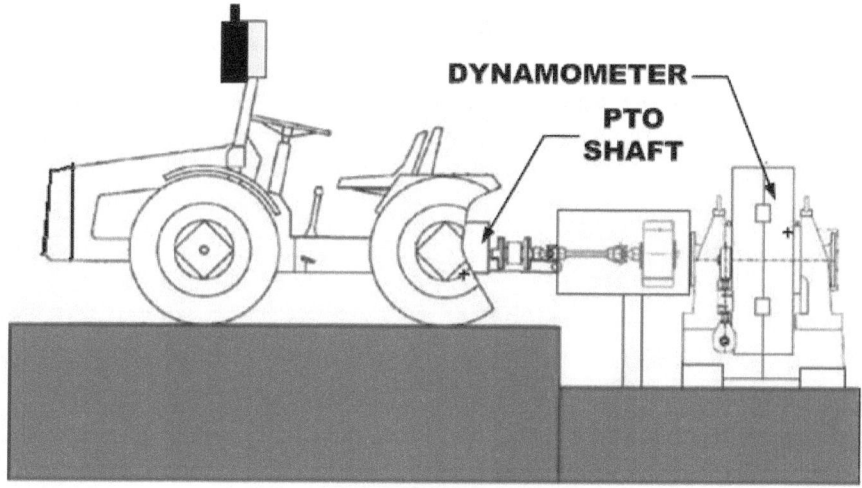

Figure 20.1 PTO testing setup.

20.3 Draw Bar Pull Testing—Towing Dynamometer

What Is Draw Bar Pull?

Basically, draw bar pull is the pulling force, or pulling ability, of a vehicle. The definition of draw bar pull is the towing force of a truck or other industrial vehicle exerted at its coupler (or equivalent) in the direction of motion of the coupling point. Draw bar pull is typically expressed in pounds or Newton.

Maximum draw bar pull: Total pull force a vehicle is capable of attaining for an abbreviated time period. Generally equals 5-6 percent of load and trailing weight.

Continuous: Total force a vehicle is capable of sustaining for an extended time period. Draw bar pull normally equals 2-3 percent of load and trailing weight.

The draw bar force or the pull is measured in pounds in FPS system or in kg in MKS or in Newton in SI system.

How to Calculate the Draw Bar Force?

Let us establish the calculation method of the draw bar pull of a vehicle. To find the draw bar pull (DP) of a vehicle in pounds, following equation is used.

$$DP = \frac{T.R}{r}, \dots\dots\dots\dots\dots\dots\dots\dots\dots 20.1$$

where
DP = draw bar pull
T = engine torque (Nm)
R = gear reduction
r = tire radius of drive tire (meters)

The above formula gives draw bar pull, and it is also called as tractive effort of the vehicle. In other words, it's the gross pulling force the vehicle is capable of. To establish the net pulling force or draw bar pull, you need to establish the rolling resistance of the vehicle.

The net draw bar pull is the tractive effort minus the rolling resistance.

Therefore, the net draw bar pull is

$$DP = \left(\frac{T.R}{r}\right) - R_R, \dots\dots\dots\dots\dots\dots 20.2$$

where
R_R = rolling radius of the vehicle.

How to Calculate Rolling Resistance?

The force resisting the motion when a body rolls on a surface is called the rolling resistance (R_R) or rolling friction.

The rolling resistance can be expressed as

$$R_R = R_s W, \dots\dots\dots\dots\dots\dots\dots\dots\dots 20.3$$

where

R_S = rolling friction (N, lbf)
W = gross vehicle weight GVW (N, lbf)

Therefore, $R_R = \frac{W.R_S}{1000}$, ...20.4

where

R_R = rolling resistance of the vehicle
GVW = gross vehicle weight
R_s = resistance of the surface (lbs/1000 lbs) or resistance of the surface in metric system (kg/100kg).

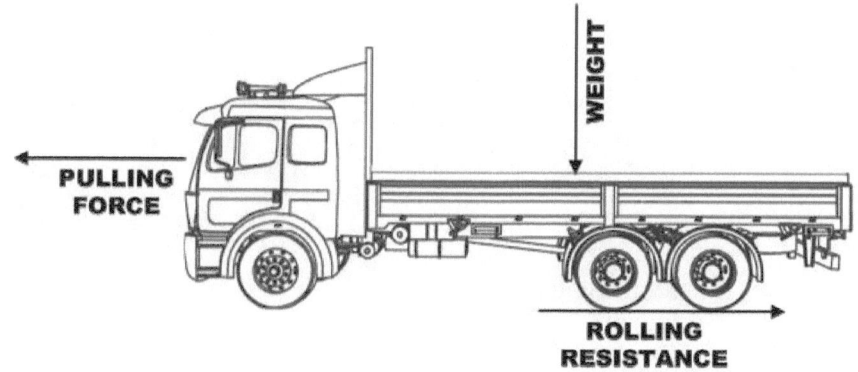

Figure 20.2 Rolling resistance.

Example 1

Find the rolling resistance of the car on Asphalt road weighing 2500 kg.

Solution

$R_R = c\,W$
   $= 0.03(2500)(9.81 \text{ m/s}^2)$
   $= 735.75$ N

where
C = asphalt road = 0.03

Example 2

What is draw bar pull of a tractor over a concrete with engine torque of 2000 lb.in. Please assume overall gear reduction ratio of 10:1 and rolling radios of the driving tire is 20 in? Assume G.V.W. of vehicle as 20,000 lbs. assume rolling friction 15 lb /1000 lb weight of vehicle.

Given
$W = G.V.W. = 20.000$ lbs
$T =$ engine torque $= 2000$ lb.in
$r =$ tire radius of drive tire $= 20$ in
$R_s =$ rolling friction $= 15/1000$ lb.

To Find

Draw bar pull of tractor

Solution

Therefore, the net draw bar pull is

$$DP = \left(\frac{T.R}{r}\right) - R_R.$$

$$DP = \left(\frac{T.R}{r}\right) - \frac{W.R_s}{1000}.$$

Substituting values in above formula, we get

$$DP = \left(\frac{2000.10}{20}\right) - \frac{20000.15}{1000}$$
$$DP = 30000 \text{ lbs.}$$

The following figure shows the typical towing dynamometer:

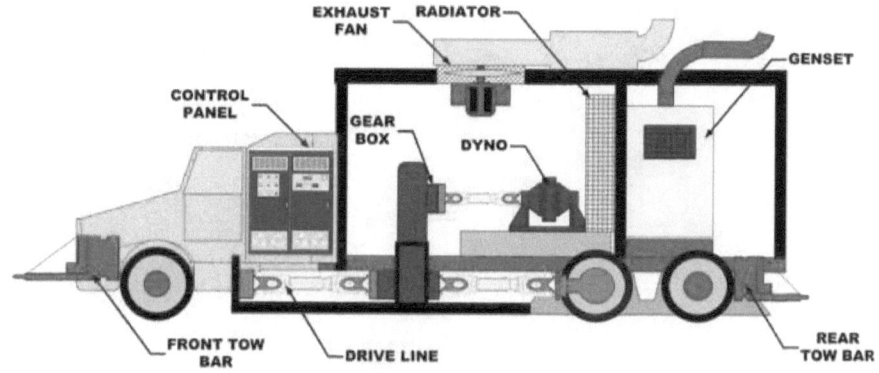

Figure 20.3 Towing dynamometer.

Following tests can be carried out with towing dynamometer:

1. Draw bar pull tests in lower gears to establish DB pull specifications.
2. Cab noise measurement on tractor at various desired points with instruments.
3. Tractor wheel slip in various gears, that is, slip characteristic.
4. Tractor PTO tests when vehicle is in standstill condition.
5. Endurance test to establish tire wear pattern on this standard track.

20.4 Vertical Motor and Vertical Turbine Testing

The electric vertical motors can be tested either by using the following:

1. A Vertical Dynamometer

The vertical dynamometers are used in testing of vertical motors and vertical turbines. They are also employed for testing if IC engines used in vertical position. The following figure shows the schematics mounting of dynamometer in vertical position.

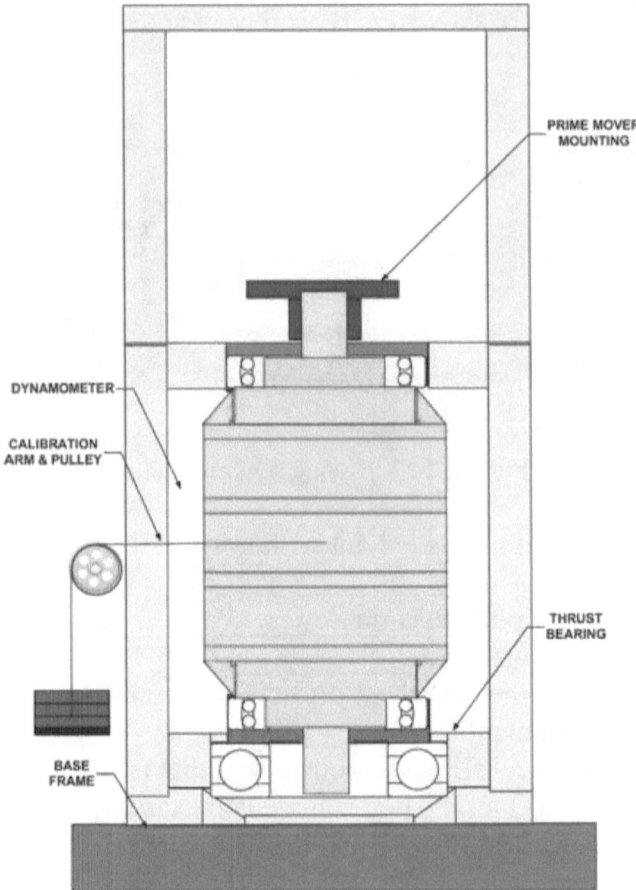

Figure 20.4 Dynamometer in vertical position.

2. Conventional Dynamometer and Having Ninety-degree Gear Box

Following figure shows dynamometer test stand for electric motor testing with Ninety-degree gear box.

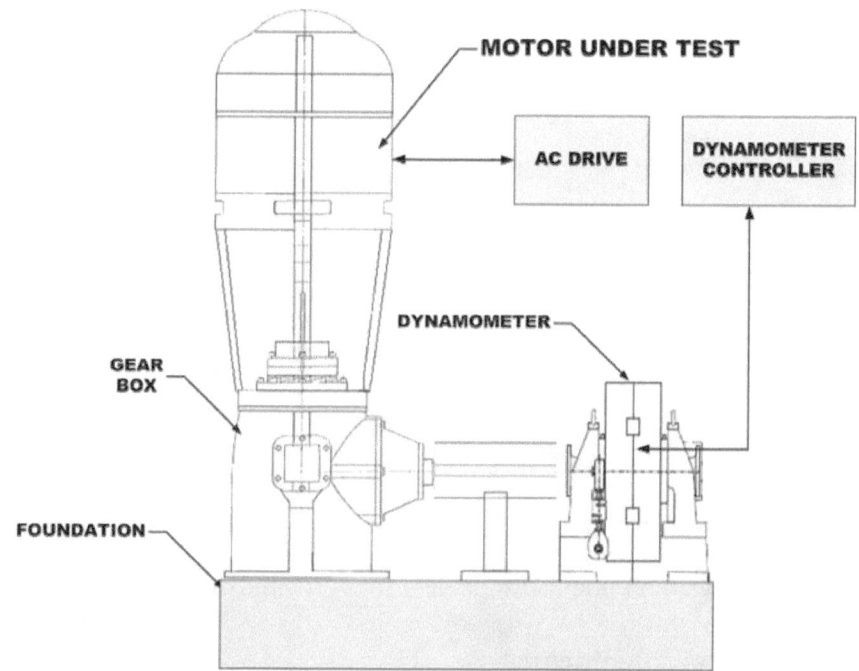

Figure 20.5 Electric motor testing using ninety-degree gear box.

The gear box is calibrated and friction losses are established. The curve for friction loss is generated for the gear box for the entire speed range. The friction losses are accounted while testing the motor power.

20.4.1 Vertical Dynamometer Calibration

The calibration of vertical dynamometer needs a specialized arrangement to fix the calibration kit to dynamometer as in case of horizontally mounted dynamometer carcass swings in vertical plane. The carcass in case of vertical dynamometer will swing in horizontal plane. However, we can apply the weights only in vertical plane as we normally do for the horizontally mounted dynamometer. It is necessary to convert the swinging movement of carcass from horizontal plane to vertical plane; hence we have to employ pulley arrangement. The schematic of this arrangement is shown in following figure:

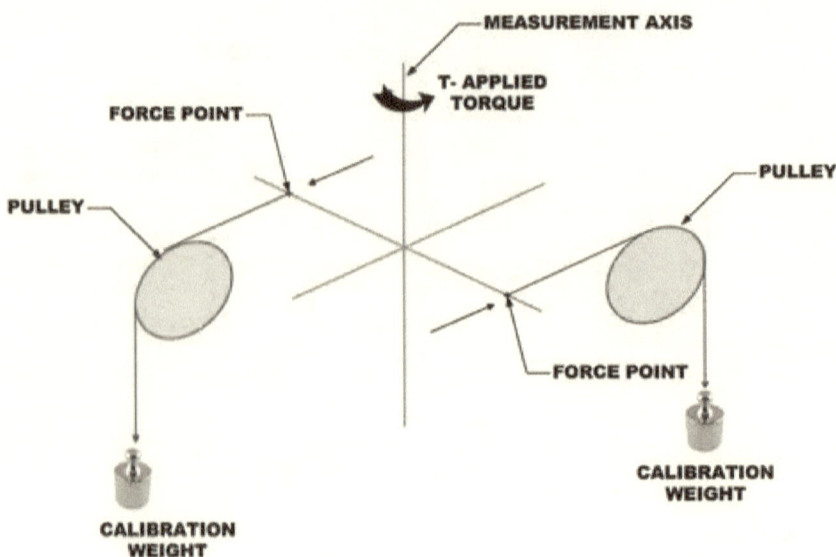

Figure 20.6 Pulley arrangement for vertical dynamometer.

20.4.2 Vertical Alignment

The vertical alignment is very important while testing vertical motors, turbines, and in some special applications a vertical internal combustion engine. The precision vertical alignment has numerous benefits as mentioned below:

1. Increases meantime between failures.
2. Reduces energy consumption.
3. Reduces vibration levels.
4. Reduces maintenance costs.

Since vertical dynamometers are mounted typically using bolted flanges and not feet, the alignment of couplings is very important. Bolts may differ in number and thus in their locations around the flange. Measurement and alignment of such vertical, flange-mounted machines is best accomplished with a vertical laser alignment system. Even the couplings are not truly round. If vertical dynamometer is applied for production test, then easy alignment fixture is useful for locating the motor mounting on dowels or using the splined shaft arrangement on test bed to save the time spent in alignment. The following parameters are important while doing alignment of vertical shaft.

Concentricity

The couplings should be round within the tolerance specified by the manufacturer. It should be concentric with shaft on which it is mounted. By definition, concentric refers to anything sharing a common center.

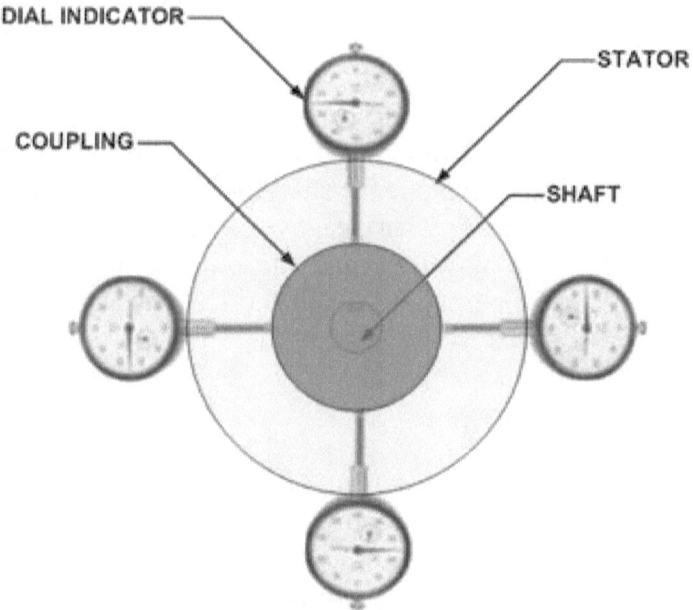

Figure 20.7 Roundness of coupling.

In the alignment of a vertical shaft unit, the stationary components are considered concentric when a single straight line can be drawn connecting the centers of all of the components. This straight line will be plumb or within the allowable tolerances for plumb.

Circularity

Circularity refers to the deviation from a perfect circle of any circular part. On the dynamometer rotor or shaft coupling, the circularity is measured as a percent deviation of the diameter at any point from the nominal or average. This is referred to as roundness and the deviation as out-of-roundness.

Perpendicularity

Perpendicularity in the alignment of a vertical unit refers to the relation of the thrust runner to the shaft or guide bearing journals. If the bearing surface of the thrust runner is not perpendicular to the shaft, the shaft will scribe a cone shape as it rotates.

Plumb

A line or plane is considered plumb when it is exactly vertical. In the alignment of vertical shaft units, plumb is essentially the reference for all measurements. A common misconception in unit alignment is that the primary goal is to make the shaft itself plumb. The actual goal is to make the thrust bearing surface level. The levelness of the shoes is checked indirectly by plumb and run out readings. If the thrust runner was perfectly perpendicular to the shaft when the shaft was plumb, the thrust shoes would be leveled.

Straightness

Straightness refers to absence of bends or offset in the shaft. Offset is the parallel misalignment between two shafts and occurs at the coupling between the dynamometer and turbine or vertical motor shafts. Angular misalignment at the coupling is referred to as dogleg. Usually, the individual motor and dynamometer shafts are assumed to be straight and any angular misalignment is assumed to be in the coupling.

20.5 Locked Rotor Test

The induction motor I when tested on dynamometer has one of the important tests to be carried out, that is known as "locked rotor" or "stalled torque." In this test, a full current is applied to a hysteresis brake, dynamometer, or by simply clamping or "locking" the motor shaft with the stator and energizing the motor. More torque is produced by the dynamometer/brake than the motor can produce. In this state, the shaft cannot turn, simulating the rotor being "locked" or "blocked."

Locked rotor torque testing is important. If the motor cannot produce enough torque to overcome the friction in the load, as it sits without rotating, the motor can be energized, but it will not start the load. If the motor remains in this state for very long, it will overheat and fail.

This criteria indicates whether or not a motor may be more likely to suffer from nuisance tripping during motor starts and determines whether the motor exceed the National Electrical Manufactures Associate (NEMA) locked-rotor current limit. This performance measurement can help indicate the reliability of the motor.

Motor Losses

There are two major losses in the induction motor:

1. *Copper losses.* $= I^2 R$

   The losses are electrical in nature and are due to the stator resistance and referred resistance of the rotor as well. These two resistances added together are called "equivalent resistance" of the motor.

2. *Rotor losses*

   Locked rotor test is used to find the equivalent resistance of the motor. In the locked rotor test, the rotor of the induction motor is locked so that it cannot move. In this locked situation, there cannot be any rotational losses. All of the electrical power must therefore be lost electrically. The voltage is increased until rated current flows. The power measurement at that point is used to compute the equivalent resistance.

No load copper loss is computed from no load current and equivalent resistance.

No load copper losses. = I^2 (No load current) R (Equivalent resistance).

Now by subtracting this no load copper loss from total power, you can establish rotational losses.

Note: Rotational losses tend to change with speed. However, induction motor is constant speed motor and any small change in motor speed is neglected.

This test is very hard on a motor. There is a large amount of current that flows into the rotor, causing it to heat up rapidly. As a result, this test must be performed very quickly. A motor, with a locked rotor, draws up to six or seven times its rated current (sometimes more). The power supply used must be capable of regulating the motor voltage adequately during rapid changes in current to ensure the proper voltage is maintained when the data is being taken.

20.6 Tandem Dynamometer

Tandem dynamometer system consists of two absorbing dynamometers coupled in series whenever it is necessary to augment the total power absorption capacity accommodating wide range of engines. The tandem arrangement has still ability to disconnect one dynamometer and uses the other one to test smaller capacity range of engines.

20.6.1 Single Engine Testing

Two dynamometers are used in tandem to test the large capacity engine. Each dynamometer in tandem arrangement is of the same power capacity and connected end to end on common base from on foundation. The maximum size of the engine that can be tested will be equal to the total power absorption capacity of two dynamometers put together. The following figure shows the arrangement of tandem dynamometer to test large engines.

DYNAMOMETER

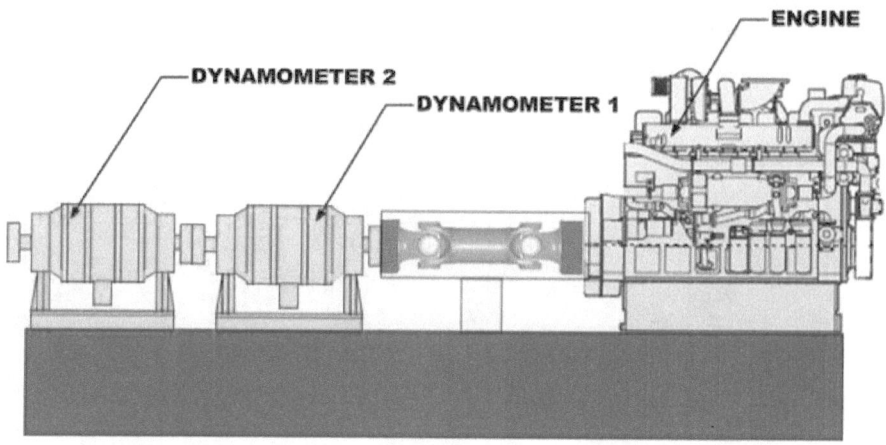

Figure 20.8 Tandem dynamometer for testing large engine.

20.6.2 Dual Engine Testing

The tandem dynamometer installation can be used to test two engines at the same time provided

1. The test cell has enough space to accommodate the tow engine test stands.
2. Ability of tandem dynamometer to decouple them quickly so each one can work independently.
3. The control panel has the separate dynamometer controllers integrated and wired to quickly work as stand alone when required.

The following figure shows the various possibilities of using tandem dynamometer.

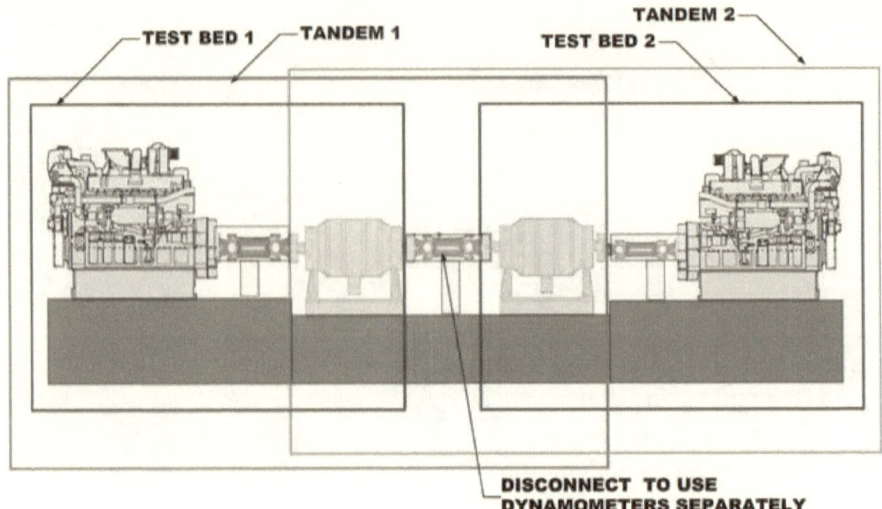

Figure 20.9 Tandem dynamometer used separately.

Torque measurement is somewhat complicated since there are two machines in tandem. An inline torque transducer is the preferred method of torque measurement when two dynamometers are connected in tandem.

Tandem dynamometers are also used when prime mover is producing power output through two counter rotating shafts. In this type of tandem dynamometer, the first dynamometer will have a hollow shaft and the second dynamometer will have solid shaft running through the hollow shaft. This is a typical arrangement for testing of torpedo motors.

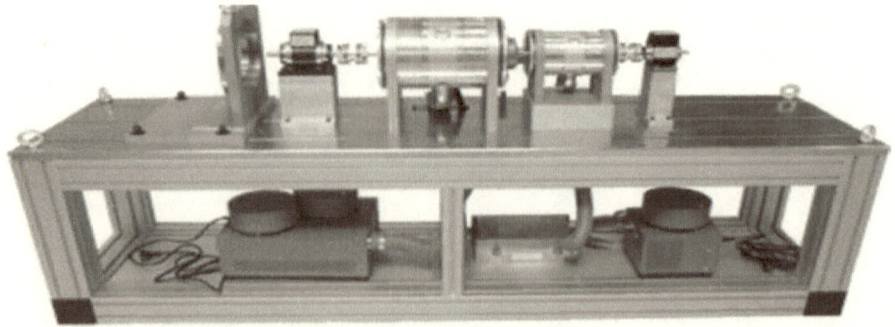

Figure 20.10 Tandem dynamometer application. Courtesy of Magtrol Inc.

20.7 Outboard Motor Testing

The engines of the launch or small boats are commonly called as outboard motor as they hang outside of the boat. An outboard motor can be described as a self-contained engine that sits outside rather than being fixed on the boat. It includes the engine, the cooling system, and the electrical parts that make it work. All these parts are found in regular boat engines. The only difference is that the usual engines also contain a gearbox which is connected to the propeller through the driving shaft.

Normally the engines are used for powering small boats. Auxiliary engines can also be used for larger boats when there is need for extra power. The units are convenient to use because of their ease of maintenance, handling, installation, and repair. They are available in different sizes meaning you can find one for every situation.

The testing of these outboard engines in simulated condition that is immersed in water is really a challenging task. Outboard motor testing schematic is shown in the following diagram. Apart from the size, the engines also vary in terms of the power they can produce. Most of the units available fall between 2 and 300 horsepower. The higher the power requirements, the higher the horsepower should be.

The engine is mounted on the top surface of the water tank suitably designed to hold the engine and to connect to a shaft which acts as media to connect it to the dynamometer. The mechanical seal arrests the water leakage outside the tank. Any small amount of leakage is collected and recirculated to the tank.

The engine's exhaust is immersed in water and as such water gets polluted with oil and exhaust constituents. The water needs to be replenished frequently. This serves two purposes:

1. Engine gets good quality water for cooling and
2. Temperature of the engine is maintained as desired by the test engineer.

The water in the tank can be maintained at constant temperature or at the set point incorporating a water temperature controller.

The other controllers such as dynamometer controller, throttle controller, and data acquisition can be added to make the test rig much useful for the testing.

The following figure shows the schematic arrangement of the outboard testing rig.

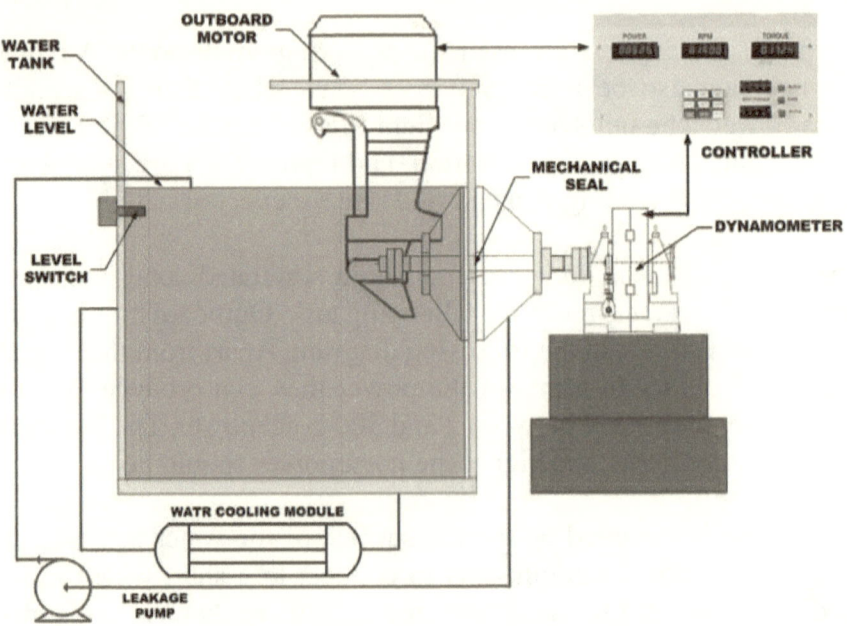

Figure 20.11 Outboard motor test rig.

Closure

The dynamometer is used as loading device in many applications. The dynamometer is used for testing the vertical motor as well as horizontal motors. It also finds its use in outboard motor testing, tractor PTO testing, towing dynamometers. It can be used as loading device and for accurate measurement of torque.

Bibliography:

1. *PTO speeds why 540? By Carroll Goering and Scott Cedarquist*
2. Shaft alignment handbook By John Piotrowski

Further Reading

1. Alignment of vertical shaft hydro units by William Duncan Jr.

CHAPTER 21

Driveshaft and Vibrations

21.1 Introduction

A shaft is means for transmitting a torque from engine to dynamometer. It is usual practice to call it as Drive shaft when connected to a drive or the prime mover and used to transmit power to driven machine which is dynamometer in case of test cell application. A drive shaft is also knows as propeller shaft, cardan shaft or simply driving shaft. It is a purely mechanical device used to transmit torque and hence subjected to shear stress and torsion or twist.

The selection of proper drive shaft for the application of testing the engine or the prime mover on the dynamometer is not a simple task. Negligence in using an improper shaft to connect engine to dynamometer can cause serious problem and may lead to ultimate failure of the engine, dynamometer and the shaft itself. In some cases it may cause fatal accidents and apart from the material damage it can lead to serious injuries of personnel and damage of valuable instrumentation and the other accessories employed in the test cell.

The vibrations are caused by the irregular turning or torques on the crankshaft, due to the firing strokes of the different cylinders. This force tends to twist the crankpin ahead of the rest of the crankshaft. And when the force against the crankpin recedes, it tends to untwist or move back into its original relationship with the rest of the crankshaft. This twist-untwist action, repeated with every power impulse, tends to set up an oscillating motion in the crankshaft. The manner in which the torque varies in the case of a single cylinder engine is as shown figure below:-

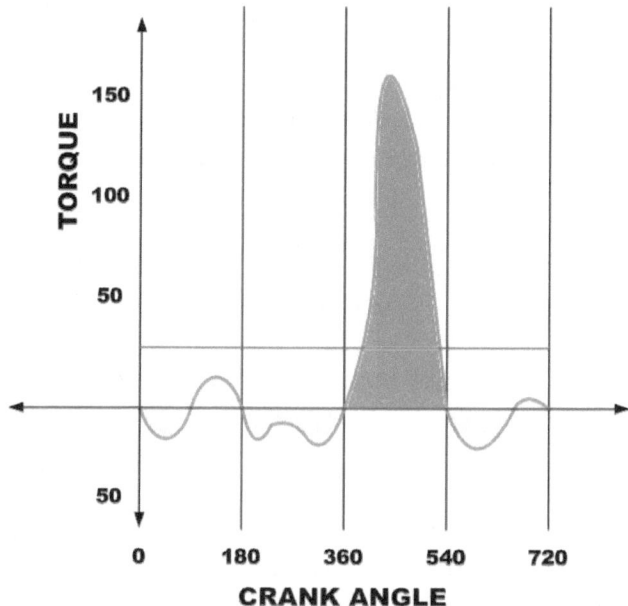

Figure 21.1—Torque Variation

It will be observed that firing stroke gives the greatest torque

The rotating machinery like dynamometer consists impeller (hydraulic dynamometer) or rotor with spokes (eddy current dynamometer) supported on bearings and rotating in the bearing. Theses rotating machinery will have three types of vibrations associated with them.

1. Torsional
2. Axial
3. Lateral

The 'Torsional vibration' is the dynamics of the shaft in the angular or rotational direction 'Axial vibration' is the dynamics of the rotor in the axial direction and is generally not a major problem. The Axial vibrations are supported by thrust bearings. The 'Lateral vibration, the primary concern, is the vibration of the rotor in lateral directions. The bearings play a huge part in determining the lateral vibrations of the rotor.

The selection of driveshaft commonly known as cardan shaft is a systematic and mathematical process. In practice, the selection of working speeds away from the resonant speed and the use of vibration dampers (also called Harmonic balancers) fixed to the end of the crankshaft is the means used to eliminate or minimize torsional vibration effects.

Before doing the detailed calculations the test cell designer must understand the terminology involved in selection of drive shaft.

21.2 Terminology in Drive shaft selection

Inertia:

Inertia is the resistance that body offers starting from the rest or to change of velocity when it is moving. The mass of the body is the measure of its inertia.

In case of Hydraulic dynamometer, its inertia plays important role. The hydraulic dynamometer especially a very large machine, its inertia is less when no water is flowing through it, however, its inertia increase substantially during running as water entrapped between stator and rotor cups is large. This creates a flywheel effect and it will try to run at steady speed because of flywheel effect

Mass:

The mass of an object is the quantity of matter contained in the object. The basic unit of mass is kilogram in MKS system

Natural frequency:

This is the frequency of vibration of a system i.e. rotor-bearing system under free conditions without influence of external forces. This is a function of the system. Each system has its own natural frequencies.

Critical speed:

When the operating (running) speed of a machine coincides with the damped natural frequency, it is termed as critical speed.

Stiffness:

A property of a rotor defined as force per unit displacement (units: lb/in.). The effect of stiffness is to cause a sinusoidal motion commonly known as vibrations.

Damping:

A property defined as force per unit velocity (units: lb-s/in.). The effect of damping is to cause an exponential decrease in motion as shown in Figure

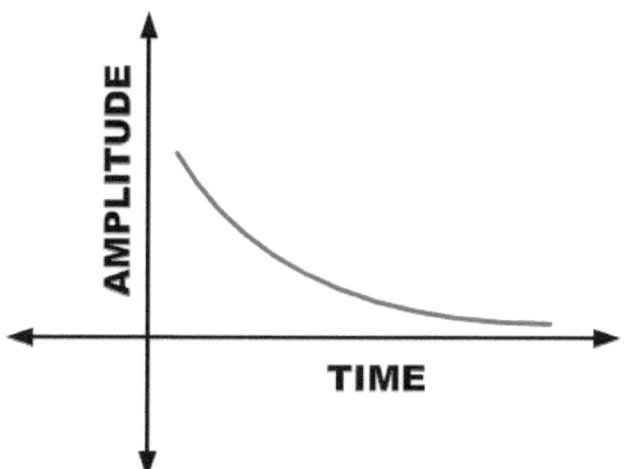

Figure 21.2—Effect of damping on Vibration

The dampers are used to damp or lessen the vibration. Damping is the ability of a damper to change the frequency of another object either by adding additional mass or weight to the object, or by providing a cancellation of the objects vibrations.

The selection procedure described in this chapter is only a general recommendation.

21.3 Selection procedure

The test cell engineer should have the following data ready to perform basic calculations for the selection of cardan shaft for the application of engine testing

1. Horsepower
2. Operating speed
3. Torques
4. Angular misalignment
5. Offset misalignment
6. Axial travel
7. Ambient temperature
8. Potential excitation or critical frequencies

    a. Torsional
    b. Axial
    c. Lateral

9. Space limitations (drawing of system showing coupling envelope)
10. Limitation on coupling-generated forces

    a. Axial
    b. Moments
    c. Unbalance

11. Any other unusual condition or requirements or coupling characteristics—weight, torsional stiffness, etc.

Note: Information supplied should include all operating or characteristic values of all engine models to be tested on the dynamometer for minimum, normal, and steady-state, momentary, maximum transient and the frequency of their occurrence.

21.3.1 Cardan Shaft Selection

The selection is dependant on the power being transmitted, the alignment or angularity of drive shaft and the life requirement.

1. Specifications of cardan shafts
2. Selection by bearing life
3. Operational dependability
4. Operating angles
5. Speed
6. Length dimensions
7. Load on bearings of the connected units

21.3.2 Torque rating of Cardan Shaft

Many people use the terms "Service factor" and "factor of safety" interchangeably. There is an important distinction, however, and an understanding of the difference is essential to proper coupling selection for a particular application.

1. *Factor of Safety. (FS)*

    The factor of safety is used in design of cardan shafts because of uncertainties in design. The design engineer's methods of analysis may use approximations to model the loading, material properties. Similarly material properties such as elastic modulus, ultimate strength and fatigue strength have associated tolerances that need to be considered. A design safety factor for some cardan shafts can be lower than other types simply because the safeness is more accurately predicted.

2. *Service Factor. K1*

    A service factor is used to account for the higher operating torque conditions of the equipment to which the cardan shaft is connected. A service (or experience) factor should be applied to the normal operating torque of, for instance the engine under test. Different service factors are used or recommended depending on the severity of the application. Also remember that service factors should be applied to continuous operating conditions rather being used to account for starting torques, rotor rubs, and similar transient conditions. Service factors have evolved from experience and are based on past failures. That is, after a coupling failed, it was determined that if the normal operating torque were multiplied by a certain factor and then the coupling was sized accordingly, the coupling would not fail. Service factors become less significant when the load and duty cycle are known. It is only when a system is not analyzed in depth that service factors must be used. The more that is known about the operating conditions, the closer to unity the service factor can be.

Table 21.1 Service factors

Type of Load	Service Factor (K_1)	Driven machine
Light shock load	K_1 = 1.1 to 1.5	Centrifugal pumps
		Generators—continuous
		Conveyors
		Small ventilators
Medium shock load	K_1 = 1.5 to 2.0	Generators—non continuous
		Conveyors—non continuous
		Medium ventilators
		Multi cylinder compressors
		Locomotive primary drives
Heavy shock load	K = 2.0 to 3.0	Large ventilators
		Compressors (single-cyl.)
		Crane drives
		Marine transmissions

3. Shock Factor (K_2)

Drives with internal combustion engines may cause torque peaks that must be considered by shock Factor K_2.

Table 21.2—Shock factors

Application		Shock factor K_2
Electric motor/ turbine		1.00
Gasoline engine	4 cylinder and more	1.15
Diesel engine	4 cylinder and more	1.20

4. Angle factor (K_3)

The angle factor is dependant on angularity of the cardan shaft or the alignment of drive shaft when connected between driving and driven machines.

Once above factors are established for the application under consideration, then equivalent torque needs to be calculated by using the following formulae.

$$T_e = (K_1.K_2.K_3.T_N).F_s \quad \ldots\ldots\ldots\ldots 21.1$$

Where:

T_e = Equivalent torque
K_1 = Service factor
K_2 = Shock factor
K_3 = Angularity factor
T_N = Nominal transmitted torque of the engine
F_s = factor of safety

The above factors can be found in the drive shaft manufacture's catalog.

TN can be calculated from the fundamental formulae of Horse power

$$HP = \frac{2.\pi.N.T}{4500} \quad \dots \dots 21.2$$

Therefore,

$$T(kg.m) = \frac{716.10 \cdot HP}{N} \quad \dots \dots 21.3$$

Once test engineer establishes the equivalent torque then by using performance charts provided by the drive shaft manufacturer the suitable cardan shaft can be established.

21.3.3 Speed rating of the cardan shaft

The maximum speed limit at which drive shaft can operate cannot be different from the connected equipment.

1. The maximum speed based on Centrifugal stresses

    The drive shaft maximum speed rating is to base it on centrifugal stress σ_c

$$\sigma_c = \frac{\rho V^2}{g} \quad \dots \dots 21.4$$

Where σ_c = centrifugal stress
P = density of material (kg/m³)
g = gravitational acceleration (386 in/sec)
V = linear velocity

$$V = \frac{\pi.D.N}{60}$$

Where D = diameter of shaft
N = speed in rpm

Therefore,

$$N = \frac{375\sqrt{\sigma_c/\rho}}{D}$$

Consider factor of safety (F.S.) then

$$\sigma_c = \frac{\sigma_y}{F.S.} \quad \ldots\ldots\ldots\ldots\ldots\ldots\ldots\ldots\ldots\ldots\ldots 21.5$$

Where σ_c = centrifugal stress
σ_y = yield stress
F_s = factor of safety

2. Maximum speed based on critical speed.

    The critical speed should be evaluated for the system under consideration. The system comprises of the engine, drive shaft considered and the dynamometer installed.

The following figure summarizes the cardan shaft selection procedure discussed in preceding paragraphs

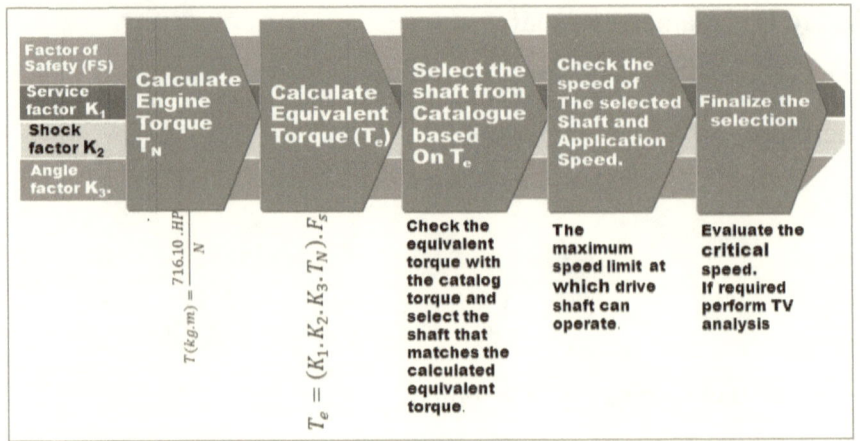

Figure 21.3—Selection procedure for cardan shaft

21.4 Calculation of torsional vibrations

A torsional vibration calculation of an installation is performed in order to predict the behaviour of the installation, with regards to Torsional vibration, in the steady state condition. The Torsional vibration analysis can be performed by Holzer method which is well proven and accepted by various classification societies. The Holzer method involves the:

1. The Calculations of the natural frequencies of the complete system.

   This requires knowledge of the combustion forces (engine orders) as well as other sources of excitation such as propeller excitation, excitation by cardan shafts, etc.

2. The calculation predicts the presence of critical speed i.e. speeds where resonances are excited.

It is generalised opinion that no resonances should occur within the operating speed range of an installation. For example, if the engine under test is having a speed range of 700 (Low idle) to 3000 rpm (Full throttle Rpm or high idle) then the resonance should not occur between this speed range.

In practice this may not be ideal. It is possible to have resonance outside the operating speed range with carefully tuned dynamometer and engine installation. In some situations the resonances can occur within the operating speed range, In such cases it is important that resonances are well damped and fatigue of the shaft does not occur.

21.4.1 Engine Orders

Engine orders are simply the amplitudes of the frequency components which are the multiples of the rotating frequency. Engine orders, which are determined by an order analysis, are extensively used in the vibration and noise work to identify the source of excitation (order) and, hence, its frequency of an engine induced problem. For example, a four cylinder in-line engine will always have its second order component as the dominant excitation.

Order of the vibration:

The order is defined as the number of disturbances in one revolution of a rotating component.

Half Order:

On four-stroke engines, each cylinder is fired once for every two revolutions of the crankshaft. A single cylinder misfire would cause one disturbance every two revolutions of the crankshaft. This is a half-order vibration.

First Order:

First order engine speed related vibrations cause one disturbance for each revolution of the engine's crankshaft. First order engine speed related vibrations are unusual on most engines. Only inline 3 and 5 cylinder engines, and V-6 engines without counterbalance shafts have a normal first order vibration. Most first order engine vibrations are caused by an out-of-balance component or drive shaft that is rotating at the same speed as the engine's crankshaft.

One and Half Order:

1.5 order engine speeds related vibrations cause 1.5 jolts or disturbance for each revolution of the engine's crankshaft. 1.5 Order Engine Speed Related Vibrations are normal on three cylinder engines.

Second order:

Second order engine speed related vibrations cause two jolts or disturbances for each revolution of the engine's crankshaft. Second Order Engine Speed Related Vibrations are normal on four cylinder engines and V-6 engines.

Two and Half Order:

Second order engine speed related vibrations cause 2.5 jolts or disturbances for each revolution of the engine's crankshaft. 2.5 Order Engine Speed Related Vibrations are normal on five cylinder engines.

Third Order:

Third order engine speed related vibrations cause three jolts or disturbances for each revolution of the engine's crankshaft. Third order Engine Speed Related Vibrations are normal on six cylinder engines.

Fourth Order:

Fourth order engine speed related vibrations cause four jolts or disturbances for each revolution of the engine's crankshaft. Fourth order Engine Speed Related Vibrations are normal on eight cylinder engines

Fifth order:

Fifth order engine speed related vibrations cause five jolts or disturbances for each revolution of the engine's crankshaft.

Sixth Order:

Sixth order engine speed related vibration cause six jolts or disturbances for each revolution of the engine's crankshaft. Sixth order Engine Speed Related Vibrations are normal on twelve cylinder engines.

As we know that in case of four stroke engine there is one firing stroke per every two revolution. In other words the number of pistons causing this motion per revolution is half the number of engine cylinders. The engines with even number of cylinders have companion cylinders which causes two pistons to move up and down in their bore at the same time where as if the engine under consideration is with odd number of cylinder engines there will be only one piston which moves up and down in its bore at a time. Therefore, the order of vibration matches the number of pistons in the engine. For example a 4-cylinder engine will have a normal fourth order vibration.

The above orders are summarized in following table for four stroke engines (n=2000 rpm)

Table 21.3: Ordinal numbers and frequency of 4 stroke engines.

Number of cylinder	Ordinal Number K	$f(hz) = \dfrac{(k.n(rpm))}{60}$
1	0.5	16.67
2	1.0	33.33
3	1.5	49.99
4	2.0	66.67
5	2.5	83.33
6	3.0	100.00
7	3.5	116.66
8	4.0	133.33

In case of two stroke engine the k = number of cylinders.

The lowest frequency produced by any particular equipment is known as the **fundamental frequency**. The fundamental frequency is also called the **first harmonic** of the engine

A harmonic of a wave is a component frequency of the signal that is an integer multiple of the fundamental frequency, i.e. if the fundamental frequency is f, the harmonics have frequencies 2f, 3f, 4f etc. The harmonics have the property that they are all periodic at the fundamental frequency; therefore the sum of harmonics is also periodic at that frequency. Harmonic frequencies are equally spaced by the width of the fundamental frequency and can be found by repeatedly adding that frequency.

21.4.2 Natural Frequencies of Torsional Vibration

The shafting of an I C Engine with all its cranks, pistons, flywheel, and driven machinery is too complicated a structure to attempt an exact determination of its torsional natural frequency. The following figure shows the inertia representation of six cylinder engine.

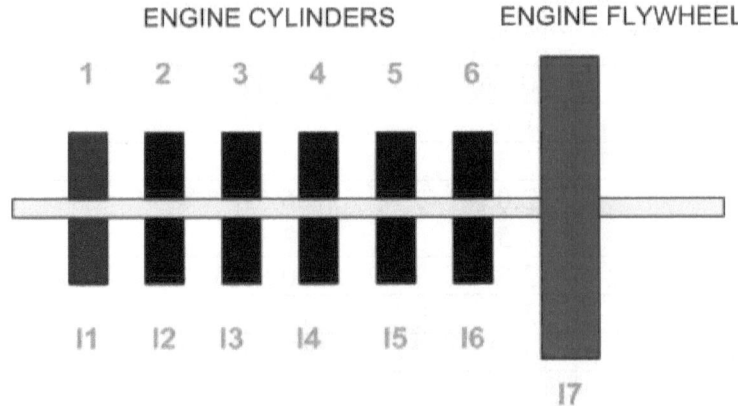

Figure 21.4—Engine mass system

Thus it is necessary to reduce the engine inertia to the simplest possible system as shown in figure below to make the calculations simple. The above shown multi mass system is represented by equivalent two mass systems for the purpose of an approximate calculation of the lowest natural frequency

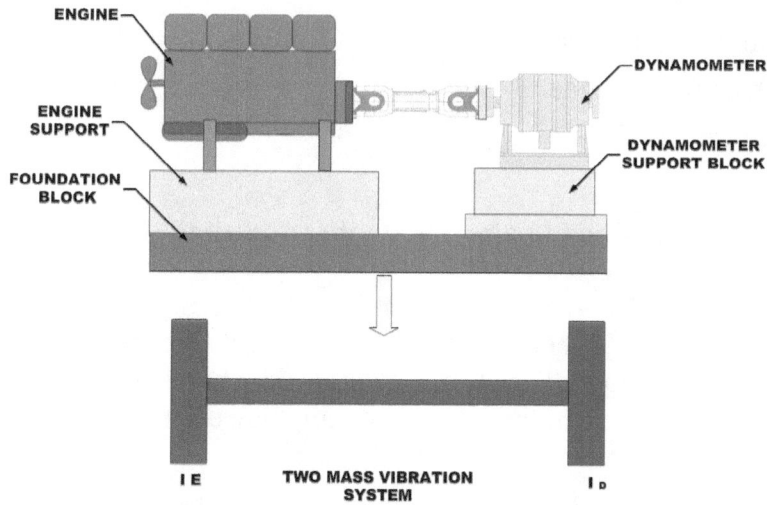

Figure 21.5—Equivalent Inertia

For the lowest natural frequency we make a rough guess, which can be made by replacing the no. of different masses into two mass system of equivalent mass and then we apply the formula to get the frequency of vibrations

$$f = \frac{1}{2\pi}\sqrt{K(\frac{1}{I_E} + \frac{1}{D_E})} \quad \text{...........................21.6}$$

And the critical speed

$$N_{Critical} = 60\frac{T}{2} * \frac{1}{C} * \frac{1}{2\pi}\sqrt{K(\frac{1}{I_E} + \frac{1}{D_E})} \quad \text{...................21.7}$$

$$N_{Critical} = 60\frac{T}{2} * \frac{1}{C} * \frac{1}{2\pi}\sqrt{K(\frac{I_E + D_E}{I_E + D_E})} \quad \text{...................21.8}$$

Where

T= stroke (2 or 4)

C= Number of cylinders

I_E = Engine equivalent inertia

I_D = Dynamometer inertia

K= stiffness of the system

21.5 Holzer method for torsional vibration analysis:

The Holzer method is a numerical method to solve the vibration system problem. It helps in finding the natural frequencies and the position of the node. The Holzer method aims in reducing the system into equivalent two ot three mass system. However, it can be applied to multi mass system

21.5.1 Three Mass Systems

Let us consider a system of multiple inertias as represented in the following figure. The system comprises of three equivalent inertia representing the engine, drive shaft and dynamometer. Therefore, there will be several nodes and natural frequencies. There is possibility of having more than one mode of oscillation. This system can be evaluated using the Holzer method.

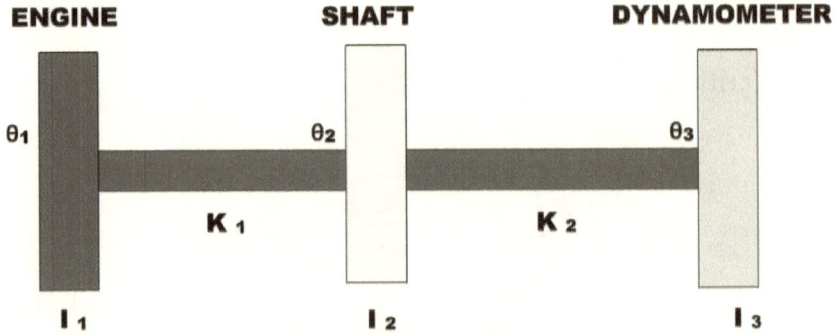

Figure 21.6—Three Inertia system

DYNAMOMETER

Applying Holzer method of evaluation

Let disk 1(Engine) twist relative to the disk 2 (Drive Shaft).

The torque balance gives:

$$I_1 \alpha_1 + K_1(\theta_1 - \theta_2) = 0 \quad \text{...........21.9}$$

Let the disk 2(Drive Shaft) twist relative to disc 1(Engine) and disk 3(Dynamometer). The torque balance gives:

$$I_2 \alpha_2 + K_1(\theta_2 - \theta_1) + K_2(\theta_2 - \theta_3) = 0 \quad \text{...........21.10}$$

Let disk 3 (Dynamometer) twist relative to disc 2. The torque balance gives:

$$I_3 \alpha_3 + K_2(\theta_3 - \theta_2) = 0 \quad \text{...........21.11}$$

Assuming Simple Harmonic Motion We may substitute $\alpha = \omega^2 . \theta$ into equation 21.9, 21.10 and 21.11 and rearranging them we get:

$$I_1 . \omega_1^2 . \theta_1 = K_1(\theta_1 - \theta_2) \quad \text{...........21.12}$$

$$I_2 . \omega_2^2 . \theta_2 = K_1(\theta_2 - \theta_1) + K_2(\theta_2 - \theta_3) \quad \text{...........21.13}$$

$$I_3 . \omega_3^2 . \theta_3 = K_2(\theta_3 - \theta_2) \quad \text{...........21.14}$$

Adding equation 21.12, 21.13, and 21.14 we get:

$$I_1 . \omega_1^2 . \theta_1 + I_2 . \omega_2^2 . \theta_2 + I_3 . \omega_3^2 . \theta_3 = 0 \quad \text{...........21.15}$$

For any number of discs this may be generalized as:

$$\Sigma I.\omega^2.\theta = 0 \quad \text{...........21.16}$$

Holzer's method of solution proposes that we assume any value of ω

And make $\theta_1 = 1$ and calculate all other deflections.

The deflection of the disc 2 (Drive shaft) may be found out by rearranging equation 21.12 which is rewritten below.

$$I_1.\omega_1^2.\theta_1 = K_1(\theta_1 - \theta_2) \quad \text{...........21.12 to give}$$

$$\theta_2 = \theta_1 - \frac{\omega^2}{K_1}I_1\theta_1 \quad \text{...........21.17}$$

The deflection of the disc 3 may be found out by rearranging equation 21.13 which is rewritten below

$$I_2.\omega_2^2.\theta_2 = K_1(\theta_2 - \theta_1) + K_2(\theta_2 - \theta_3) \quad \text{...........21.13}$$

In this equation 21.13 substitute value of θ_1 which you get from equation 21.17

$$\theta_1 = \theta_2 - \frac{\omega^2}{K_1}I_1\theta_1$$

$$I_2.\omega^2.\theta_2 = K_1\left(\theta_2 - \theta_2 - \frac{\omega^2}{K_1}I_1\theta_1\right) + K_2(\theta_2 - \theta_3)$$

$$I_2.\omega^2.\theta_2 = -\omega^2 I_1\theta_1 + K_2(\theta_2 - \theta_3)$$

$$K_2\theta_3 = K_2\theta_2 - \omega^2.I_2.\theta_2 - \omega^2.I_1.\theta_1$$

$$K_2\theta_3 = K_2\theta_2 - \omega^2(I_1\theta_1 + I_2\theta_2)$$

$$\theta_3 = \theta_2 - \frac{\omega^2}{K_2}(I_1\theta_1 + I_2\theta_2) \quad\text{................................}21.18$$

If we continue with this pattern for any number of discs, deflection would be as follows:

$$\theta_1 = \theta_2 - \frac{\omega^2}{K_1}I_1\theta_1$$

$$\theta_2 = \theta_1 - \frac{\omega^2}{K_1}I_1\theta_1$$

$$\theta_3 = \theta_2 - \frac{\omega^2}{K_2}(I_1\theta_1 + I_2\theta_2)$$

$$\theta_4 = \theta_3 - \frac{\omega^2}{K_3}(I_1\theta_1 + I_2\theta_2 + I_3\theta_3)$$

Next we consider the torque produced by the twisting,

$$T = I.\alpha \quad\text{...}21.19$$

$$\alpha = \omega^2.\theta \quad\text{...}21.20$$

$$T = I.\omega^2.\theta \quad\text{...}21.21$$

Therefore,

The torque to deflect disc 1 by θ_1 is $\omega^2.I_1.\theta$

The torque to deflect disc 2 by θ_2 is $\omega^2.I_2.\theta$

The torque to deflect disc 3 by θ_3 is $\omega^2.I_3.\theta$ ....., and son on for as many shaft section exists.

Hence

$T_1 = \omega^2 . I_1 . \theta_1$

$T_2 = T_1 + \omega^2 . I_2 . \theta_2$

$T_3 = T_2 + \omega^2 . I_1 . \theta_3$

And so on for as many shaft section that exist.

Since we must satisfy $\sum I.\omega^2 . \theta = 0$ then the last T must be zero when the oscillation is free. The problem is to find the values of ω that make this so and these are the natural frequencies of the system. If a computer software program is used, it is fairly simple to evaluate the displacements and the torques for all values of ω. This is easily achieved with Microsoft excel.

21.5.2 Two Inertia systems

Let us consider a shaft with two inertia I_1 and I_2 having stiffness K_1. The shaft is free to rotate in bearings. The torsional vibration will cause both end to twist, however, some point in between will not be twisting. This point is called node (zero displacement)

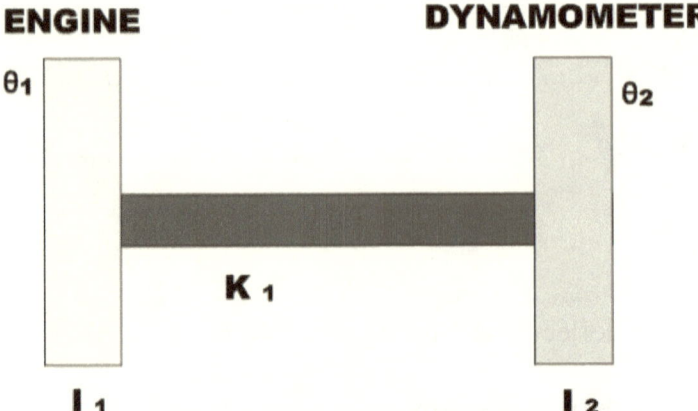

TWO INERTIA SYSTEM

Figure 21.7—two Inertia system

The natural frequencies can be derives using the work in preceding paragraphs. In this example of two rotors $T_2 = 0$

$$\theta_2 = \theta_1 - \frac{\omega^2}{K_1} I_1 \theta_1 \quad \text{...............21.22}$$

$$T_1 = \omega^2 . I_1 . \theta_1 \quad \text{...............21.23}$$

$$T_2 = T_1 + \omega^2 . I_2 . \theta_2 \quad \text{...............21.24}$$

$$T_2 = 0 = T_1 + \omega^2 . I_2 . \theta_2 = \omega^2 . I_1 . \theta_1 + \omega^2 . I_2 . \theta_2 \quad \text{...............21.25}$$

Substituting for θ_2 from equation 1 in to equation 4, we get

$$0 = \omega^2 . I_1 . \theta_1 + \omega^2 . I_2 . \left(\theta_1 - \frac{\omega^2}{K_1} I_1 \theta_1 \right) \quad \text{...............21.26}$$

Simplifying, we get

$$\omega^2 = K_1 \left(\frac{I_1 + I_2}{I_1 . I_2} \right) \quad \text{...............21.27}$$

The node will in between two rotors

Let us solve a practical example by using Holzer method.

EXAMPLE 1: THREE INERTIA SYSTEMS

The engine under test is connected to a dynamometer with a cardan shaft. Assume this system as three inertias on a shaft.

Data given: Engine inertia 2 kgm²
Shaft Inertia 4 kgm²
Dynamometer inertia 2 kgm²

The shaft connecting the first disk and second disk has stiffness of 3×10^6 m/radian and shaft connecting the last two disc has a stiffness of 2×10^6 m/radian. Ignore the inertia of the shafts.

Find:

1. Natural frequencies of the system.
2. Determine the nodal points

Solution:

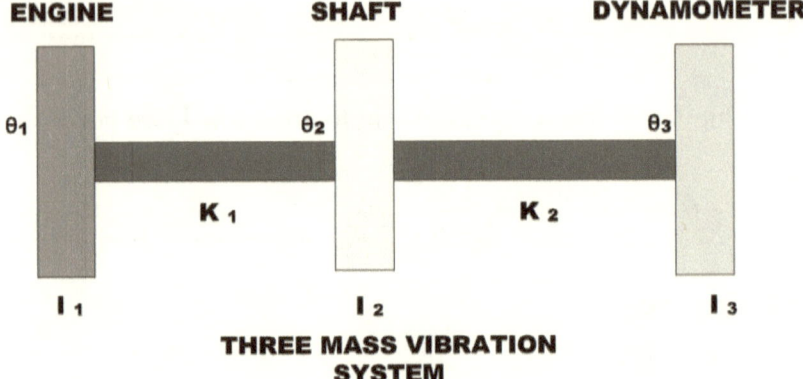

Figure 21.8—Representation of rotors

The solution by Holzer method suggests that we assume any ω and make $\theta_1 = 1$ and calculate all other deflections.

$$\theta_1 = 1$$

$$\theta_2 = \theta_1 - \frac{\omega^2}{K_1} I_1 \cdot \theta_1 = 1 - \frac{2\omega^2}{3.10^6}$$

$$\theta_3 = \theta_2 - \frac{\omega^2}{K_2}(I_1 \cdot \theta_1 + I_2 \cdot \theta_2) = \theta_2 - \frac{\omega^2}{2.10^6}(2 \times 1 + 4\theta_2)$$

$$T_1 = \omega^2 I_1 \theta_1 = 2\omega^2$$

$$T_2 = T_1 + \omega^2 I_2 \theta_2 = 2\omega^2 + 4\omega^2 \theta_2$$

$$T_3 = T_2 + \omega^2 I_3 \theta_3 = 2\omega^2 + 4\omega^2 \theta_2 + 2\omega^2 \theta_3$$

These values should be evaluated for all values of ω, *and* T_3 and plotted against ω

The point where $T_3 = 0$ gives the natural frequencies. The above equations can be solved by using the computing power of the Microsoft Excel. In the excel spread sheet start with the large values of ω and narrowing it down to the points where T_3 changes from positive to negative value.

The snap shot of spreadsheet is presented below.

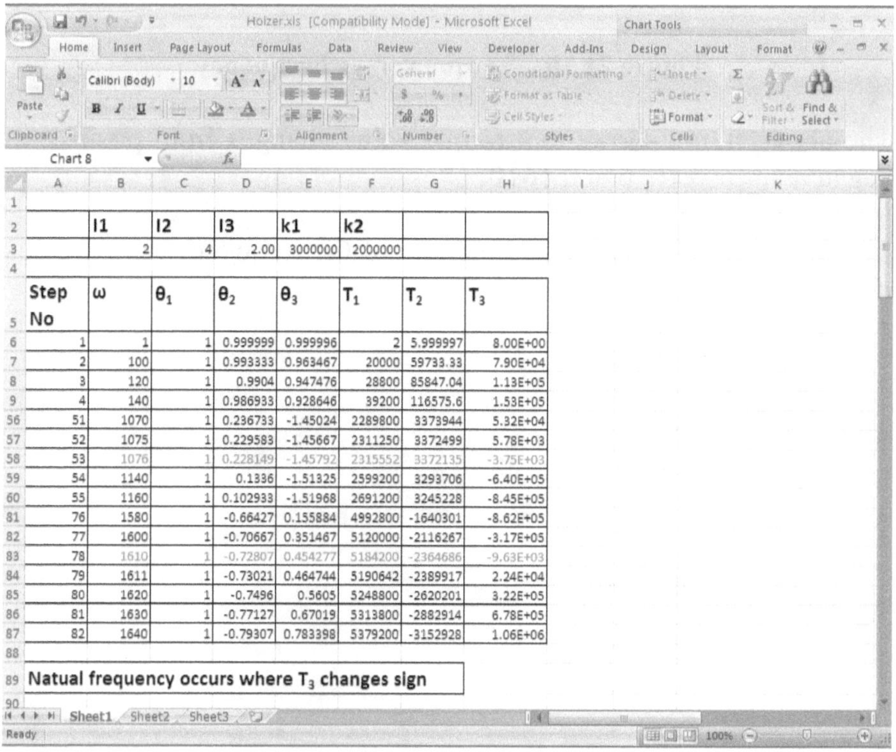

		A	B	C	D	E	F	G	H
2		I1	I2	I3	k1	k2		
3		2	4	2.00	3000000	2000000		

	Step No	ω	θ_1	θ_2	θ_3	T_1	T_2	T_3
6	1	1	1	0.999999	0.999996	2	5.999997	8.00E+00
7	2	100	1	0.993333	0.963467	20000	59733.33	7.90E+04
8	3	120	1	0.9904	0.947476	28800	85847.04	1.13E+05
9	4	140	1	0.986933	0.928646	39200	116575.6	1.53E+05
56	51	1070	1	0.236733	-1.45024	2289800	3373944	5.32E+04
57	52	1075	1	0.229583	-1.45667	2311250	3372499	5.78E+03
58	53	1076	1	0.228149	-1.45792	2315552	3372135	-3.75E+03
59	54	1140	1	0.1336	-1.51325	2599200	3293706	-6.40E+05
60	55	1160	1	0.102933	-1.51968	2691200	3245228	-8.45E+05
81	76	1580	1	-0.66427	0.155884	4992800	-1640301	-8.62E+05
82	77	1600	1	-0.70667	0.351467	5120000	-2116267	-3.17E+05
83	78	1610	1	-0.72807	0.454277	5184200	-2364686	-9.63E+03
84	79	1611	1	-0.73021	0.464744	5190642	-2389917	2.24E+04
85	80	1620	1	-0.7496	0.5605	5248800	-2620201	3.22E+05
86	81	1630	1	-0.77127	0.67019	5313800	-2882914	6.78E+05
87	82	1640	1	-0.79307	0.783398	5379200	-3152928	1.06E+06

89 Natual frequency occurs where T_3 changes sign

Figure 21.9—Excel snap shot

The following figure shows the graph of frequency versus torque:

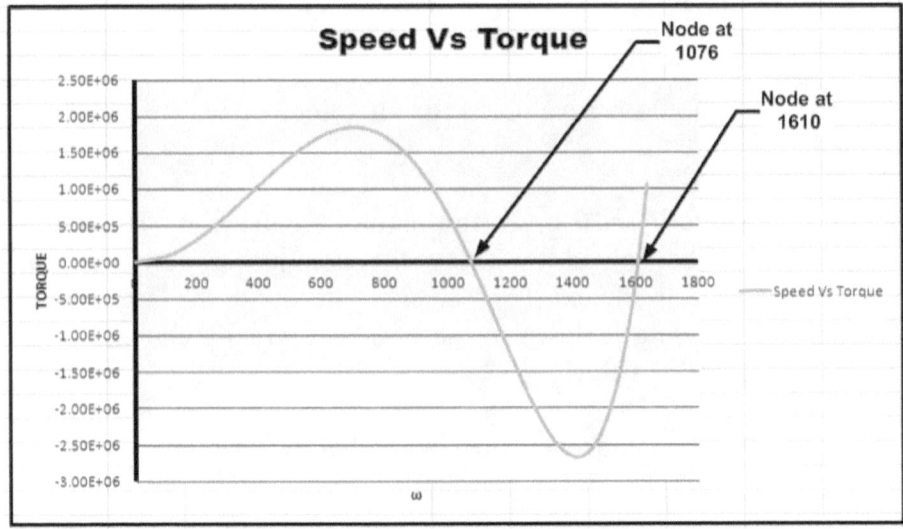

Figure 21.10—Torque versus Frequency

The above figure shows that torque value changes from positive to negative at ω = 1016 and 1610

The following figure shows the graph of the frequency (ω_r) versus displacement θ

DYNAMOMETER

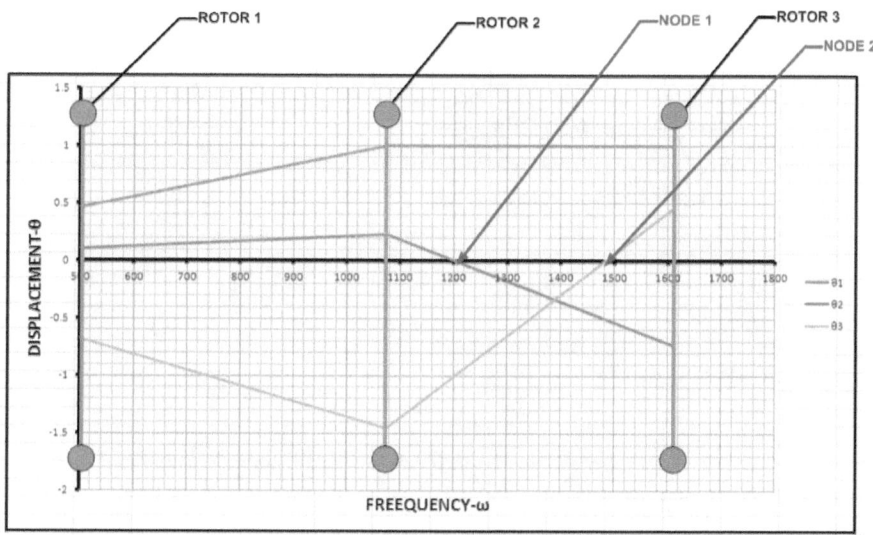

Figure 21.11—Displacement versus Frequency

The graph of frequency versus displacement clearly shows that nodes are between rotor 2 and 3 and one node is closer to rotor 2 and one closer to rotor 3.

The system of engine dynamometer may be further simplified as two mass system of 2 disk system. The calculations will be further simplified; however it will be an approximation of the true results.

EXAMPLE 2: TWO INERTIA SYSTEMS

A shaft rotating in bearing carries a disc having $I_1 = 2$ kgm² at one end and disk having $I_2 = 4$ kgm² on other end. The shaft connecting them has a stiffness of $K_1 = 4 \times 10^6$ m/rad. Determine the natural frequency and the position of node.

Solution:

$$\omega^2 = K_1\left(\frac{I_1 + I_2}{I_1 \cdot I_2}\right) = 4 \times 10^6 \left(\frac{2+4}{2 \times 4}\right)$$
$$= 3 \times 10^6$$

$$\omega = 1732 \frac{rad}{s}$$

Or $fn = 275.6\ Hz$

If we consider the node as a fixed point, each rotor will have the same natural frequency about that point. For a single rotor system

$$\omega^2 = \frac{K}{I}$$

For the first rotor

$$\omega^2 = \frac{K_1}{I}$$

$$3 \times 10^6 = \frac{K_1}{2}$$

$$K_1 = 6 \times 10^6$$

For the second rotor

$$\omega^2 = \frac{K_2}{I}$$

$$3 \times 10^6 = \frac{K_2}{4}$$

$$K_2 = 12 \times 10^6$$

The difference in stiffness is due to difference in length of the shaft

$$K = \frac{GJ}{l}$$

and GJ is same for the both sections.

Now

$$\frac{K_1}{K_2} = \frac{L_2}{L_1} = \frac{6}{12}$$

Therefore,

$$L_2 = \frac{L_1}{2}$$

$$L = L_1 + L_2 = L_1 + \frac{L_1}{2}$$

$$L = \frac{3}{2}L_1 \text{ or } L_1 = \frac{2L}{3}$$

And

$$L_2 = \frac{L}{3}$$

Displacements: Let $\theta_1 = 1$

$$\theta_2 = \theta_1 - \frac{\omega^2}{K_1}I_1\theta_1 = 1 - \frac{3 \times 10^6}{4 \times 10^6}2 \times 1 = 1 - 1.5 = -0.5$$

The node position is shown in following figure:

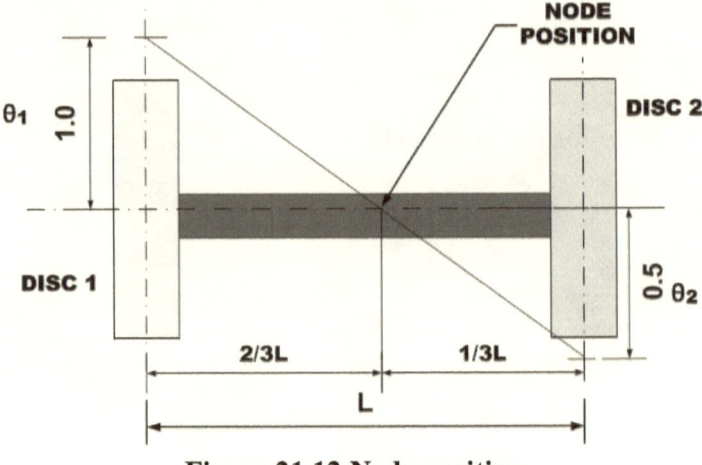

Figure 21.12 Node position

21.6 Engine and dynamometer foundation:

The dynamometer and engine foundation plays important role in reducing the overall vibrations of the entire installation. The engine and dynamometer can be installed on soft or hard foundations. The following figures show the various ways of installing the engine and dynamometer test bed. These installation possibilities are general schematics. In actual practice the full torsional vibration analysis is recommended for a particular test bed. Test cell design engineer can choose the appropriate scheme and do the full evaluation of the torsional vibration.

21.6.1 Schematic 1—Engine and dynamometer firm on hard foundation

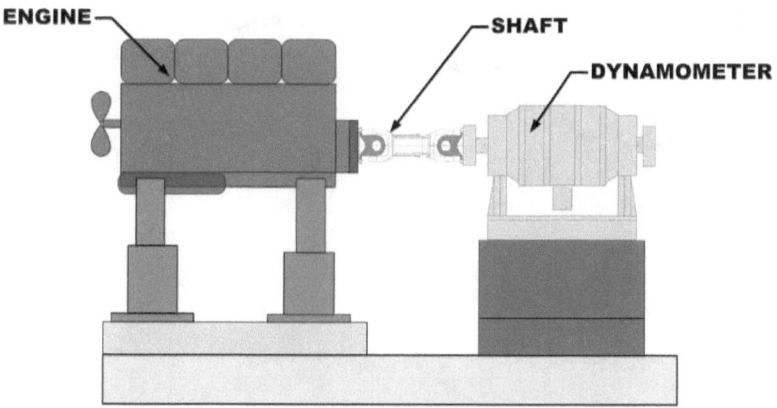

Figure 21.13—Engine and dynamometer hard on foundation

21.6.2 Schematic 2—Engine flexible mounted on hard foundation and dynamometer firm on hard foundation.

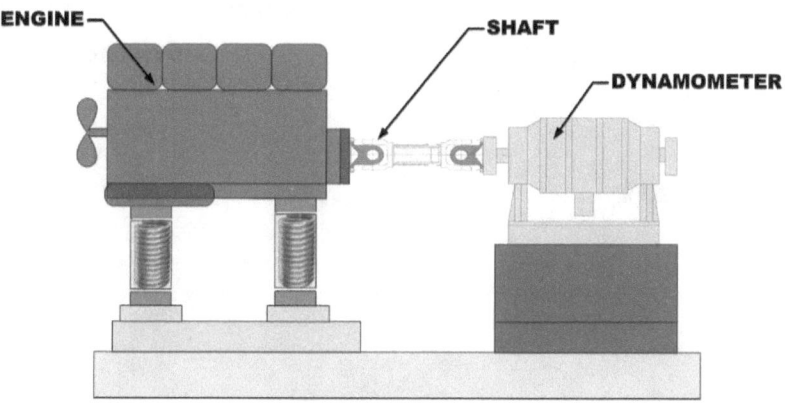

Figure 21.14-Engine flexible mounted on hard engine stand

21.6.3 Schematic 3—Engine Hard mounted on flexible Engine stand and dynamometer firm on hard foundation.

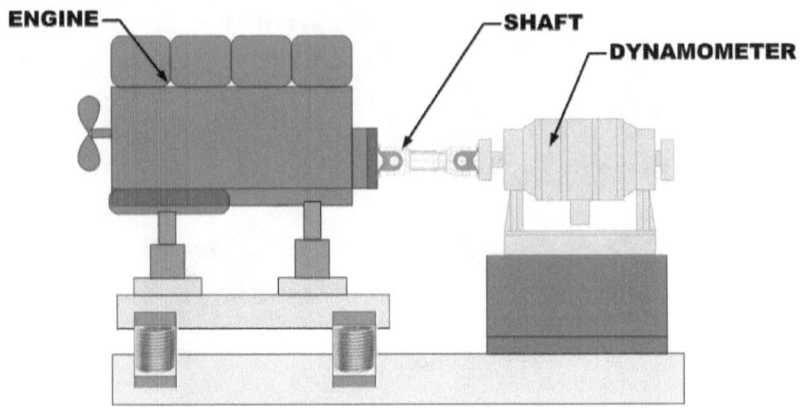

Figure 21.15—Engine hard on flexible engine stand

21.6.4 Schematic 4:—Engine and Dynamometer hard on flexible foundation.

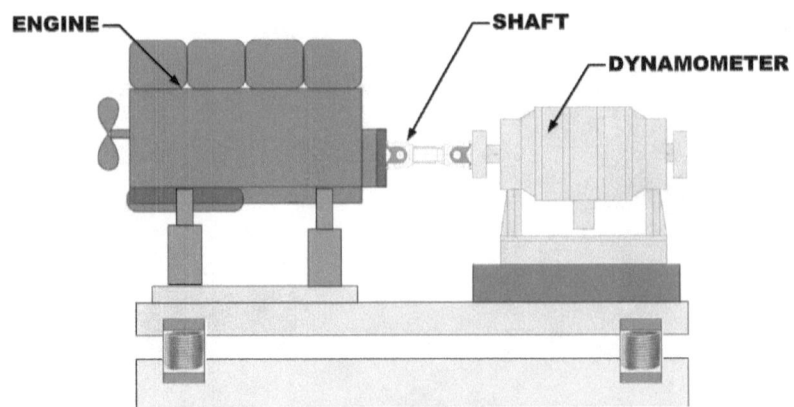

Figure 21.16—Engine and dynamometer hard on soft foundation

21.6.5 Schematic 5:—Engine flexible mounting and Dynamometer hard and both flexible mounted on common foundation.

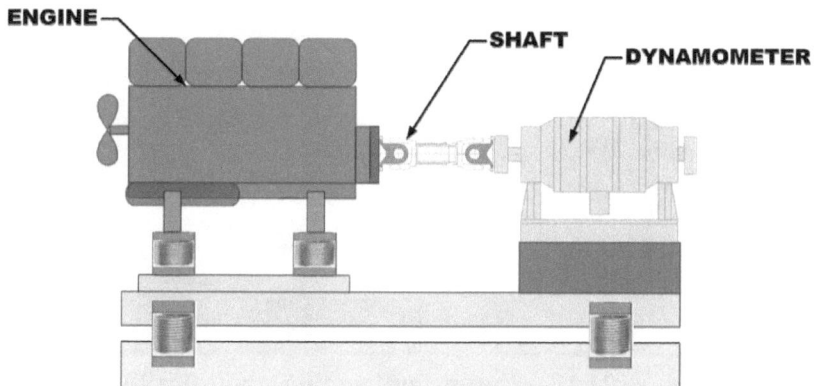

Figure 21.17 Engine flexible on stand which is on flexible foundation

21.6.6 Schematic 6:—Engine stand flexible mounting and engine hard on engine stand. Dynamometer hard on its own foundation. Engine stand and dynamometer foundation both flexible mounted on foundation.

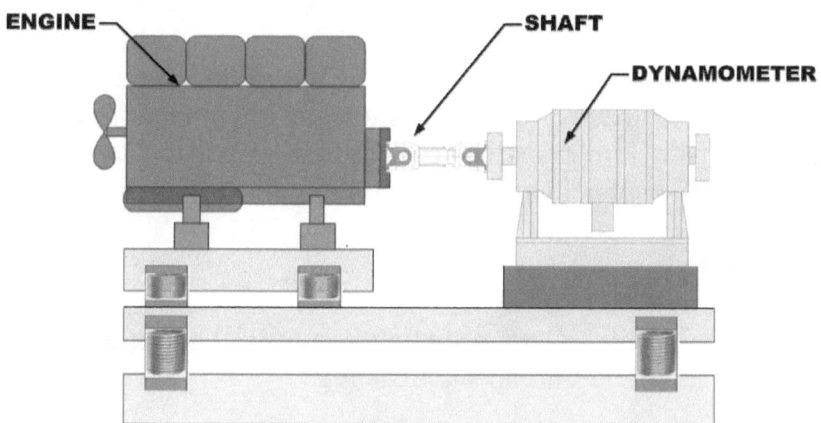

Figure 21.18—Engine stand flexible on flexible foundation and dyno hard

21.6.7 Schematic 7:—Engine flexible on engine stand which is flexible on foundation. Dynamometer firm on its own foundation

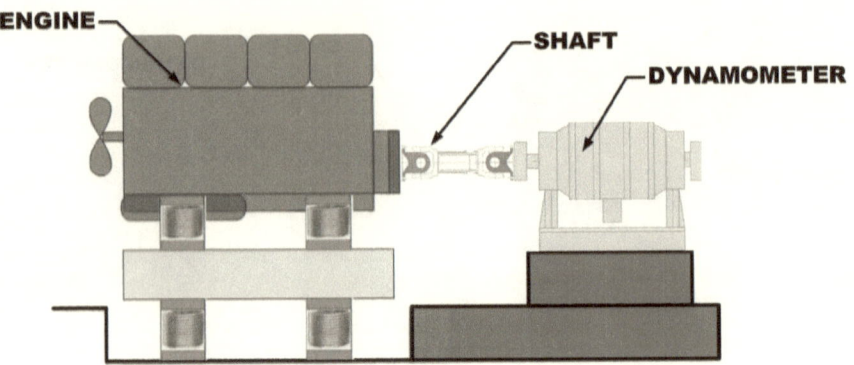

Figure 21.19—Engine and engine stand flexible

Closure:

Selection procedure of the cardan shaft is important and needs to be implemented properly with care so that proper most suitable cardan shaft is selected for the application under test. Right selection of the drive shaft is very important for the safe and trouble free running of the test cell. The torsional vibration analysis can be performed using various methods such as Holzer's method, to ensure that the selected cardan shaft is right match for the testing application. The other factors such as correct installation and alignment of the cardan shaft are equally important to avoid undue vibration and to ensure the safety of the personnel working in the test cell.

Bibliography:

1. Dana Spicer—Cardan shafts for industrial applications.
2. Cardan India Cardan shaft catalog.

3. Mechanical Vibrations—Theory and Applications—second edition by Francis S Tse, Ivan E. Morse and Rolland T. Hinkkle. Published by Allyn and Bacon, Inc. 1978
4. Practical solution of torsional vibration problems: with examples from marine, electrical, and automobile engineering practice / by W. Ker Wilson.

Further reading:

1. Coupling and joints—Design, selection and application by Jon A. Mancuso Pulished by Marcel Decker, Inc, 1999

CHAPTER 22

Air Intake and Exhaust Extraction Systems

I. Air Intake System

22.1 Introduction

Air intake conditions are very important for the engine to produce the correct designed power. The combustion in the cylinder of an internal combustion engine is mainly dependent upon:

1. Air
2. Fuel
3. Ignition

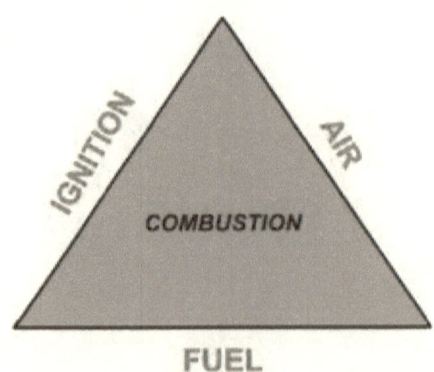

Figure 22.1 Essentials of combustion process.

The requirement of fuel is of prime importance for ignition to happen. The ignition is also equally important and needs to ignite it. For proper combustion, sufficient air is required and many times its importance is ignored. If any one of the three elements is removed, combustion process stops.

Combustion air pressure, temperature, and humidity play an important role in influencing the power output and the exhaust emissions of the internal combustion engine.

The most fundamental requirement of any engine test cell is the delivery of air to the test unit, that is, the engine under testing, at the required conditions of pressure, temperature, air mass flow, etc. In particular, the air must be supplied to the test cell with a dew point which is low enough to obviate the formation of ice.

As engines have become much more complex with sophisticated air handling, fuel systems, and integrated exhaust aftertreatment systems, the required dynamometer with independent control of the intake manifold air temperature and engine coolant temperature has become increasingly important as engine performance and combustion recipes are tailored to meet low emission standards.

The internal combustion engine has an air inlet for supplying intake airflow for combustion and an exhaust outlet for letting out an exhaust flow exiting from the internal combustion engine. The modern engine testing method includes the steps operating the internal combustion engine while both of the air inlet and the exhaust outlet are subjected to the simulated atmospheric pressure. The exhaust pressure controller is used for maintaining the exhaust outlet of the internal combustion engine which simulates the predetermined exhaust pressure during operation of the internal combustion engine. Similarly air intake pressure controller is used for maintaining the air inlet of the internal combustion engine substantially equal to a determined intake pressure during operation of the internal combustion engine.

22.2 The Purpose of Combustion Air Handling Unit—CAHU

The combustion air handling unit is used to simulate the various atmospheric conditions. It is primarily used for the following purpose:

1. Simulate various altitude conditions.
2. Simulate sea level conditions.
3. Simulate the turbocharger behavior.
4. To manipulate inlet temperature and pressure as desired by test engineer and development team.
5. To achieve a good reproducibility of measuring results within the test cell.

22.3 Combustion Air Handling Unit—CAHU

The Combustion Air Handling Unit also known as CAHU is used for the engine combustion air temperature, relative humidity, and atmospheric pressure control. The most consistent engine combustion air conditions have been 75 degrees dry bulb, 50 percent relative humidity, and about .04" hg pressure above ambient pressure. It has been found experimentally that by maintaining these combustion air criteria that the dynamometer correction factor stabilizes and dynamometer repeatability is assured.

The following figure shows the block schematic diagram of the combustion air handling unit. The incoming air is processed as it passes through various stages of the combustion air handling unit.

Stage 1

The combustion air handling unit has a centrifugal blower unit which pumps in the outside air and processes it to the test conditions demanded by the test engineer. This stage consists of the convergent incoming nozzle and the air filter. The filter prohibits the suspended particles in the air entering the system. The converging nozzle increases the velocity of the incoming air. The blower also increases the air pressure. The blower is driven by AC motor with variable speed drive.

Stage 2

The filtered air enters the cooling chamber where it is cooled by the cooling coil which carries the chilled water supply mixed with anti-freeze supplied by the chiller unit. The cooling section has condensate drain which drains out the condensed excess vapor content in the air.

Stage 3

This is humidifying stage. The humidity of the processed air is increased when passed through this section. The steam is injected into the air flow to increase the humidity of the incoming air.

Stage 4

The air after the humidifying section passes through the heating section. This sections consists the battery of heating elements. The elements are switched as required by the system demand to increase the temperature of incoming air.

Stage 5

The secondary cooling coil helps maintain the temperature of the processed air. It helps to remove the excess heat from the air added by the heating section and thus helps in achieving a fine control and maintaining the demanded temperature of the process air.

Stage 6

The processed air then is delivered to the delivery chamber and the flow and the pressure is regulated by the butterfly valves. One valve is used to regulate the inlet flow to the delivery chamber and the other valve regulates the excess flow to the atmosphere. This chamber helps to remove the pressure and flow ripple and helps delivering smooth flow to the engine.

Stage 7

The processed and conditioned air as per the requirement is delivered to the engine through the metering flow meter. This is the actual mass of the air finally delivered to the engine. The feedback signal from the flow meter is used by the controller to control the flow as per demand.

This option allows you not only to control and adjust the amount of volume of air but also to control the barometric pressure at the engine, at a predetermined barometric pressure selected above outside ambient barometric pressure.

The vacuum or negative pressure in the system is prevented by an under pressure relief valve which activates the flap and allows atmospheric air into the system in the event of failure of centrifugal blower or the inlet blockage. In case of the blockage in the system, the engine under test continues to draw air from the combustion handling unit and at this moment the "under pressure relief valve" opens and allows the atmospheric air to rush in to the unit. This situation prevents the engine from starving for lack of air, though the air may not be conditioned as per requirement.

The following figure shows the block schematic of combustion air handling unit:

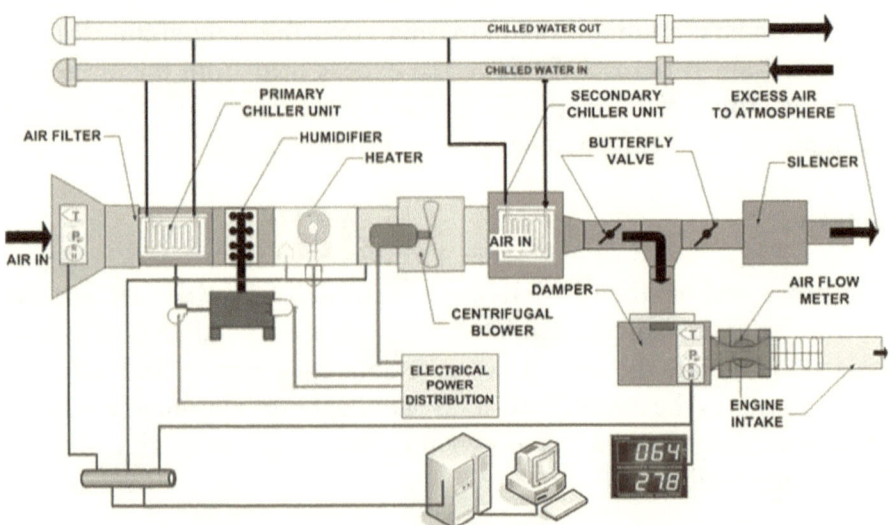

Figure 22.2 Combustion air handling unit.

22.4 Air-Fuel Ratio

Air-fuel ration is defined as mass ratio of air to fuel present in an internal combustion engine.

$$Air\ Fuel\ Ratio\ (AFR) = \frac{m_{Air}}{m_{fuel}} \quad \quad \quad 22.1$$

If the engine under test is provided with enough air to completely burn all of the fuel, the ratio is known as the stoichiometric mixture. Air-fuel ration is an important measure for emission and performance measurements. Stoichiometric or theoretical combustion is the ideal combustion process where fuel is burned completely.

The word "stoichiometric" is derived from the Greek words stoicheion, meaning element and metron, meaning measure.

Today's engines with engine management systems employ fuel injection system in place of carburetor and attempt to achieve stoichiometric mixture moderate load cruise situations. The gasoline fuel have stoichiometric air-fuel ratio as approximately 14.7, that is the approximate mass of air is 14.7 mass of fuel.

The air-fuel mixture less than 14.7 to 1 is considered to be a rich mixture and more than 14.7 to 1 to be a lean mixture.

22.5 Lambda

Lambda (λ) is the ratio of actual air-fuel ratio to stoichiometric ratio for a given mixture. Lambda of 1.0 is at stoichiometric and Lambda greater than 1.0 is commonly known as rich mixture, whereas Lambda less than 1.0 is known as lean mixture.

$$\lambda = \frac{AFR_{AIR}}{AFR_{Stoichimetric}} \quad \quad \quad 22.2$$

Above formula clearly indicates a direct relationship between lambda and AFR. To calculate AFR from a given lambda, multiply the measured lambda by the stoichiometric AFR for that fuel used for the engine under testing.

22.6 What Is the Effect of Changing the Air-Fuel Ratio?

1. To investigate at what air-fuel ratio knocking occurs.
2. If the mixture is weakened, the flame speed is reduced, consequently less heat is converted to mechanical energy, leaving heat in the cylinder walls and head, potentially inducing knock.
3. To investigate at what air-fuel ration the thermal efficiency is maximum.

II Engine Exhaust Extraction

22.7 Introduction

The exhaust extraction from test cell is very important, whether it is engine test cell or vehicle test cell. The small leakage of the exhaust gas in to the cell will deteriorate the engine performance. Besides this it is also harmful to humans inhaling these fumes. Normally the exhaust is extracted out of the test cell. The quick connecting flanges and locking clamps are normally employed to facilitate the faster engine changes in case of production cells.

The function of the exhaust system is to convey engine exhaust safely outside the test cell or the test house and to disperse the exhaust fumes, soot, and noise away from testing personnel and buildings. The exhaust system must be designed to minimize back pressure on the engine. Excessive exhaust restriction will result in increased fuel consumption, abnormally high exhaust temperature, and failures related to high exhaust temperature as well as excessive black smoke.

The test engineer and the test cell designer should have the basic knowledge of the exhaust handling, extraction methods, and its implications on the testing performance.

22.8 What Is the Exhaust?

Once the combustion process is complete and high pressure generated has done the useful work to turn the crankshaft during the expansion stroke, the products of the combustion commonly called as exhaust gases need to be removed from the engine cylinder to make room for the fresh air fuel charge for the next cycle.

The exhaust process takes place in two stages:

1. Exhaust blowdown.
2. Exhaust process.

The entire exhaust flow is discharged into the exhaust manifold and in turn it is delivered to exhaust pipe. In case of four-stroke engine, exhaust stroke occurs once in to revolutions. Therefore, the resulting flow passing through the exhaust pipe is nonsteady and pulsating.

22.9 Exhaust Blowdown

When the exhaust valve starts to open toward the end of the power stroke, the exhaust blow down occurs. This happens when the piston has almost reached bottom dead center on its travel during power stroke. At this instant, the pressure in the cylinder is much higher than the atmosphere and temperature of the exhaust gases is still very high

to the tune of 1000-1200°K, whereas the pressure in the exhaust pipe is about atmospheric. This pressure difference causes the rapid flow of exhaust gases into exhaust extraction system. The rush of flow of exhaust gases is called the exhaust blowdown.

As piston still continues to travel toward BDC (bottom dead center), the pressure in the cylinder decreases during the blowdown process, the gas leaving will have progressively lower velocity and kinetic energy.

22.10 Exhaust Stroke

When the blowdown process is near completion, the piston passes through the BDC (bottom dead center) and begins its travel toward TDC (Top dead Center) during the exhaust stroke. During this stage, the exhaust valve remains open and the pressure in the cylinder is above the pressure in the exhaust pipe as the piston is forcing the exhaust gases through the exhaust valve which is open during the exhaust stroke.

Theoretically, at the end of the exhaust stroke when the piston reaches TDC, all the exhaust gases are removed from the cylinder and the exhaust valve closes. When the exhaust valve is finally closed, there is still a residual of exhaust gases trapped in the clearance volume of the cylinder. The higher the compression ratio of the engine, the less clearance volume exists to trap this exhaust residual.

22.11 The Exhaust Backflow

Ideally the intake valve should be totally open at TDC when the intake stroke starts and the exhaust valve should be closed. However, due to the finite time required to open the inlet valve, it must start to open 10°-25° before TDC. There is, therefore, a period of 15°-50° of engine rotation when both intake and exhaust valves are open. This is called valve overlap.

During valve overlap, as both inlet and exhaust valves are open for a small duration of time, there can be some reverse flow of exhaust gas back into the intake system. When the intake process starts, this exhaust is drawn back into the cylinder along with the air-fuel charge. This results in a larger exhaust residual during the rest of the cycle.

This backflow of exhaust gases is a greater problem at low engine speeds, being worst at idle conditions. At most low engine speeds, the intake throttle is at least partially closed, creating low pressure in the intake manifold. This creates a greater pressure differential, forcing exhaust gas back into the intake manifold. Cylinder pressure is about one atmosphere, while intake pressure can be quite low. In addition, real time of valve overlap is greater at low engine speed, allowing more backflow. Some engines are designed to use this small backflow of hot exhaust gas to help vaporize the fuel that has been injected directly behind the intake valve face.

Once the exhaust gases leave the cylinders by passing through the exhaust valves, they pass through the exhaust manifold, a piping system that directs the flow into one or more exhaust pipes.

The orientation of the engine's air intake and exhaust pipes in relation to the cell's fresh air and exhaust ducts are crucial to successfully keeping all exhaust gases out of the engine's fresh air supply. Exhaust manifolds are usually made of cast iron and are sometimes designed to have close thermal contact with the intake manifold. This is to provide heating and vaporization in the intake manifold.

22.12 The Importance of Correct Back Pressure

Exhaust gases are passed through the muffler to reduce the noise of engine combustion. At the same time, back pressure causes exhaust gases to remain in the engine cylinder after the exhaust stroke. While a certain amount of back pressure is vital to optimum performance, too much back pressure can result in loss of horsepower and excessive engine/turbocharger operating temperatures. When this happens, performance and fuel economy suffer. It may not take much to alter the balance or to affect engine operation.

22.13 Exhaust Extraction from the Engine Test Cell

The exhaust extraction systems employed in the test cell are of the following types with each having its own advantages and disadvantages.

22.13.1 Underground Extraction System

This method of exhaust extraction is most popular. The underground system extraction is highly efficient, cost-effective, and trouble-free. The most important factor of this method of extraction is un-obtrusive and plays a very important role in test cell design. While routing the exhaust piping underground care should be taken not to provide the exhaust exit stream with excessive resistance due to the pipe work system. The following figure shows the schematic arrangement of underground exhaust extraction system. An exhaust system must reduce engine noise and discharge exhaust gases safely away from the test cell. An efficient exhaust system can improve engine performance.

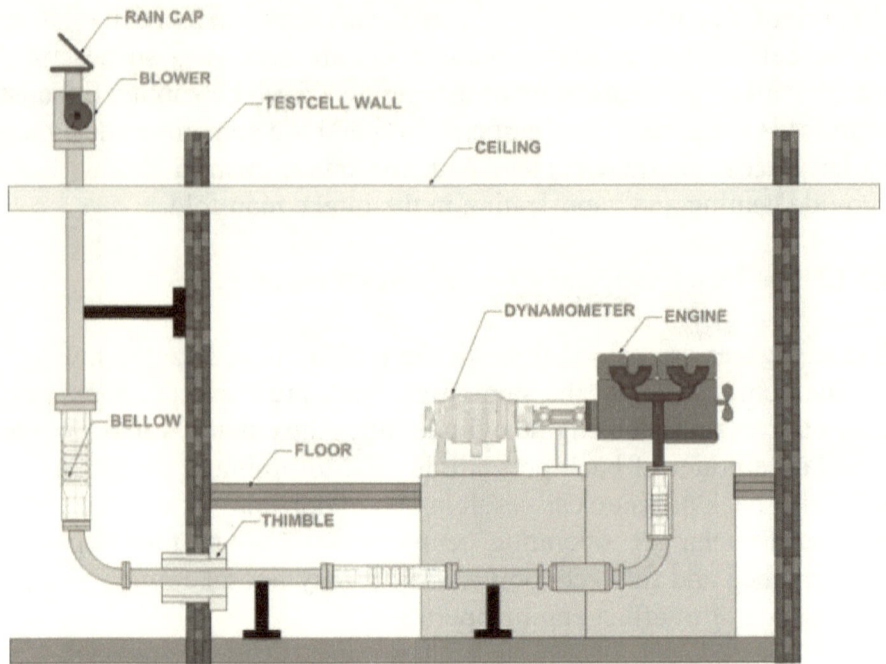

Figure 22.3 Underground exhaust extraction system.

22.13.2 Overhead Extraction Systems

The overhead systems are sometimes called as suspended systems, as exhaust piping is suspend from the ceiling of the test cell. The suspended exhaust system in test cell extracts the exhaust via ducting which is suspended to the ceiling and taken out of the test cell. Some outlets are provided for the smoke measurement purpose. Routing of exhaust piping at least 6 to 8 feet (2 to 3 meters) above the floor will also help to prevent accidental contact with the exhaust system.

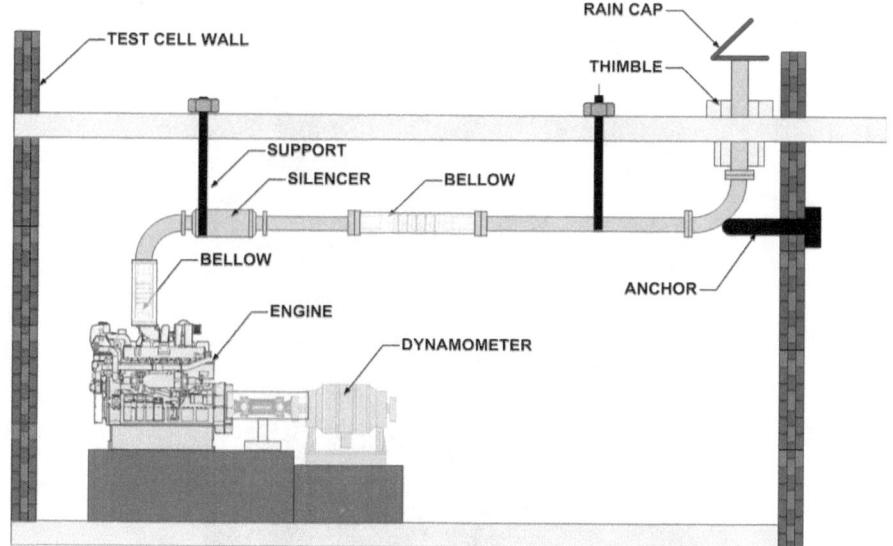

Figure 22.4 Overhead exhausts extraction system.

22.14 Exhaust Extraction from the Vehicle Test Cell

The vehicle testing in a closed cell needs a proper extraction system. The exhaust from the test cell needs to be extracted for the safety of the people and its proper disposal into atmosphere.

22.14.1 Telescopic System

Telescoping design eliminates the need for winches, ropes and pulleys, and hose reel. The exhaust pipes are telescopic and extend from one another telescopically. The following figure shows the telescopic exhaust hose and its application to extract exhaust from the exhaust pipe of a car.

Figure 22.5 Telescopic exhaust.
Courtesy of Crushproof Tubing Company, USA.

22.14.2 Common Rail System

The following figure shows the common rail system. In this type of system, there are multiple exhaust extraction stations in a service shop. These individual exhaust extraction stations are connected to common hollow rail.

Figure 22.6 Common rail systems. Courtesy of Fume-A-Vent.

The common rail system provides the following advantages:

1. The common rail traverses over the entire length of the shop providing mobility and quiet efficient exhaust extraction.
2. Exhaust carriage carrying the exhaust hose moves over the rail (similar to overhead electric hoist). The carriage can be brought to the point of use.
3. This design offers clean aesthetically pleasing design.

22.15 The Essential Elements of Exhaust Extraction

1. Exhaust capture nozzle,
2. Crushproof hose,
3. Fan/motor assembly,
4. Exhaust to atmosphere ductwork,
5. Hose reel, manual or motorized,
6. Spring balancer for hose.

22.15.1 Exhaust Capture Nozzles

The exhaust capture nozzles are available in different shapes and material of construction. They are classified as follows:

A. Material of construction

Neoprene Rubber

Neoprene nozzles are resistant to exhaust temperatures up to 125-150°C at normal use. Neoprene reduces the risk of damaging vehicle tail pipe surfaces.

Figure 22.7 Neoprene nozzle. Courtesy of Fume-A-Vent.

Stainless Steel

These nozzles have spring-loaded lid which secures the nozzle in place on the mouth of exhaust pipe. A steel nozzle is recommended in conjunction with high exhaust temperature.

B. Shape of Nozzle

1. Circular/Conical

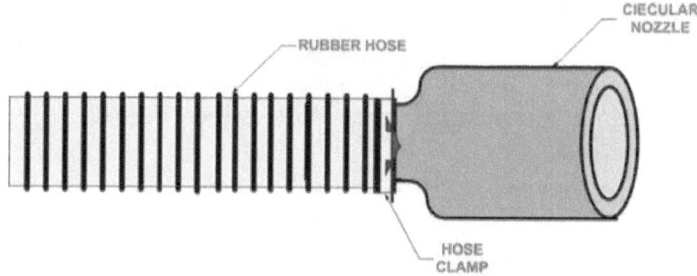

Figure 22.8 Circular nozzle.

The conical nozzle will have a conical opening which can accommodate wide range of tail pipe diameters.

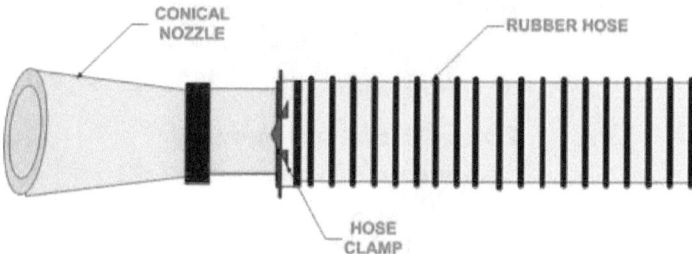

Figure 22.9 Conical nozzle.

2. Angular

These types of nozzles are used for connecting the exhaust tail pipe of two wheelers (e.g., motorcycles).

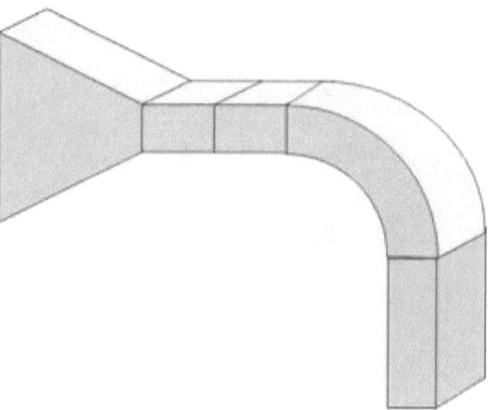

Figure 22.10 Angular nozzles.

C. Type of Fitting

The exhaust nozzles are classified based on their fitting method on exhaust tail pipe as described below.

- **Slip on type**

These can easily slide off the exhaust pipe and can only be used with exhaust pipes that are protruding from the vehicle; no ambient air is fed into the nozzle for temperature reduction.

Figure 22.11 Round nozzles.
(Courtesy of Fumeavent)

- **Clamp on type**

The clamp on type exhaust nozzles are used to connect the exhaust extraction hose pipe quickly on the tail pipe of the vehicles. This is very useful especially when you need to connect the exhaust extraction hose on tail pipe of motorcycle being tested on chassis dynamometer.

Figure 22.12 Clamp on nozzles. Courtesy of Fume-A-Vent.

The following illustration shows the clamp on nozzle mounted on an exhaust hose on reel. This type of exhaust extraction system is highly suitable for the production testing of the two wheelers where vehicle testing turnover is very high. The quick connection to the tail pipe of motorcycle/scooter is of very important which saves total testing cycle time.

Figure 22.13 Exhaust Hose Reel. Courtesy of Fume-A-Vent.

- **Spring Loaded**

This type of nozzle is equipped with a spring loaded shutoff flap which will seal off intake into the nozzle when not in use.

Normally the exhaust extraction nozzles are zinc coated to prevent the corrosion, and it has built-in provision to mount exhaust probe for the emission analysis.

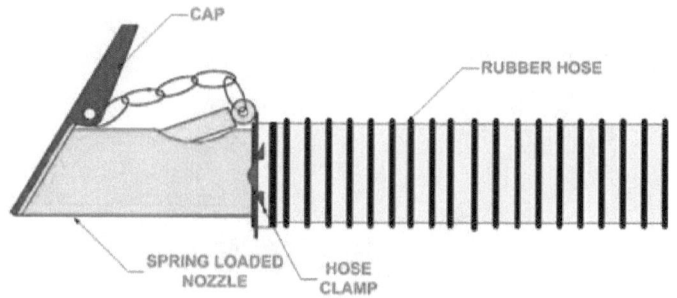

Figure 22.14 Spring loaded nozzles.

The test cell designer is normally required to adopt the nozzle from available size and type from the market or sometimes he may have to go for a custom-made nozzle to meet his requirement. It is advisable to use safety gloves while handling the nozzles as they will be hot due to contact with exhaust pipe.

Commercially various sizes of the exhaust nozzles are available to suit the individual requirement of engine as well as vehicle testing. The following illustration shows the commercially available exhaust extraction nozzles. These exhaust extraction nozzles are also known commercially as tail pipe adapters.

Figure 22.15 Exhaust extraction nozzles. Courtesy of Fume-A-Vent.

22.16 Exhaust Crush-proof Hoses

It is sometimes convenient to use flexible hose pipes which are resistant to high temperature and pressure. They are also known as crush-proof hoses pipes. They are highly suitable for vehicle testing. Crush-proof hoses are strong enough to sustain the medium-sized vehicle weight. It happens many times that vehicles are brought to chassis dynamometer or service center and are run over the exhaust hoses. The crush-proof hoses can tolerate the weight of the vehicles without getting crushed under them.

Figure 22.16 Crush-proof exhaust hose.
Courtesy of Crushproof Tubing Company, USA.

22.17 Fan Motor Assemblies

This assembly consists of blower assembly with aluminum or steel fans which are designed for high-efficiency exhaust extraction. The exhaust extraction fan along with butterfly valve in exhaust outlet helps maintain the exhaust back pressure.

Figure 22.17 Typical exhaust handling blower. Courtesy of Fume-A-Vent.

22.18 Hose Reel

The hose reel is a hose mounted on a winch drum. The hose is wound around the drum. At the time of testing, the hose is unwound and connected to the vehicle exhaust pipe. Once testing is done, the hose is wound back on the winch drum. The operation of winding and unwinding can be done manually or it can be motorized. Once the hose is wound on drum after the testing, it leaves the test area free of hanging hoses.

Figure 22.18 Hose reel with integral exhaust fan.
Courtesy of Fume-A-Vent.

22.19 Spring Balancer for Hose

This system is particularly suitable for the small engines used for scooters and motorcycles production test beds. In this type of system, the engines are changed rapidly especially when engines are checked for fixed load or fixed test point such as Go—No Go test. The operator can just pull the hose and clamp on the exhaust of the engine. Once the engine is tested, the operator needs to unclamp the exhaust hose which will retract automatically and will be out of the way of vehicle handling.

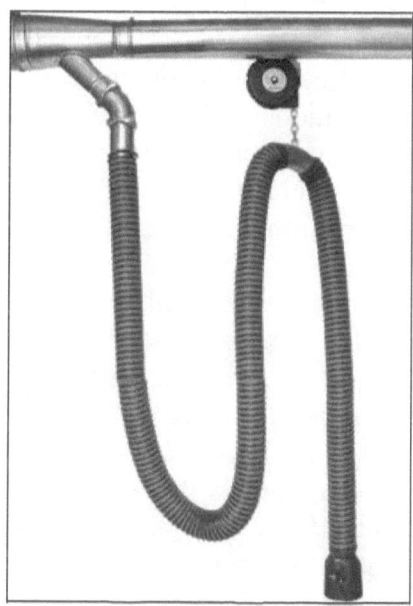

Figure 22.19 Self-retracting spring balancer with exhaust hose. (Courtesy of Fume-A-Vent.)

The Details of Spring Balancer

The spring balancer is shown in Figure 22.20. The spring balancer helps to retract the exhaust hose automatically and helps in reducing the manual effort.

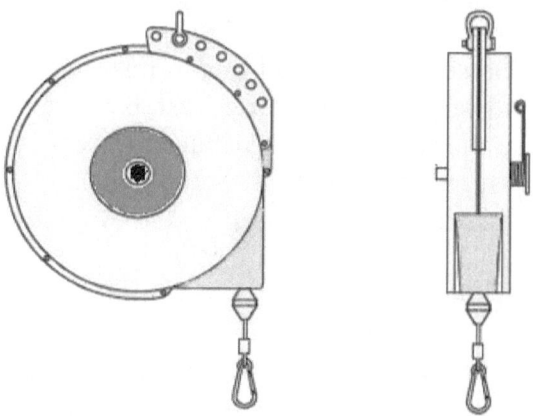

Figure 22.20 Spring balancer. Courtesy of Fume-A-Vent.

22.20 Multipoint System

The multiple exhaust extraction system with common exhaust blower is suitable for the applications such as service stations. At a suitable point, a single large extraction fan/motor assembly is provided. Nozzles would normally include a shut-off valve/lid so that when not in use, the reel is isolated from the general exhaust system.

However, for the dynamometer test cell, the exhaust of one test cell is not normally connected to common header in which other test cells are discharging the exhaust as this becomes the conflicting issues on the power measurement and performance testing. Moreover, the exhaust back pressure also plays an important role on power measured by dynamometer. This also applies to the test cell installed with chassis dynamometer. Normally for the engine test cells, the independent exhaust extraction system is provided.

22.21 Design Considerations in Exhaust Extraction System

While designing the exhaust extraction system for the engine test cell, the test cell designer should consider the following aspects.

22.21.1 Exhaust Back Pressure

Exhaust piping should be of sufficient diameter to limit exhaust backpressure to a value within the rating of the specific engine under test. Exhaust backpressure must not exceed the allowable backpressure specified by the engine manufacturer. Excessive exhaust backpressure reduces engine power and engine life and may lead to high exhaust temperatures and smoke. Engine exhaust backpressure should be estimated before the layout of the exhaust system is finalized, and it should be measured at the exhaust outlet under full-load operation before the set is placed in service.

The test cells are normally built to test engines of varying sizes. Therefore, it is evident that the engines on a particular test bed will have different exhaust sizes and different backpressure limitations. Piping of smaller diameter than the exhaust outlet must never be used. Piping that is larger than necessary is more subject to corrosion due to condensation than smaller pipe. Piping that is too large also reduces the exhaust gas velocity available for dispersing the exhaust gases up and into the outdoor wind stream.

22.21.2 Pipes and Flexible Pipes

Iron pipe may be used for exhaust piping, or alternatively, prefabricated stainless steel exhaust pipes can be used. It is common practice to connect flexible corrugated stainless steel exhaust bellows to the engine exhaust outlet to allow for thermal expansion and engine movement and vibration whenever the engine is mounted on vibration isolators.

Test cell engineer should be aware of the fact that weight on the engine exhaust outlet can cause damage to the engine exhaust manifold and can cause vibration from the engine test bed to be transmitted into the building structure.

The use of exhaust mounts with isolator further limits vibration from being induced into the building structure.

To reduce corrosion due to condensation, a muffler should be installed as close as practical to the engine so that it heats up quickly. Locating the silencer close to the engine also improves the sound attenuation of the muffler. Pipe bend radii should be as long as practical.

A test cell should not be connected to an exhaust system of other test cell. Soot, corrosive condensate, and high exhaust gas temperatures can damage idle equipment served by a common exhaust system.

22.21.3 Insulation for Pipes

Exhaust tubing and piping should be of the same nominal diameter as the engine exhaust outlet (or larger) throughout the exhaust system.

The entire engine exhaust system components should be covered with insulation or have barriers to prevent dangerous accidental contact.

Thermally insulated exhaust system will

1. Avoid fires due to accidentally spilled fuel on exhaust system.
2. Prevent burns from accidental contact.
3. Prevent activation of fire detection devices and sprinklers.
4. Reduce corrosion due to condensation.
5. Reduce the amount of heat radiated to the test cell.

22.21.4 Expansion Joints and Bellows

Expansion joints, engine exhaust manifolds, and turbocharger housings, unless water cooled, must never be insulated. Insulating exhaust manifolds and turbochargers can result in high material temperatures that can destroy the manifold and turbocharger, particularly in applications where the engine will be tested for a large number of hours or especially in the event of endurance test.

Flexible exhaust tubing must not be used to form bends. Isolated noncombustible hangers or supports should be adopted to support mufflers and piping.

Exhaust pipe, especially those made of steel, expands as exhaust temperature rise above ambient temperature. Therefore, it is required that exhaust expansion joints be used to take up expansion in long, straight runs of pipe.

Expansion joints should be provided at each point where the exhaust changes direction. Exhaust piping must be routed sufficiently away keeping enough space from combustible construction. It is in practice to use appropriately designed thimbles where exhaust piping must pass through combustible walls or ceilings.

22.21.5 Exhaust to Atmosphere

The exhaust system outlet direction should also be carefully considered. Exhaust should never be directed toward the roof of a building or toward combustible surfaces. Exhaust from a diesel engine is hot and will contain soot and other contaminants that can adhere to surrounding surfaces. It should be vent out to open atmosphere.

Locate the exhaust outlet and direct it away from the ventilation air intakes.

If noise is a factor, direct the exhaust outlet away from critical locations. Mufflers/silencers should be used to reduce the noise.

The rain cap should be provided to prevent rain from entering the exhaust system outlets which are vertical in nature. This might include a rain cap or exhaust trap. Horizontal exhaust outlets may be cut off at an angle and protected with bird screen.

Rain caps can freeze closed in cold environments, disabling the engine, so other protective devices may be best for those situations.

22.21.6 Condensate Extraction

The exhaust pipes get cooled during the night time when engine testing is not done, and moisture in exhaust condensates in the pipe. The condensate in the pipe may react with the sulfur from unburned fuel and form the sulfuric acid leading to the corrosion of pipes and exhaust manifolds. Therefore, it is essential to drain off the system for accumulated condensate.

Horizontal runs of exhaust piping should slope downward, away from the engine, to the outdoor or to a condensate trap.

A condensate drain trap and plug should be provided where piping turns to rise vertically. Condensate traps may also be provided with a silencer.

Closure

The engine intake and exhaust systems play vital role in establishing the engine performance in terms of power torque and fuel consumption. The engine air intake system plays an important role in maintaining the standard intake conditions and simulates the altitude. Exhaust handling especially in the test cell also plays an important role in establishing the engine performance. The test cell designer needs to plan in detail while designing the test cells and should consider the right air intake and exhaust handling system for the application whether it is test cell or a service station.

References

1. Image curtsey Crushproof hoses.
2. Image curtsey of Fume a vent.

APPENDIX -1

Power calculation formulae

Sr. No.	Torque (M)	Power (P)	Speed (N)
1	M in kgm	$P = \dfrac{M.N}{716.20} \ (ps)$	N in rpm
2	M in kgm	$P = \dfrac{M.N}{973.76} \ (Kw)$	N in rpm
3	M in Nm	$P = \dfrac{M.N}{9549.305} \ (Kw)$	N in rpm
4	M in Nm	$P = \dfrac{M.N}{159.16} \ (ps)$	N in rps
5	M in lbf.ft	$P = \dfrac{M.N}{5252.1} \ (Kw)$	N in rpm
6	M in lbf.ft	$P = \dfrac{M.N}{7043.2} \ (Kw)$	N in rpm
7	Force (F) in Kg for lever arm length 716.20 mm	$P = \dfrac{F.N}{1000} \ (ps)$	N in rpm
8	Force (F) in Kg for lever arm length 973.76 mm	$P = \dfrac{F.N}{1000} \ (Kw)$	N in rpm
9	Force (F) in Newton (N) for lever arm length 954.93 mm	$P = \dfrac{F.N}{1000} \ (Kw)$	N in rpm
10	Force (F) in Newton (N) for lever arm length 1591.6 mm	$P = \dfrac{F.N}{100} \ (Kw)$	N in rps
11	Force (F) in lbf for lever arm length 2.6261 mm	$P = \dfrac{F.N}{2000} \ (hp)$	N in rpm
12	Force (F) in lbf for lever arm length 3.5216 mm	$P = \dfrac{F.N}{2000} \ (Kw)$	N in rpm

All above formulae are derived by using fundamental formula for:

1. $HP = \frac{2.\pi.N.T}{4500}$ T in Kgm and N in rpm

2. $KW = \frac{2.\pi.N.T}{60000}$ T in Nm and N in rpm

3. $HP = \frac{2.\pi.N.T}{33000}$ T in lbft and N in rpm

APPENDIX II

Engine terminology

Spark Ignition (SI)
An engine in which the combustion process in each cycle is started by use of a spark plug.

Compression Ignition (CI)
An engine in which the combustion process starts when the air-fuel mixture self-ignites due to high temperature in the combustion chamber caused by high compression. CI engines are often called **Diesel** engines, especially in the non-technical community

Top-Dead-Center (TDC)
Position of the piston when it stops at the furthest point away from the crankshaft. Top because this position is at the top of most engines (not always), and dead because the piston stops at this point. Because in some engines top-de ad-center is not at the top of the engine (e.g., horizontally opposed engines, radial engines, etc.), some Sources call this position

Head-End-Dead-Center (HEDC).
When an occurrence in a cycle happens before TDC, it is often abbreviated bTDC or bTe. When the occurrence happens after TDC, it will beabbreviated aTDC or aTe. When the piston is at TDC, the volume in the cylinder is a minimum called the *clearance volume.* This is also know as top center.

Bottom-Dead-Center (BDC)
Position of the piston when it stops at the point closest to the crankshaft. Some sources call this **Crank-End-Dead-Center** because

587

it is not always at the bottom of the engine. Some sources call this point **Bottom-Center**. During an engine cycle things can happen before abBotet.om-dead-center, bBDC or bBC, and after bottom-dead-center,

Direct Injection (DI)
Fuel injection into the main combustion chamber of an engine. Engines have either one main combustion chamber (open chamber or a divided combustion chamber made up of a main chamber and a smaller connected secondary chamber.

Indirect Injection (IDI)
Fuel injection into the secondary chamber of an engine with a divided combustion chamber.

Bore
Diameter of the cylinder or diameter of the piston face, which is the same minus a very small clearance.

Stroke
Movement distance of the piston from one extreme position to the other: TDC to BDC or BDC to TDC.

Clearance Volume
Minimum volume in the combustion chamber with piston at TDC.

Displacement or Displacement Volume
Volume displaced by the piston as it travels through one stroke. Displacement can be given for one cylinder or for the entire engine (one cylinder times number of cylinders). Some literature calls this *swept volume*.

Wide-Open Throttle (WOT)
Engine operated with throttle valve fully open when maximum power and/or speed is desired.

Ignition Delay.
It is the time interval between ignition initiation and the actual start of combustion.

Air-Fuel Ratio
Ratio of mass of air to mass of fuel input into engine.

Fuel-Air Ratio (
Ratio of mass of fuel to mass of air input into engine.

Brake Maximum Torque
The speed at which maximum torque occurs.

Overhead Valve
Valves mounted in engine head.

Overhead Cam
Camshaft mounted in engine head, giving more direct control of valves which are also mounted in engine head.

APPENDIX III
Pressure Units

Since 1 Pa is a small pressure unit, the unit hectoPascal (hPa) is widely used, especially in meteorology. The unit kiloPascal (kPa) is commonly used design of technical applications like HVAC systems, piping systems and similar.

- *1 hectoPascal = 100 Pascal = 1 millibar*
- *1 kiloPascal = 1000 Pascal*

Some Alternative Units of Pressure

- *1 bar - 100,000 Pa*
- *1 millibar - 100 Pa*
- *1 atmosphere - 101,325 Pa*
- *1 mm Hg - 133 Pa*
- *1 inch Hg - 3,386 Pa*

A **torr** (torr) is named after Torricelli and is the pressure produced by a column of mercury *1 mm* high - equals to *1 / 760th* of an atmosphere.

- *1 atm = 760 torr = 14.696 psi*

Pounds per square inch (psi) was common in U.K. but has now been replaced in almost every country except in the U.S. by the SI units. Since atmospheric pressure is 14.696 psi - a column of air on a area of one square inch area from the Earth's surface to the space - weights 14.696 pounds.

The **bar** (bar) is common in the industry. One bar is 100,000 Pa, and for most practical purposes can be approximated to one atmosphere even if *1 Bar = 0.9869 atm*

There are 1,000 **millibar** (mbar) in one bar, a unit common in meteorology.

1 millibar = 0.001 bar = 0.750 torr = 100 Pa

Pressure Units

	Pascal (Pa)	Bar (bar)	Technical Atmosphere (at)	Atmosphere (atm)	torr (Torr)	pound-force per square inch (psi)
1 Pa	$\equiv 1$ N/m^2	10^{-5}	1.0197×10^{-5}	9.8692×10^{-6}	7.5006×10^{-3}	145.04×10^{-6}
1 bar	100,000	$\equiv 10^6$ dyn/cm^2	1.0197	0.98692	750.06	14.5037744
1 at	98,066.5	0.980665	$\equiv 1$ kgf/cm^2	0.96784	735.56	14.223
1 atm	101,325	1.01325	1.0332	$\equiv 1$ atm	760	14.696
1 torr	133.322	1.3332×10^{-3}	1.3595×10^{-3}	1.3158×10^{-3}	$\equiv 1$ Torr; ≈ 1 mmHg	19.337×10^{-3}
1 psi	6.894×10^3	68.948×10^{-3}	70.307×10^{-3}	68.046×10^{-3}	51.715	$\equiv 1$ lbf/in^2

Example reading: 1 Pa = 1 N/m^2 = 10^{-5} bar = 10.197×10^{-6} at = 9.8692×10^{-6} atm, etc.

PRESSURE VS. ALTITUDE

Altitude (Feet)	Pressure		
	in of Hg	mm of Hg	psia
-1,000	31.02	787.9	15.25
-500	30.47	773.8	14.94
0	29.921	760	14.7
500	29.38	746.4	14.43
1,000	28.86	732.9	14.18
1,500	28.33	719.7	13.9
2,000	27.82	706.6	13.67
2,500	27.31	693.8	13.41
3,000	26.81	681.1	13.19
3,500	26.32	668.6	12.92
4,000	25.84	656.3	12.7
4,500	25.36	644.2	12.45
5,000	24.89	632.3	12.23
10,000	20.58	522.6	10.1
15,000	16.88	428.8	8.28
20,000	13.75	349.1	6.75
30,000	8.88	225.6	4.36
40,000	5.54	140.7	2.72
50,000	3.426	87.3	1.689
60,000	2.132	54.15	1.048
70,000	1.322	33.59	0.649
80,000	0.82	20.83	0.403

APPENDIX IV

Temperature Conversion

Celcius	Fahrenheit
°C = (°F - 32) / 1.8	°F = 1.8C + 32
TEMPERATURE CONVERSION	
°C	°F
-200	-328
-180	292
-160	-256
-140	-220
-120	-184
-100	-148
-90	-130
-80	-112
-70	-94
-60	-76
-50	-58
-40	-40
-30	-22
-20	-4
-10	14
0	32
5	41
10	50
15	59
20	68

25	77
30	86
40	104
50	122
60	140
70	158
80	176
90	194
100	212
120	248
140	284
160	320
180	356
200	392
212	414
220	428
250	482
300	572
350	662
400	752
450	842
500	932
600	1112
700	1292
800	1472
900	1652
1000	1832

APPENDIX V

Torque Conversion

Torque Conversion Chart English to Metric

Foot Pounds (ft. lbs.)	Newton Meters (N-m)	Foot Pounds (ft. lbs.)	Newton Meters (N-m)	Foot Pounds (ft. lbs.)	Newton Meters (N-m)	Foot Pounds (ft. lbs.)	Newton Meters (N-m)	Foot Pounds (ft. lbs.)	Newton Meters (N-m)
1	1.3558	32	43.3856	63	85.4154	94	127.4452	125	169.475
2	2.7116	33	44.7414	64	86.7712	95	128.801	126	170.8308
3	4.0674	34	46.0972	65	88.127	96	130.1568	127	172.1866
4	5.4232	35	47.453	66	89.4828	97	131.5126	128	173.5424
5	6.779	36	48.8088	67	90.8386	98	132.8684	129	174.8982
6	8.1348	37	50.1646	68	92.1944	99	134.2242	130	176.254
7	9.4906	38	51.5204	69	93.5502	100	135.58	131	177.6098
8	10.8464	39	52.8762	70	94.906	101	136.9358	132	178.9656
9	12.2022	40	54.232	71	96.2618	102	138.2916	133	180.3214
10	13.558	41	55.5878	72	97.6176	103	139.6474	134	181.6772
11	14.9138	42	56.9436	73	98.9734	104	141.0032	135	183.033
12	16.2696	43	58.2994	74	100.3292	105	142.359	136	184.3888
13	17.6254	44	59.6552	75	101.685	106	143.7148	137	185.7446
14	18.9812	45	61.011	76	103.0408	107	145.0706	138	187.1004
15	20.337	46	62.3668	77	104.3966	108	146.4264	139	188.4562
16	21.6928	47	63.7226	78	105.7524	109	147.7822	140	189.812
17	23.0486	48	65.0784	79	107.1082	110	149.138	141	191.1678
18	24.4044	49	66.4342	80	108.464	111	150.4938	142	192.5236
19	25.7602	50	67.79	81	109.8198	112	151.8496	143	193.8794
20	27.116	51	69.1458	82	111.1756	113	153.2054	144	195.2352
21	28.4718	52	70.5016	83	112.5314	114	154.5612	145	196.591
22	29.8276	53	71.8574	84	113.8872	115	155.917	146	197.9468
23	31.1834	54	73.2132	85	115.243	116	157.2728	147	199.3026
24	32.5392	55	74.569	86	116.5988	117	158.6286	148	200.6584
25	33.895	56	75.9248	87	117.9546	118	159.9844	149	202.0142
26	35.2508	57	77.2806	88	119.3104	119	161.3402	150	203.37
27	36.6066	58	78.6364	89	120.6662	120	162.696	151	204.7258
28	37.9624	59	79.9922	90	122.022	121	164.0518	152	206.0816
29	39.3182	60	81.348	91	123.3778	122	165.4076	153	207.4374
30	40.674	61	82.7038	92	124.7336	123	166.7634	154	208.7932
31	42.0298	62	84.0596	93	126.0894	124	168.1192	155	210.149

UNITS TO CONVER	AMERICAN			S.I.			METRIC		
	ozf.in	lbf.in	lbf.ft	mN.m	cN.m	N.m	gf.cm	kgf.cm	kgf.m
ozf.in	1	0.0625	0.005	7.062	0.706	0.007	72.007	0.072	0.0007
lbf.in	16	1	0.083	113	11.2985	0.113	1152.1	1.152	0.0115
lbf.ft	192	12	1	1356	135.6	1.3558	13,286	13.83	0.138
mN.m	0.142	0.009	0.0007	1	0.1	0.001	10.2	0.01	0.0001
cN.m	1.416	0.088	0.007	10	1	0.01	102	0.102	0.001
N.m	141.6	8.851	0.738	1000	100	1	10,197	10.2	0.102
gf.cm	0.014	0.0009	0.00007	0.098	0.01	0.0001	1	0.001	0.00001
kgf.cm	13.89	0.868	0.072	98.07	9.807	0.098	1000	1	0.01
kgf.m	1389	86.8	7.233	9807	980.7	9.807	100,000	100	1
inf.g				0.2491	0.0249				
daN.m	14.1611	0.8851	0.0738	100	10	0.001	1.0197	1.02	0.0102

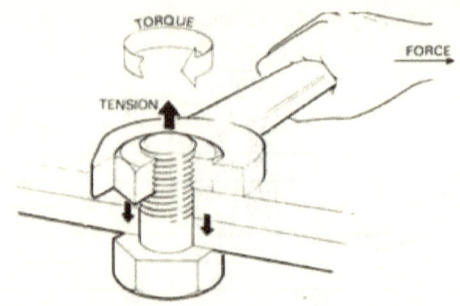

Torque = T   Force = F
Torque = Force by Turning or Twisting.
Tension = Power by Creating a Straight Pull.

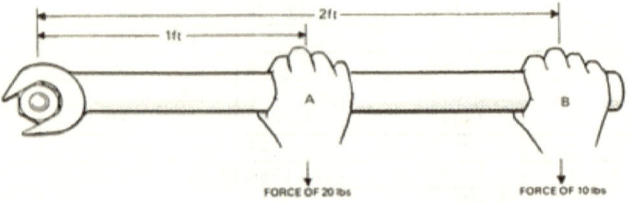

Torque = Force x Distance
T = F x L so that:
 T1 = F1 x L1 = 10 lbs x 2 ft = 20 ftlbs
T2 = F2 x L2 = 20 lbs x 1 ft = 20 ftlbs
kgf·cm = 0.098066N·m=0.098N·m
1N·m = 10.1972kgf·cm=10.2kgf·cm

Image courtesy of Maryland Metrics, USA

APPENDIX VI

Permissions and Approvals

APPENDIX-PERMISSIONS AND APPROVALS

Sr. No.	Name of the Company	Communication
1	ABB Ltd.	Permission through email
2	Acromag, Incorporated	Permission through email
3	Air Cleaning Specialist Inc. USA	Permission through email
4	Associated Electrodyne Industries Pvt. Ltd	Permission through email
5	Blue-White Ind., USA	Permission through email
6	Crushproof Tubing Company, USA	Permission through email
7	DAVCO Technology, USA	Permission through email
8	Dyne systems. Inc, USA	Permission through email
9	Do India, Pune, India	Permission through email
10	Horiba Ltd. Germany	Permission through email
11	Hyper Physics by Rod Nave, Georgia State University	Permission through email
12	Magtrol Inc. USA	Permission through email
13	Maryland Metrics	Permission through email
14	Moore Industry California, USA	Permission through email
15	Omniscient Engineering India	Permission through email
16	Pico Technology, Cambridge shire, UK	Permission through email
17	Piper Dynamometers, UK	Permission through email
18	Power test, USA	Permission through email
19	Power Testing Intrument Co., Ltd, Hong Kong	Permission through email
20	Rice Lake Weighing System	Permission through email
21	Saj Test Plant Pvt. Ltd.	Permission through email
22	Siemens Industries, Inc.	Permission through email
23	Taylor dynamometer USA	Permission through email
24	Tectos gmbh, Graz, Austria	Permission through email
25	Technogerma GmbH, Germany	Permission through email

INDEX

A

absolute humidity, 298
absolute pressure, 276
AC (alternating current)
    dynamometer, 59-150, 167, 213
  accuracies of, 157-58
  compared with DC dynamometer, 155, 173, 175-80
  inertia of, 157
  operating principle of, 150, 158-66
  power and torque envelopes of, 156
  rating of, 150
acquisition channels, 235, 335-36, 338, 344, 348, 354, 364-71, 503, 522
air-fuel ration, 562-63, 564
Air Intake System, 558
alarm annunciator, 349
alarm trips
  audiovisual indications of, 356
  choice of, 353-54
  configuration of, 352
  types of, 349-52
ambient conditions, 218-19
analysis of variance, 200
Aneroid Barometer, 297-98
ANOVA method, 200
armature choke, 147
atmospheric pressure, 275, 296-97
automation, 303-4
  classification of, 306
    auxiliary equipment, 324
    control system, 307-10
    engine handling, 312-23
  levels of, 304
  strategies for, 306

B

Babbage, Charles, 31
barometer, 297
barometric pressure, 296
barometric sensitivity, 266
Bernoulli's equation, 413, 429-30
BHP Concepts, 37
Bourdon mechanical gauge, 279
brinelling effect, 35, 196, 201, 496
bump-less transfer, 192

C

CAHU (combustion air handling unit), 559, 559-62
calibration, 202-4
calibration curve, 203, 210, 266, 269-70
Cardan shaft. *See* driveshaft

cavitation, 82
  avoiding, 85
  effect of, 86
  mechanism of, 83-84, 150
cavitation inception, 82, 84
chassis dynamometer, 30
CJC (Cold Junction Compensation), 241, 243-44, 251-52
Collins, E. V., 31
constant capacity range, 130
constant torque range, 130
correction factors, 218
  and horsepower, 218
  kinds of, 223
    DIN 70012, 225, 229
    ISO 1585, 231-32
    JIS D 1001, 230-31
    SAE J1349, 224-29
  use of, 220

D

data acquisition
  definition of, 336
  function of, 336
  terminology of, 340-44
data acquisition systems, 336-37
  with AC Dynamometer, 367
  displaying data, 369
  with eddy-current dynamometer, 366-67
  elements of, 337
  executing the automated test, 369
  managing and analyzing data, 370
  reading and recording data, 369
  with transmission test system, 368
  types of, 344
    plug-in boards, 344, 347

serial communication, 345
USB, 347
wireless, 344-45
  uses of, 336
Davco slow flow meter, 445-46
Davy, Humphrey, 246
dBA
  definition of, 377
  interpreting, 379
DC (direct current) dynamometer, 59, 126-27, 147
  accuracies of, 134
  compared with DC dynamometer, 155, 173, 175-80
  construction of, 130-32
  control system of, 134-41
  direction of rotation of, 141
  emergency breaking, 143
  in engine testing, 145
  inertia of, 133
  operating principle of, 127-29
  power and torque curves of, 133
  rating of, 129
  in speed control mode, 143-44
decibels (dB), 377
de Prony brake, 31, 43-45
Diaphragm Pressure Gauge, 280-81
draw bar pull, 507-8, 510-11
driveshaft, 491, 524, 526, 556
  selecting, 528-34
  terminology in selecting, 528
dry bulb temperature, 223
dynamometer
  accuracy of, 37
  applications of, 504
    draw bar pull testing, 507-11
    locked rotor testing, 516-18
    outboard motor testing, 521-22
    tandem dynamometer, 518-20

tractor PTO testing, 504-6
vertical motor and vertical turbine testing, 511-16
definition of, 29-30, 33
etymology of, 32
features of, 32
history of, 30-32
measurement of torque and power of, 196-97
  accuracy, 197-98
  measurement error, 202
  precision, 198
  repeatability, 198-99
  reproducibility, 199-200
  resolution, 201
  sensitivity, 200-201
mounting of, 34-36
operation modes and torque envelope of, 187-89
operation modes of, 182
  open loop, 183-85
  power law, 186
  speed constant, 185-86
  torque constant, 186
selecting the right operation mode, 189-90
selection of, 86-87, 168-69, 169-70
speed measurement of, 32
and torque measurement, 32
types of, 41
dynamometer constant, 34

E

EDACS (Engine Data Acquisition and Control System), 308, 364-65

eddy-current dynamometer, 106, 113, 170
eddy-current dynamometers
  compared with hydraulic dynamometer, 170
  construction of, 95-98
  electrical laws on, 93-94
  history of, 92
  inertia of, 98
  operating principle of, 92-93
  power and torque curves of, 99-102
  types of
    dry gap, 58, 104
    dry gap twin coil, 105
    wet gap, 59, 103-4
eddy currents, 91-92, 100
encoder, 293
engine dynamometer, 30
engine exhaust extraction, 564-83
engines
  classifications of, 449
  definition of, 449
  measurements
    airflow, 426-39
    blow by, 443
    fuel consumption, 409
    oil consumption, 444-48
  testing, 468-86
EOTC (engine oil temperature control), 327-30
EWTC (Engine Water Temperature Controller), 331

F

fan brake dynamometer, 50-54
Faraday, Michael, 91
flexure support, 36

forced air cooling, 151
Foucault, Jean B. L., 58, 91
Foucault currents. *See* eddy currents
four-quadrant design. *See* regeneration
Froude, William, 31, 68-69, 89, 206
Froude number, 68
fundamental frequency, 538

G

gauge pressure, 275
gravimetric fuel consumption meter, 424

H

horsepower
  definition of, 30
  and torque, 221
hot wire anemometer, 426-28
hydraulic dynamometers, 31, 54, 68, 89
  application of, 88
  cavitation in, 82-84
  compared with eddy-current dynamometer, 170-73
  direction of rotation of, 78
  inertia of, 73-74
  load control mechanism of, 79
  power and torque curves of, 74-78
  servo controls
    electrical, 80
    electrohydraulic, 81, 260
  types of
    pressure controlled / variable fill, 55-57, 70-71, 80
    sluice gate controlled / constant fill, 57, 69-70, 79
  water movement inside, 72-73
hysteresis dynamometers, 107, 110-14
  accuracy of, 114
  application of, 114
  construction of, 111
  operating principles of, 110
  power curve of, 112
  torque envelope of, 112

I

internal combustion engines
  basics of, 458
  components of, 450-56
  heat balance of, 479
  testing classification for, 466
  testing procedure for, 466-68
  types of
    four strokes, 460
    two strokes, 460, 465-66
International Prototype Kilogram, 204, 212

J

Johnston, Edward A., 504

L

lambda, 563
load bank, 146
  types of
    capacitive, 146-47
    reactive, 147
    resistive, 146
load cell, 259

classification of, 261-65
operating principles of, 260-61
terminology of, 265-74
theory of operation of, 260
Lord Kelvin, 45

M

Mach number, 68
magnetic pickup, 289
magnetic powder dynamometer,
   accuracy of, 114
magnetic powder dynamometers,
   107-10, 113
  application of, 114
  power curves of, 108-9
  torque curves of, 109
manometer, 276
  inclined tube, 279
  U tube, 277
MAP (manifold absolute pressure)
   sensor, 191
measurands, 196
measurement error, 202
moisture, 436
moment of a force. *See* torque

N

noise, 377
NRC (Noise reduction coefficient),
   377
NTP (normal temperature and
   pressure), 219

O

orifice meter, 412

P

PID (proportional, integral,
   derivative) controller, 192-95,
   326
Pitot, Henri, 428
Pitot tube, 428-29, 431
portable dynamometers, 115-16,
   124
  advantages of, 122
  applications of, 121-24
  construction of, 116
  cross section of, 117
  power curve of, 118-20
power
  definition of, 221
  effect of altitude on, 221-22
  effect of humidity on, 222
  effect of temperature on, 223
pressure, definition of, 274
pressure transducers, 274, 281
process value, 327
Prony, Gaspard Clair François
   Marie Riche de, 31, 43
PTO (power take-off) shaft, 504
PTO horsepower, 506
PWM (pulse width modulation),
   163-65

R

regeneration, 137, 142, 144
relative humidity (RH), 223, 232,
   294, 298-99, 301, 436-37, 503,
   560
relays, types of, 354-55
repeatability error, 198
resolution error, 201
rolling resistance, 508-9

rope brake dynamometer, 41, 45-50
rotameter, 415-19
rotary torque transducer, 282
  calibration of
    electrical, 287
    mechanical, 287
  coupling requirements for, 286
  designs of
    electrical, 284
    mechanical, 283
  installation schematics of, 286
  theory of operation, 282-83
RTD (Resistance Temperature Detector), 246-52, 302
  advantages of, 249
  compared with thermocouple, 250-53
  disadvantages of, 249
  linearization of, 248-49
  theory of operation of, 248
  tolerance or accuracy for, 249

S

SAE (Society of Automotive Engineers), 224-25, 234
Seebeck, Thomas Johann, 239
Seebeck effect, 239
sensor, 338
sensor break detect, 325
sequence generator, 369
shunt calibration, 287-88
signals, 338
sling psychrometer, 299-301
standard atmospheric pressure, 276
STC (Sound transmission class), 377
stoichiometric mixture, 563
STP (standard temperature and pressure), 219
strain gauge, 253
  attaching, 256-58
  factors affecting selection of, 255-56
  signal conditioning for, 259
  strain measurement of, 258-59
  theory of operation of, 253-55

T

tachogenerator, 292
tachometer, 29
temperature, measurement of, 246

test cells
  environmental measurements of, 294
  essentials of, 487-503
  noise in, 376
  role of engine and dynamometer foundation in, 399
  systems of
    communication, 406
    compressed air, 394-97
    engine exhaust handling, 386
    engine oil supply, 398-99
    firefighting, 406-8
    fuel, 397
    lighting, 405-6
    ventilation, 379-85
    water supply, 387-94
  types of, 376
theoretical torque, 208
thermocouples, 239
  accuracy of, 240, 241
  compared with RTD, 250-53

precautions and considerations for using, 244-46
theory of operation of, 239-40
types of, 241-42
using CJC (cold junction compensation), 243-44
thermocouple wire, 240-41
thermoelectric effect, 239
Thoma's number, 84-85
thyristor, 134
thyristor bridge, 134, 139
torque, 32-33
torque measurement, 205-16
Torricelli, Evangelista, 297
Torricelli's vacuum, 277
torsional vibrations
  calculation of, 534-40
  Holzer method for analysis of, 540-52
  role of engine and dynamometer foundation in, 552-56
Traceability, 204-5
transmission dynamometer, 61
  belt, 64
    Tatham, 64-65
    Von Hefner, 65-67
  epicyclical train, 62-63
tribology, 168
trunnion, 34-35
trunnion bearings, 33-35

U

USB (Universal Serial Bus), 339, 346

V

volumetric efficiency, 439, 478
volumetric fuel consumption, 419, 421-23

W

Walker, W. G., 50
water brakes. *See* hydraulic dynamometer
Watt (unit of power), 38
Watt, James, 30, 38
weight, 213
wet bulb temperature, 223
Winther, Anthony, 31, 92
Winther, Martin, 31, 92
work, equation for, 37

www.ingramcontent.com/pod-product-compliance
Lightning Source LLC
Chambersburg PA
CBHW020718180526
45163CB00001B/17